Advances in Waste Materials' Valorization

Advances in Waste Materials' Valorization

Guest Editor

Natalia Howaniec

Basel • Beijing • Wuhan • Barcelona • Belgrade • Novi Sad • Cluj • Manchester

Guest Editor
Natalia Howaniec
Department of Energy
Efficiency, Environmental
Monitoring and Alternative
Fuel Technologies
Central Mining
Institute—National Research
Institute
Katowice
Poland

Editorial Office
MDPI AG
Grosspeteranlage 5
4052 Basel, Switzerland

This is a reprint of the Special Issue, published open access by the journal *Materials* (ISSN 1996-1944), freely accessible at: https://www.mdpi.com/journal/materials/special_issues/704IXLHDMC.

For citation purposes, cite each article independently as indicated on the article page online and as indicated below:

Lastname, A.A.; Lastname, B.B. Article Title. *Journal Name* **Year**, *Volume Number*, Page Range.

ISBN 978-3-7258-8291-5 (Hbk)
ISBN 978-3-7258-8292-2 (PDF)
https://doi.org/10.3390/books978-3-7258-8292-2

Contents

materials

Editorial

Advances in Waste Materials' Valorization

Natalia Howaniec

Department of Energy Efficiency, Environmental Monitoring and Alternative Fuel Technologies, Central Mining Institute—National Research Institute, Pl. Gwarków 1, 40-166 Katowice, Poland; n.howaniec@gig.eu

1. Introduction

Sustainable management of natural resources remains one of the principal economic and environmental challenges. Considerable efforts are being undertaken worldwide, ranging from the development and implementation of cost-effective and more resource- and energy-efficient production technologies to the recovery, reuse and recycling of waste materials. This is of particular importance in the energy sector, fueling entire economies and having a direct impact on related fundamental challenges of energy security and the green transition [1,2]. The cement production sector is another major economic area and a source of carbon dioxide emissions, focusing the attention of scientists and engineers active in the field and involved in the development of waste-derived cementitious materials [3]. These trends were clearly reflected in the dominant thematic area of this collection.

2. Insight Into Recent Advances in Waste Valorization

The compiled works covered, in particular, research devoted to the evaluation of carbon fiber (CF) recovery methods; the application of CF, recovered from retired wind turbine blades, with oil shale ash in cement-based composites, or with steelmaking dust and coal-derived fly ash in the production of geopolymer-based granulates; the use of blast-furnace slag, low-grade calcined clay, limestone and lava in cementitious materials development; the application of H_2S, potentially derived from biotechnological processing of organic waste, in the synthesis of selenium sulfide; the development of metal oxide nanoparticle-functionalized antibacterial textiles using recycled cotton and polyester yarns; and the preparation and characterization of new layered composite materials based on silicone implant waste, biochar–sulfur composites, and mesoporous silica from bottle-derived soda–lime glass.

Industrial demand for CF components is growing and reflects their increasing use in the production of lightweight and durable materials, in particular for automotive, aerospace and wind turbine applications. This is inherently combined with the need for cost-effective, environmentally friendly and energy-efficient methods for managing CF-derived waste materials [4]. In Contribution 1, by Poranek et al., the results of laboratory tests and a life cycle assessment (LCA) of two solvolysis-based CF composite recovery methods were presented. The first one consisted of ambient pressure solvolysis in 0.5 and 2 L batch reactors connected to a reflux condenser and an inert gas supply tank, using a solution of ethylene glycol and potassium hydroxide [5]. The second one used plasma-enhanced nitric acid solvolysis to increase the CF recovery rate [6]. LCA results showed that the second option, which is still at the development stage, has the highest negative environmental impact compared to CF production from virgin materials, but also the potential for improved performance through optimization. In contrast, the first option offers environmental benefits compared to virgin fiber production.

Another example of the valorization of short carbon fibers (SCFs), widely used in a variety of industrial applications [7], is provided by Kalpokaitė-Dičkuvienė et al. in Contribution 2. SCFs recovered from the pyrolysis of retired wind turbine blades were employed to modify the functional properties of mortar composites, while oil shale combustion fly ash was tested as a partial cement substitute. The effects on hydration phase composition, pore size distribution, specific surface area, compressive and flexural strength, freeze–thaw cycling, and water absorption of composites over a period of up to 120 days were observed, and their environmental performance was assessed. The use of ash resulted in a refined pore structure and an increased volume of reaction products, primarily calcium silicate hydrates (C–S–H), which are critical for cement strength. The mineralogical composition of the modified binder was reported to be the dominant factor in achieving a dense microstructure that enhanced resistance to water ingress and improved mechanical performance under long-term hydration and freeze–thaw exposure. LCA demonstrated that combining complex binders with SCF may mitigate the environmental impact of cement production in terms of global warming potential, terrestrial ecotoxicity, human non-carcinogenic toxicity and fossil resource scarcity. Both waste materials tested were proven suitable for recycling in the development of cementitious materials for selected applications. Considering the potential of the cement and wind turbine markets [8,9], as well as the CO_2 emissions associated with the power sector and cement industry [3,10], these findings may provide a foundation for the development of next-generation binder systems.

Composites recovered from retired wind turbine blades were also tested as a primary aluminosilicate precursor, while steelmaking dust and coal-derived fly ash were used as additives in the production of geopolymer-based aggregates by Cempa et al. in Contribution 3. The study contributed to current research focused on recycling selected waste materials from the energy sector for the development of construction materials [3,11]. Reactive solids and alkaline activator liquids were mixed and granulated in a single operation using an intensive counter-current mixer. The materials developed were cold-cured at room temperature and characterized after 28 days in terms of grain size distribution, crushing resistance, water absorption, abrasion, and leaching properties. Scanning electron microscopy imaging and energy-dispersive X-ray spectroscopy analysis were also performed. The results indicated that the additives improved the mechanical performance of the aggregates compared to commercial lightweight aggregates and enhanced their microstructure, while effective immobilization of several heavy metals was also achieved.

The next case of waste valorization as a supplementary cementitious material [12] is Contribution 4 by Etcheverry et al., who tested cement blends with ground-granulated blast-furnace slag, low-grade calcined clay, limestone or lava. The reactivity of the blends was tested using isothermal calorimetry. Results showed slightly lower 28-day strength for blends with clay than for those with blast-furnace slag, as well as a strength increase at later ages for blends with lava due to delayed pozzolanic activity. Concrete produced using low-grade calcined clay and lava was reported to achieve compressive strength comparable to that of the reference cement III/A, but exhibited higher capillary porosity, leading to increased water absorption, drying shrinkage, and reduced freeze–thaw resistance.

Contribution 5 by Gong et al. is a comprehensive review on the recycling of another industrial waste, soda residue, whose unique structural properties and cementitious activity make it a promising mineral admixture for cementitious materials, contributing to the development of the green construction sector [13]. The authors concluded that the physicochemical properties of soda residue, its influence on microstructure, and its effect on macro performance enhance the compactness and early strength of cementitious materials through physical filling and chemical activation. However, its chloride content and the associated salt crystallization phenomenon weaken workability and durability. Its

https://doi.org/10.3390/ma19122459

successful employment requires precise dosage control, chemical modification, and the use of mineral admixtures. The identified future research needs include the development of composite binder systems, innovations in solid waste-based material preparation, and advancements in desalination technologies to enable the large-scale and environmentally friendly use of the waste concerned.

Solid products derived from the thermochemical processing of apple wood chips under a carbon dioxide atmosphere were tested in the production of biochar–sulfur composites by Syguła et al. in Contribution 6. The effects of the preparation method (using a muffle furnace or electric burner) and sulfur content (in the range of 60–80%) on the mechanical performance of the composites, as potential durable and sustainable materials for emerging applications [14,15], were determined. The results showed that an increase in sulfur content led to an improvement in the ultimate strength of the composites and that the use of an electric burner resulted in composites exhibiting superior strength and reduced brittleness compared to those fabricated in a muffle furnace. The latter effect was attributed to improved sulfur distribution and more effective infiltration of liquid sulfur into the porous biochar structure under electric burner conditions.

The use of bottle-derived soda–lime glass as a sustainable silica source for the synthesis of mesoporous silica, as an alternative to conventional silica precursors, was tested by de Oliveira Lima et al. in Contribution 7. Glass powder was dissolved in sodium hydroxide to form silica gel, which was subsequently applied as a precursor in porous silica synthesis [16]. The structural, morphological, surface, and textural properties of the materials tested were determined. X-ray diffraction analysis confirmed the amorphous nature of all samples, while Fourier-transform infrared spectroscopy spectra indicated successful silica network formation with modifications in bond connectivity. Scanning electron microscopy imaging revealed spherical particles with average diameters of approx. 0.19 ± 0.02 μm for silica gel and 0.15 ± 0.03 μm for mesoporous silica. Zeta potential measurements indicated a negative surface charge and good colloidal stability in aqueous media. On the basis of the specific surface area, it was concluded that the development of porous structure was limited by the solubility of silica gel in acidic media, which prevented ideal condensation on the surface of the surfactant micelles. While the waste glass derived silica gel was assessed as a promising precursor, the lack of a highly ordered mesostructure indicated the need for more effective control of precursor solubility and pH.

Following the increased interest in developing more sustainable methods of chemical production for the pharmaceutical and cosmetic industries [17,18], Safinazlou et al. presented in Contribution 8, a preliminary feasibility study devoted to the synthesis of selenium sulfide using H_2S, which can be produced via a biotechnological route from organic waste sources (e.g., fermentation juice from vegetables, spoiled milk, or compost) and naturally occurring mineral SO_4^{2-} sources (such as mineral water from gypsum-rich soils), as an alternative to the conventional and more complex chemical method [19]. The preliminary results of the study were considered to be promising; however, further testing, optimization, and upscaling are needed before the ecological benefits can be confirmed and economic value realized.

A study contributing to reports on the sustainable management of breast implant waste [20] was reported in Contribution 9 by Kozłowska et al., who examined its potential use as a component of layered composites. Experimental data acquired using attenuated total reflection Fourier-transform infrared spectroscopy, differential scanning calorimetry, and simultaneous thermogravimetric/derivative thermogravimetric analysis coupled with Fourier-transform infrared spectroscopy confirmed that radiation did not affect the structure, thermal properties, or oxidative decomposition behavior of the shell and gel layers of breast silicone implants. The application of a urethane–methacrylate matrix in composite

 https://doi.org/10.3390/ma19122459

materials affected the crystallization and melting behavior of the waste-derived silicone phase, whose presence, in turn, increased the thermal stability of the composite while decreasing the glass transition temperature, storage modulus, and hardness compared to the matrix alone. The type of implant waste layer used (shell or gel) did not significantly affect the melting and glass transition temperatures, thermal stability, or oxidative decomposition behavior of the developed materials. The potential practical applicability of the layered composites developed and tested within the study was demonstrated.

An experimental work contributing to the research area devoted to the application of recycled materials in functionalized textiles [21] was presented by Faheem et al. in Contribution 10. A preliminary study on the development of metal oxide nanoparticle-functionalized antibacterial textiles using recycled cotton and polyester yarns was performed. Antibacterial finishes were applied at both the yarn and fabric stages to analyze differences in performance. Agar plate qualitative antibacterial analyses proved the presence of antibacterial properties before and after commercial laundering. Thermophysiological comfort properties were reported to be dependent on recycled yarn types and finishing stages. The air permeability of fabric-finished specimens was about 47% lower, while the overall moisture management capability index was 21% higher, than that of yarn-finished specimens. Bursting strength was observed to decrease since the finishing treatments made the yarns stiffer. A more detailed characterization of nanoparticles deposited onto the textile material and the determination of the effects of specific functional nanoparticles on antibacterial performance were mentioned among the potential subjects for future research.

3. Summary

This Special Issue provided a platform for communicating recent insights into the implementation of conventional and emerging material-processing techniques and strategies for the recovery and recycling of waste materials of various origin. The majority of the works collected covered the valorization of energy sector-derived waste, largely in the development of cementitious materials. The remaining studies concerned the recycling of a wide range of wastes with the aim of developing functional materials for defined or emerging purposes in the cosmetic, medical, glass or construction industries. The studies presented included material recovery methods, preparation strategies, and assessments of properties determined experimentally using a variety of analytical techniques. The active engagement of researchers representing various scientific disciplines, reflected in the coverage of this Special Issue, proves that the subject remains both topical and promising.

Funding: This research received no external funding.

Acknowledgments: I sincerely appreciate the active contributions of authors, reviewers, as well as the editorial office and editorial board members, to the creation of this Special Issue.

Conflicts of Interest: The author declares no conflicts of interest.

List of Contributions:

1. Poranek, N.; Pikoń, K.; Generowicz-Caba, N.; Mańka, M.; Kulczycka, J.; Marinis, D.; Farsari, E.; Amanatides, E.; Lewandowska, A.; Sajdak, M.; et al. Comparative LCA Analysis of Selected Recycling Methods for Carbon Fibers and Socio-Economic Analysis. *Materials* **2025**, *18*, 2660. https://doi.org/10.3390/ma18112660.
2. Kalpokaitė-Dičkuvienė, R.; Stasiulaitiene, I.; Baltušnikas, A.; Yousef, S. Transforming Waste into Value: The Role of Recovered Carbon Fibre and Oil Shale Ash in Enhancing Cement-Based Structural Composites. *Materials* **2025**, *18*, 5636. https://doi.org/10.3390/ma18245636.
3. Cempa, M.; Korol, J.; Klupa, A. Assessment of the Potential for Producing Geopolymer-Based Granulates as a Substitute for Natural Aggregates. *Materials* **2025**, *18*, 5275. https://doi.org/10.3390/ma18235275.

4. Kozłowska, A.; Worzakowska, M.; Tarasiuk, B.; Piziorska, M.; Podkościelna B. Study and Reuse of Silicone Implants After Radiotherapy. *Materials* **2026**, *19*, 1798. https://doi.org/10.3390/ma19091798.
5. Syguła, E.; Słupska, M.; Radziemska, M.; Białowiec, A. Mechanical Properties of Biochar–Sulfur Composites. *Materials* **2026**, *19*, 549. https://doi.org/10.3390/ma19030549.
6. Faheem, M.S.; Jamshaid, H.; Mishra, R.K.; Abbas, A.; Muller, M.; Urban, J.; Penc, M. Preliminary Assessment of Commercial Antibacterial Finishes on Knitted Fabrics from Recycled Cotton and Polyester. *Materials* **2025**, *18*, 5319. https://doi.org/10.3390/ma18235319.
7. Etcheverry, J.M.; Detemmerman, L.; Degezelle, K.; Grigorjev, V.; De Brabandere, L.; De Belie, N. Durability Considerations in Replacing Blast Furnace Slag with Low-Grade Calcined Clay and Natural Pozzolan in Quaternary Cements. *Materials* **2025**, *18*, 5048. https://doi.org/10.3390/ma18215048.
8. Safinazlou, S.; Abdin, A.Y.; Tiganescu, E.; Lilischkis, R.; Schäfer, K.-H.; Fink-Straube, C.; Nasim, M.J.; Jacob, C. Valorizing Organic Waste: Selenium Sulfide Production Mediated by Sulfate-Reducing Bacteria. *Materials* **2025**, *18*, 2784. https://doi.org/10.3390/ma18122784.
9. Gong, Y.; Zhao, K.; Ba, Y.; Wu, Y.; Liu, G.; Liu, Z.; Yang, Y. Research progress on the application of soda residue in cementitious materials. *Materials* **2026**, *19*, 2228. https://doi.org/10.3390/ma19112228.
10. De Oliveira Lima, A.M.; Fernandes Graça, M.P.; Hammami, I.; Mathias Macêdo, A.A. Waste-Glass-Derived Porous Silica: Synthesis and Structural Characterization. *Materials* **2026**, *19*, 601. https://doi.org/10.3390/ma19030601.

References

1. Zdeb, J.; Howaniec, N.; Smoliński, A. Utilization of carbon dioxide in coal gasification—Experimental study. *Energies* **2019**, *12*, 140. [CrossRef]
2. Smoliński, A.; Howaniec, N.; Gąsior, R.; Polański, J.; Magdziarczyk, M. Hydrogen rich gas production through co-gasification of low rank coal, flotation concentrates and municipal refuse derived fuel. *Energy* **2021**, *235*, 121348. [CrossRef]
3. Zdeb, J.; Howaniec, N.; Smoliński, A. Experimental Study on Combined Valorization of Bituminous Coal Derived Fluidized Bed Fly Ash and Carbon Dioxide from Energy Sector. *Energy* **2023**, *265*, 126367. [CrossRef]
4. Rani, M.; Choudhary, P.; Krishnan, V.; Zafar, S. A review on recycling and reuse methods for carbon fiber/glass fiber composites waste from wind turbine blades. *Compos. Part B-Eng.* **2021**, *215*, 108768. [CrossRef]
5. Muzyka, R.; Mumtaz, H.H.; Sobek, S.; Werle, S.; Adamek, J.; Semitekolos, D.; Charitidis, C.A.; Tiriakidou, T.; Sajdak, M. Solvolysis and oxidative liquefaction of the end-of-life composite wastes as an element of the circular economy assumptions. *J. Clean. Prod.* **2024**, *478*, 143916. [CrossRef]
6. Marinis, D.; Markatos, D.; Farsari, E.; Amanatides, E.; Mataras, D.; Pantelakis, S. A Novel Plasma-Enhanced Solvolysis as Alternative for Recycling Composites. *Polymers* **2024**, *16*, 2836. [CrossRef] [PubMed]
7. Agarwal, N.; Rangamani, A.; Bhavsar, K.; Virnodkar, S.S.; Fernandes, A.A.A.; Chadha, U.; Srivastava, D.; Patterson, A.E.; Rajasekharan, V. An Overview of Carbon-Carbon Composite Materials and Their Applications. *Front. Mater.* **2024**, *11*, 1374034. [CrossRef]
8. Taherinezhad Tayebi, S.; Sambucci, M.; Valente, M. Waste Management of Wind Turbine Blades: A Comprehensive Review on Available Recycling Technologies with A Focus on Overcoming Potential Environmental Hazards Caused by Microplastic Production. *Sustainability* **2024**, *16*, 4517. [CrossRef]
9. Cavalett, O.; Watanabe, M.D.B.; Voldsund, M.; Roussanaly, S.; Cherubini, F. Paving the way for sustainable decarbonization of the European cement industry. *Nat. Sustain.* **2024**, *7*, 568–580. [CrossRef]
10. Barbhuiya, S.; Kanavaris, F.; Das, B.B.; Idrees, M. Decarbonising cement and concrete production: Strategies, challenges and pathways for sustainable development. *J. Build. Eng.* **2024**, *86*, 108861. [CrossRef]
11. Zdeb, J.; Howaniec, N. Energy Sector Derived Combustion Products Utilization—Current Advances in Carbon Dioxide Mineralization. *Energies* **2022**, *15*, 9033. [CrossRef]
12. Snellings, R.; Mertens, G.; Elsen, J. Supplementary cementitious materials. *Rev. Mineral. Geochem.* **2012**, *74*, 211–278. [CrossRef]
13. Wu, P.; Zhang, T.; Geng, Y.; Huang, S.B.; Wang, W.X.; Lv, X.J. Research progress on the physical and chemical properties and applications of soda residue. *China Powder Sci. Technol.* **2022**, *28*, 35–42.
14. Robinson, P.R. Sulfur Removal and Recovery. In *Springer Handbook of Petroleum Technology; Springer Handbooks*; Hsu, C.S., Robinson, P.R., Eds.; Springer: Cham, Switzerland, 2017.

15. Yaashikaa, P.R.; Kumar, P.S.; Varjani, S.; Saravanan, A. A critical review on the biochar production techniques, characterization, stability and applications for circular bioeconomy. *Biotechnol. Rep.* **2020**, *28*, e00570. [CrossRef] [PubMed]
16. Liu, Y.; Shi, C.; Zhang, Z.; Li, N. An Overview on the Reuse of Waste Glasses in Alkali-Activated Materials. *Resour. Conserv. Recycl.* **2019**, *144*, 297–309. [CrossRef]
17. Heath, R.S.; Ruscoe, R.E.; Turner, N.J. The Beauty of Biocatalysis: Sustainable Synthesis of Ingredients in Cosmetics. *Nat. Prod. Rep.* **2022**, *39*, 335–388. [CrossRef] [PubMed]
18. Wynendaele, E.; Furman, C.; Wielgomas, B.; Larsson, P.; Hak, E.; Block, T.; Van Calenbergh, S.; Willand, N.; Markuszewski, M.; Odell, L.R.; et al. Sustainability in Drug Discovery. *Med. Drug Discov.* **2021**, *12*, 100107. [CrossRef]
19. Zhu, S.H.; Zhu, S.M.; Zhu, L. Production Method of Selenium Sulfide. CN 201010571362, 25 November 2010.
20. Ahmed, Z.; Zargaran, A.; Marudamuthu, D.; Sousi, S.; Zargaran, D.; Hamilton, S.; Mosahebi, A. Sustainability in breast surgery: An eco-audit of the breast implant pathway. *J. Plast. Reconstr. Aesthetic Surg.* **2025**, *105*, 219–229. [CrossRef] [PubMed]
21. Huang, X.; Tan, Y.; Huang, J.; Zhu, G.; Yin, R.; Tao, X.; Tian, X. Industrialization of open- and closed-loop waste textile recycling towards sustainability: A review. *J. Clean. Prod.* **2024**, *436*, 140676. [CrossRef]

 https://doi.org/10.3390/ma19122459

materials

MDPI

Article

Comparative LCA Analysis of Selected Recycling Methods for Carbon Fibers and Socio-Economic Analysis

Nikolina Poranek [1,*], Krzysztof Pikoń [1], Natalia Generowicz-Caba [2], Maciej Mańka [2], Joanna Kulczycka [3], Dimitrios Marinis [4], Ergina Farsari [4], Eleftherios Amanatides [4], Anna Lewandowska [5], Marcin Sajdak [6], Sebastian Werle [7] and Szymon Sobek [8]

1 Department of Technologies and Installations for Waste Management, Faculty of Energy and Environmental Engineering, Silesian University of Technology, Konarskiego 18, 44-100 Gliwice, Poland; krzysztof.pikon@polsl.pl

2 Mineral and Energy Economy Research Institute Polish Academy of Sciences, Wybickiego 7A Street, 31-261 Krakow, Poland; ngenerowicz@min-pan.krakow.pl (N.G.-C.); mmanka@min-pan.krakow.pl (M.M.)

3 Faculty of Management, AGH University of Science and Technology, Mickiewicza 30, 30-059 Krakow, Poland; kulczycka@min-pan.krakow.pl

4 Department of Chemical Engineering, University of Patras, 26504 Patra, Greece; marinis@chemeng.upatras.gr (D.M.); efarsari@chemeng.upatras.gr (E.F.); lef@chemeng.upatras.gr (E.A.)

5 Institute of Management, Poznan University of Economics and Business, al. Niepodległości 10, 61-875 Poznan, Poland; anna.lewandowska@ue.poznan.pl

6 Department of Air Protection, Silesian University of Technology, Stanisława Konarskiego 22B, 44-100 Gliwice, Poland; marcin.sajdak@polsl.pl

7 Department Thermal Technology, Silesian University of Technology, Stanisława Konarskiego 22, 44-100 Gliwice, Poland

8 Department of Heating, Ventilation and Dust Removal Technology, Silesian University of Technology, Stanisława Konarskiego 20, 44-100 Gliwice, Poland; szymon.sobek@polsl.pl

* Correspondence: nikolina.poranek@polsl.pl

Abstract: Carbon fiber is essential in many industries. Since primary production is highly energy-intensive, recycling technologies are being sought. A goal of the research was to develop at a laboratory scale a chemical recycling method aimed at recovering carbon fiber. Two variants of the method have been established and environmentally compared with a primary production version. Methods: The life cycle assessment methodology has been used to assess and quantify the environmental impacts. The cradle to gate analysis was performed with the functional unit defined as a production of 1 kg of carbon fiber. Results: The best environmental option turned out to be a developed chemical recycling technology named Scenario 1. It is a solvolysis performed using an ambient-pressure-operated batch reactor connected to a reflux condenser and an inert gas supply tank, using an ethylene glycol and potassium hydroxide solution. The worst case appeared to be the second variant of the chemical recycling, named Scenario 2 (plasma-enhanced nitric acid solvolysis). Conclusions: In Scenario 1, a production of the ethylene glycol was recognized as a key environmental driver, while in Scenarios 2 and 3 the energy-related impact was the most influential.

Keywords: carbon fiber; environment; life cycle management; solvolysis; waste management

1. Introduction

The research and development of zero-carbon energy sources have become top priorities for countries worldwide since the beginning of the 21st century, largely due to the ongoing use of fossil fuels and significant greenhouse gas emissions [1]. According to the Ellen MacArthur Foundation report, the implementation of a circular economy (CE) could result in material savings of up to USD 630 billion annually by 2025 [2]. Transitioning to a CE could contribute to a reduction of global greenhouse gas emissions by 9.3 billion tons annually by 2050, which is equivalent to about a quarter of global annual emissions [3].

The production of carbon fiber components is experiencing significant growth, driven by increasing demand for lightweight and durable materials across various industries. In 2023, the global carbon fiber market was valued at approximately USD 7.1 billion and is projected to reach $23.2 billion by 2033, reflecting a compound annual growth rate (CAGR) of 12.6% [4–6].

This upward trend is particularly evident in sectors such as aerospace, automotive, and wind energy, where the emphasis is on reducing structural weight while maintaining high strength and energy efficiency. For instance, in the aerospace industry, carbon fiber's superior strength-to-weight ratio makes it ideal for manufacturing aircraft components, contributing to reduced fuel consumption and emissions. Similarly, in the automotive sector, the use of carbon fiber enhances vehicle performance and fuel efficiency. In the wind energy industry, carbon fiber is utilized in the production of turbine blades, benefiting from its strength and lightweight properties [7].

Overall, the growing environmental awareness and pursuit of technological innovation are further propelling the expansion of the carbon fiber market. Carbon fiber finds widespread applications in the market. It is used, among other things, to manufacture a torque transmission shaft, which serves as a research material [8]. A torque transmission shaft is a mechanical component designed to transfer rotational force (torque) from one part of a system to another, commonly found in machinery, vehicles, and industrial equipment where power needs to be transmitted efficiently and reliably [9]. Its primary functions include transferring power from a source such as an engine or motor to another component like a gearbox or wheels, absorbing and distributing loads without excessive deformation or failure, and ensuring smooth rotation to prevent vibrations and mechanical issues. There are various types of torque transmission shafts, including solid shafts made of steel or alloy for high strength and durability, hollow shafts that reduce weight while maintaining strength, flexible shafts used in systems with misalignment, and splined shafts with grooves or teeth for precise torque transfer without slippage. These shafts must be made of strong, fatigue-resistant materials such as carbon steel, alloy steel, stainless steel, or composite materials for lightweight applications. Key design considerations include diameter and length to withstand torque and bending forces, surface finish to minimize friction and wear, and proper balancing and alignment to prevent vibrations and mechanical failures. Torque transmission shafts are widely used in automotive applications (as drive shafts, axles, and propeller shafts), in industrial machinery (like conveyor belts, pumps, and compressors), in aerospace engineering (where lightweight and strong materials are crucial), and in robotics for precision movement control. A well-designed torque transmission shaft ensures efficient power transfer, longevity, and reliability in mechanical systems [10].

Carbon fiber components are valued for their exceptional strength-to-weight ratio, corrosion resistance, and durability. They consist of 92–99% carbon atoms, and their carbon fiber-reinforced polymers (CFRPs) exhibit high tensile strength (up to 7 GPa) and a tensile modulus ranging from 230 to 600 GPa, making them significantly stronger yet lighter than steel [11,12]. These materials can reduce component weight by up to 70% while maintaining excellent fatigue resistance and thermal stability, withstanding temperatures up to 2000 °C in non-oxidizing environments. However, carbon fiber production is highly energy-intensive, consuming 20–30 times more energy per unit mass than steel production [12,13]. Additionally, recycling CFRPs remains a challenge due to their thermoset resin matrix—current methods such as pyrolysis, solvolysis, and mechanical grinding face efficiency and cost barriers. Due to their exceptional durability, these materials are difficult to recycle, especially carbon fibers embedded in resin, which complicates their recovery and reuse. Since the use of carbon fiber components is becoming increasingly desirable, the amount of waste generated will continue to grow in the market, making it crucial to find a recycling method that provides the greatest environmental relief [13,14]. Table 1 shows examples of different limitations in carbon fiber recycling.

Table 1. Examples of limitations of carbon fiber recycling for selected technologies.

Recycling Method	Description	Key Limitations	Source
Mechanical Recycling	Grinding/shredding CFRP into small particles for fillers or low-grade reuse	Fibers are short and damaged; Loss of mechanical properties; Low-value recycled product	[15]
Thermal Recycling	Pyrolysis or fluidized bed to burn off resin and recover fibers	High energy demand; Potential fiber surface degradation; Emission of VOCs without control	[16]
Chemical Recycling (Solvolysis)	Use of solvents under heat/pressure to dissolve matrix and recover clean fibers	Complex solvent handling; High process cost and time; Challenges in scaling up and purifying recovered oligomers	[17]

CFRP (Carbon Fiber Reinforced Polymer) is a relatively new material on the market. In 2000, global CFRP production was approximately 10,000 tons per year. By 2010, this figure had increased to around 40,000 tons annually. Forecasts suggest that by 2025 production will exceed 200,000 tons per year [18–20].

The lifespan of CFRP products depends on the sector. For example, the estimated service life is 10–20 years in the automotive industry, 20–30 years in aviation, and 20–25 years in the wind energy sector. It is estimated that more than 172,500 tons of CFRP waste are generated [18,21].

Carbon fiber components are built from multi-layer laminates made of carbon fibers and a thermosetting polymer matrix. Adhesives or foams, used to partially fill certain parts, may also be involved in their assembly. Resin is a thermosetting material that undergoes a permanent chemical transformation, unlike thermoplastics, which presents a major challenge for recycling. Parts of boats, cars, and airplanes are made from these highly durable composite materials. Wind turbines account for only 5–10% of global composite material consumption. Therefore, research into the recycling of materials containing carbon fiber will also contribute to the development of recycling in other industries [22].

The waste management hierarchy includes options ranked from least to most preferred: disposal → energy recovery → recycling → repurposing for other uses, reuse → reducing waste generation and prevention. Figure 1 presents a diagram of the available methods for processing wasted carbon fiber components [23].

As indicated in Figure 1, the wasted carbon fiber components may be reused (after preparing for re-use), recycled, or be a subject of energy recovery. A valuable option is a primary (closed loop) recycling where the wasted components are mechanically processed to obtain a material with the same function and with no losses in quality. Secondary (open loop) recycling is also a possible scenario. In this case, the waste is mechanically transformed into materials with different properties and functions. The carbon fiber components can be shredded and used as fillers, for instance, in concrete mixtures. One ton of carbon fiber component waste can reduce CO_2 emissions by 110 kg and save 461 kg of raw materials compared to standard cement production. However, this approach wastes the structural properties of the material. Cut materials can also serve as fillers for vehicles or bicycles (reducing weight and lowering CO_2 emissions) [24,25]. Tertiary recycling is another solution where the wasted materials are used as a feedstock in a process aimed at creating chemicals or fuels. An example of tertiary recycling is pyrolysis. This process involves heating the material to a temperature of 400–700 degrees Celsius in a low-oxygen environment. This breaks down the resin into simpler substances, allowing the fibers to be recovered and reused. Research shows that after heating the strength of the fibers can decrease by as much as 50%. Additionally, the pyrolysis process is energy-intensive and does not allow for resin recovery [26,27]. However, pyrolysis is being developed further. A Dutch team has developed a technology for extracting fibers from blades by briefly introducing oxygen into the process. Before oxygen is introduced, the resin and fibers are separated through oxygen-free pyrolysis. These recovered fibers can be used in automotive parts. Fibers recovered using the two-stage technology demonstrated 19% better tensile strength and 43% better fracture toughness compared to fibers recovered through single-step high-temperature pyrolysis [27,28].

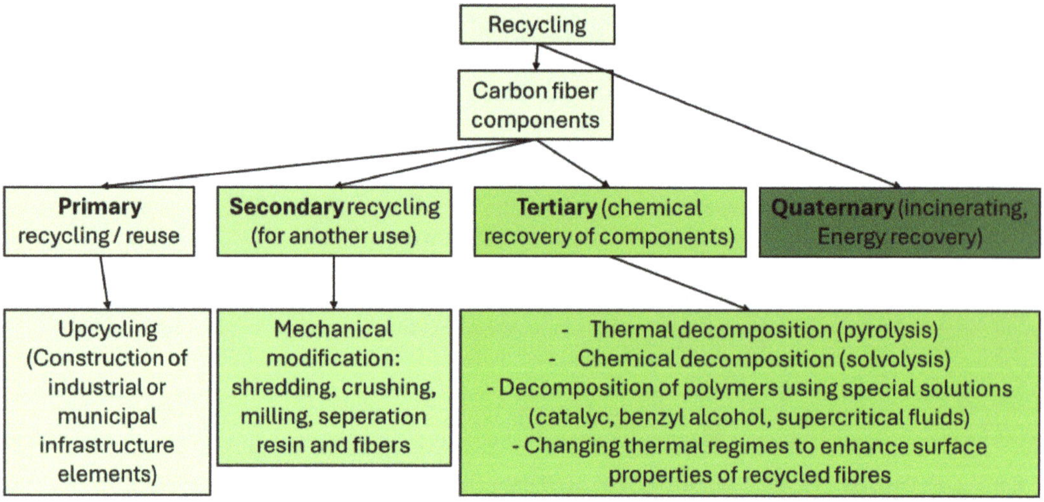

Figure 1. Possible recycling methods: Reuse, repurpose, decomposition, energy recovery [23].

Another tertiary recycling method is microwave processing. In this process, glass fibers can lose about 25% of their strength. An ideal recycling process should enable the recovery of both fibers and resin to produce new products. At the same time, the strength of the fibers should be comparable to that of virgin fibers. Chemical decomposition is another option for tertiary recycling. A review of the literature suggests that the solvolysis process may hold potential for recycling of the carbon fiber components. In the solvolysis process, the polymer matrix of the composite is swelled using volatile, high-boiling, low-molecular-weight solvents or solvent mixtures. Depending on the type of resin used in the composite, a catalyst may also be required. This catalyst, aided by the loosened composite matrix, can more effectively penetrate deep into the material and facilitate its degradation. The process occurs at an elevated temperature, restricted by the solvent's boiling point, within a nitrogen-rich atmosphere at atmospheric pressure [29,30].

Chemical recycling has the potential to provide cleaner fibers while preserving more strength due to the lower temperatures involved compared to pyrolysis. Various solutions can be employed to break down polymers, such as catalytic processes, nitric acid, ammonia, glycol, and ethanol. In solvolysis processes, the phenomenon of the critical temperature can be utilized. This is the temperature above which the difference in density between the gaseous and liquid states of a substance disappears, making it impossible to liquefy the gas despite increasing pressure.

Solvolysis processes can also be combined with other techniques, such as microwaves or pyrolysis. Microwave-assisted chemical recycling processes can be used to recover glass fibers from composite waste, with the recovered fibers retaining 93–99% of the strength of virgin fibers. The recovered glass fibers demonstrated 99% tensile strength, 93% Young's modulus, and 95% strain-to-failure compared to virgin fibers [28]. Different technologies consume varying amounts of materials, reagents, and energy needed for processing. This affects not only costs but also the environmental impact [28–30].

The aim of this study is to perform a life cycle assessment (LCA) comparing two recycling methods for carbon fiber-reinforced composites with the production of components made from virgin carbon fiber. The recovered carbon fibers originate from the recycling of a torque transmission shaft, as described in a later section of the article. The data for virgin carbon fiber were sourced from the Ecoinvent database (entry: Carbon fiber reinforced plastic, injection molded {GLO} | carbon fiber reinforced plastic, injection molded | Cut-off, U). The analysis was carried out using SimaPro software (PRé Sustainability, Stationsplein 121, Amersfoort, The Netherlands).

2. Materials and Methods

Environmental analyses are the foundation for a rational, evidence-based decision-making process. This is an important issue, closely linked to the economy. Proper decision-making processes determine vast financial flows directed towards investments. They also influence the mitigation of negative environmental impacts, which can be avoided or minimized in a much more effective way than by implementing many advanced technologies [29,30].

Environmental costs are difficult to estimate due to the lack of market pricing, and thus they are considered typical external costs that occur when one party's profit or utility is impaired by the actions of another, without compensation. For this reason, purely economic approaches, which require the monetization of all effects, are often debated. A significant challenge in estimating environmental effects is the considerable time gap between the occurrence of an event and its environmental consequences. For example, the effects (and thus the environmental costs) of phenomena such as global climate change will be felt by many generations to come [31–33].

The life cycle assessment (LCA) is a sophisticated ecodesign tool and an element of environmental management systems [34,35]. The group of ISO 14040s standards include guidelines and requirements for performing LCA [36,37]. The LCA can be characterized by the following features: life cycle oriented, input–output related, environmentally multicriterial, impact assessment based, quantitative, scientifically grounded, standardized, and widely recognized. The LCA enables assessing and quantifying potential environmental impacts in the life cycle of goods and services. Additionally, it may be used to evaluate technologies and organizations. The analysis includes four phases: a goal and scope definition, a life cycle inventory (LCI), a life cycle impact assessment (LCIA), and an interpretation. More information about life cycle assessment methodology can be found in the literature [38,39].

2.1. Goal and Scope Definition

A main goal of the presented research was to develop in a laboratory scale a chemical recycling method aimed at recovering from wasted material a reclaimed carbon fiber. The method is based on the solvolysis and has been assumed as a potentially environmentally friendly alternative for a primary production. The wasted materials investigated for the different routes of the recycling were the torque transmission shafts (B&T Composites, Florina, Greece) cut to 20 cm length and 8 cm outer diameter comprising carbon fiber epoxy composite matrix. The samples were processes in the "as received" state, without any further cutting, milling, or other pretreatment methods.

Two variants of the recycling method have been assumed. In Scenario 1, a solvolysis process for the waste composite recycling was performed using a 0.5 and 2 L ambient pressure operated batch reactors connected to a reflux condenser and an inert gas supply tank (N2), using an ethylene glycol (EG) and potassium hydroxide (KOH) solution. The process was presented and discussed in detail in [27]. In Scenario 2, solvolysis was performed using plasma-enhanced nitric acid solvolysis of carbon fiber component composite scrap. This process synergistically combines the traditional nitric acid solvolysis with plasma aiming at the significant increase in the recovery rate of CFs. More details of the process can be found in [40].

The specific goals of the comparative LCA study are as follows:

- To evaluate the potential environmental impact and identify key drivers for the developed recycling processes (Scenario 1 and 2);
- To compare the recycling of the secondary carbon fiber with the primary production.

The potential audience of the results are all parties interested in ecodesign, carbon fiber production, and waste management.

The function of the recycling processes can be perceived from two perspectives: as disposing of waste (the end of life) and as generating secondary materials (the cradle). In our case study, there is a focus on the second function and the functional unit (FU) has been defined as a production of 1 kg of carbon fiber. The multifunctionality of recycling calls for the allocation procedure [41]. The developed recycling process combines two life cycles: the previous life cycle where the primary carbon fiber was incorporated in the torque transmission shafts and the subsequent life cycle where the reclaimed secondary carbon fiber will be used in similar or other applications. The carbon fiber included in the transmission shafts was based on a primary material, and for this reason, a virgin production was previously used for this purpose. This virgin material was present in the wasted shafts that were a subject of recycling. A heart of allocation is the following question, "how to allocate the environmental burdens related to the virgin production and to the recycling process between the previous and subsequent life cycles?". There can be different answers, as many allocation approaches exist. In our case study, the cut-off approach has been used. This means that 100% virgin production and 0% recycling has been allocated to the previous life cycle (transmission shafts), and 0% virgin production and 100% recycling has been allocated to the subsequent life cycle (where the reclaimed carbon fiber will be used). No credits (avoided burdens) have been accounted for the recycling.

The LCA analysis has been scoped from the cradle to gate. Figure 2 presents general system boundaries. More detailed information is presented in process diagrams (Figures 3 and 4). A transport of the wasted material to the laboratory has been included while a quality assessment between the primary and the reclaimed carbon fibers has been excluded from the analysis.

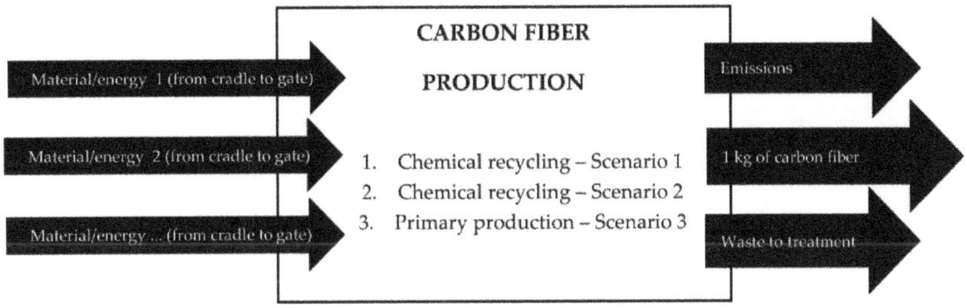

Figure 2. The system boundaries of LCA.

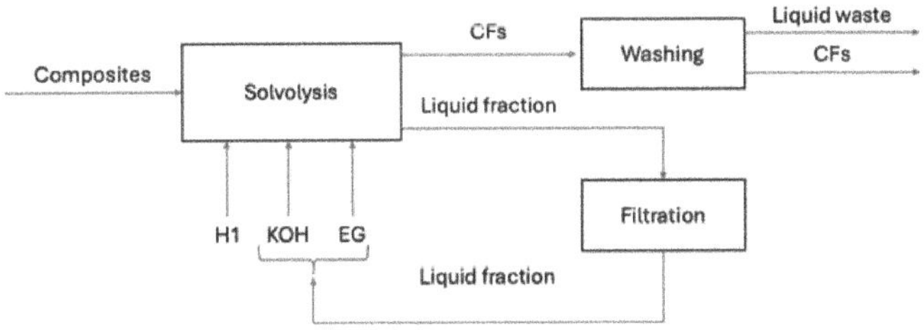

Figure 3. Schematic diagram of the solvolysis process for Scenario 1.

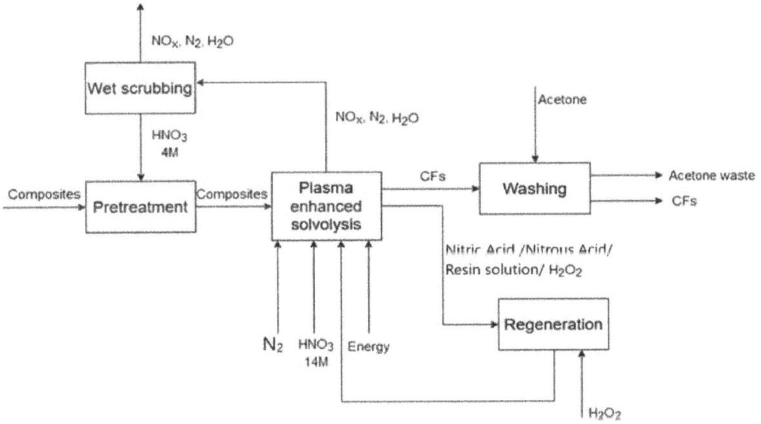

Figure 4. Schematic diagram of the solvolysis process for Scenario 2.

2.2. Life Cycle Inventory (LCI)

In the presented case study, two data sources have been used. The research presented in this study concerns torque transmission shafts, which were used as the representative waste material for evaluating various recycling methods. For both variants of recycling process (Scenarios 1 and 2) activity data have been modeled with primary information based on stoichiometric calculations and reactions made in the laboratory. The data can be recognized as good quality in terms of temporal, geographical, and technological scope [42]. It means that they are consistent with the goal of the study. A main limitation lies in their reliability and completeness, as the data have been calculated based on activities made in a laboratory, not on a commercial scale. It makes it probable that the activity data for Scenarios 1 and 2 could be overestimated. The activity data for the primary production (Scenario 3) have been taken from the ecoinvent dataset carbon fiber reinforced plastic, injection molded (GLO). All background processes have been modeled with the ecoinvent v3.10 secondary data. Schematic process diagrams and inventory results for Scenarios 1 and 2 are presented in Figures 3 and 4 and Tables 2 and 3.

Table 2. Inventory results for recycling process-Scenario 1 (per functional unit).

Name of Input/Output	Amount	Unit
Inputs		
Wasted carbon fiber component (B&T Composites)	1.44	kg
Transport of wasted carbon fiber component, lorry 16–32, Euro 5 (100 km)	1.44×10^2	kgkm
Solvent (closed looped circulation)	5.74	kg
Ethylene glycol	8.97×10^{-1}	kg
Potassium Hydroxide	8.97×10^{-2}	kg
Electricity (low voltage, at grid, Poland)	2.16	kWh
Outputs		
Reclaimed carbon fiber	1.00	kg
Solvent (closed looped circulation)	5.74	kg
Liquid waste	9.87×10^{-1}	kg

Table 3. Inventory results for recycling process-Scenario 2 (per functional unit).

Name of Input/Output	Amount	Unit
Inputs		
Wasted carbon fiber component (B&T Composites)	1.67	kg
Transport of wasted carbon fiber component, lorry 16–32, Euro 5 (100 km)	1.67×10^2	kgkm
Nitric Acid 65%/H_2O_2 30% (closed looped circulation)	2.77×10^1	kg
Nitrogen gas	1.63×10^1	kg
Acetone	4.00	kg
Electricity (low voltage, at grid, Poland)	7.17×10^1	kWh
Outputs		
Reclaimed carbon fiber	1.00	kg
NO_X gasses that enter the wet scrubber	6.83	kg
NO_X gasses that escape the wet scrubber	5.00	kg
Wet scrubbing solution (HNO_3)	1.83	kg
Nitric Acid/Nitrous Acid/Resin solution/H_2O_2 (closed looped circulation)	2.08×10^1	kg
Nitrogen gas	1.63×10^1	kg
Acetone/resin residuals	4.00	kg

2.3. *Life Cycle Impact Assessment (LCIA)*

The life cycle impact assessment calculations have been made with the Environmental Footprint EF 3.1 method (adapted) v3.10, available in the SimaPro software. The LCIA method has been developed by the European Commission as an element of Environmental Footprints methodology [43]. With this method, the environmental impact is to be assessed within 16 impact categories reflecting various environmental problems: Acidification, Climate change, Ecotoxicity-freshwater, Particulate matter, Eutrophication-marine, Eutrophication-freshwater, Eutrophication-terrestrial, Human toxicity-cancer, Human toxicity-non-cancer, Ionizing radiation, Land use, Ozone depletion, Photochemical ozone formation, Resource use-fossils, Resource use-minerals and metals and Water use. LCIA results are presented in Section 3.

3. Results

Two sorts of the results are presented in Section 3: characterized (where each impact category has its own unit) and weighted (all impact categories have the same unit-milipoints). In both cases the same interpretation rule works. The higher the value of the indicator, the higher the negative environmental impact is. Table 4 and Figure 5 show the weighted results for production of 1 kg of carbon fiber with three processes: the chemical recycling (Scenario 1 and Scenario 2) and primary production (Scenario 3). The cumulated impact (single sore) is 0.33 mPt for Scenario 1, 21.44 mPt for Scenario 2, and 5.85 mPt for Scenario 3 (the primary production). The results clearly show that Scenario 1 has the lowest environmental impact while Scenario 2 is the worst option. In Table 4, the cells marked in gray indicate the most relevant impact categories (the categories with the highest contribution).

Table 4. The environmental impact per functional unit-weighted results for Scenarios 1–3.

Impact Category	Scenario 1		Scenario 2		Scenario 3 (Virgin)	
	mPt	%	mPt	%	mPt	%
Single Score	**3.31×10^{-1}**	**100%**	**2.14×10^{1}**	**100%**	**5.85**	**100%**
Climate change	1.43×10^{-1}	43%	2.38	11%	2.27	39%
Resource use, fossils	8.77×10^{-2}	27%	1.36	6%	1.25	21%
Particulate matter	2.09×10^{-2}	6%	1.38	6%	9.24×10^{-1}	16%
Ecotoxicity, freshwater	1.79×10^{-2}	5%	8.24×10^{-2}	0.4%	6.21×10^{-2}	1.1%
Acidification	1.58×10^{-2}	5%	3.41	16%	5.02×10^{-1}	9%
Photochemical ozone formation	1.58×10^{-2}	5%	6.85	32%	2.71×10^{-1}	5%
Water use	1.07×10^{-2}	3%	1.60×10^{-1}	0.7%	1.04×10^{-1}	2%
Eutrophication, terrestrial	6.50×10^{-3}	2%	2.38	11%	1.55×10^{-1}	3%
Eutrophication, marine	4.20×10^{-3}	1.3%	3.04	14%	1.03×10^{-1}	2%
Human toxicity, non-cancer	2.70×10^{-3}	0.8%	1.45×10^{-1}	0.7%	7.58×10^{-2}	1.3%
Eutrophication, freshwater	2.20×10^{-3}	0.7%	1.67×10^{-1}	0.8%	5.53×10^{-2}	0.9%
Ionizing radiation	1.30×10^{-3}	0.4%	1.96×10^{-2}	0.1%	2.86×10^{-2}	0.5%
Land use	9.00×10^{-4}	0.3%	1.45×10^{-2}	0.1%	2.13×10^{-2}	0.4%
Human toxicity, cancer	8.00×10^{-4}	0.2%	3.63×10^{-2}	0.2%	1.89×10^{-2}	0.3%
Resource use, minerals and metals	4.00×10^{-4}	0.1%	3.60×10^{-3}	0.02%	5.60×10^{-3}	0.1%
Ozone depletion	1.00×10^{-4}	0.03%	7.00×10^{-4}	0.00%	5.00×10^{-4}	0.0%

Table 5 and Figure 6 present the environmental impact as the characterized results. It can be observed that Scenario 1 is the best option for all impact categories.

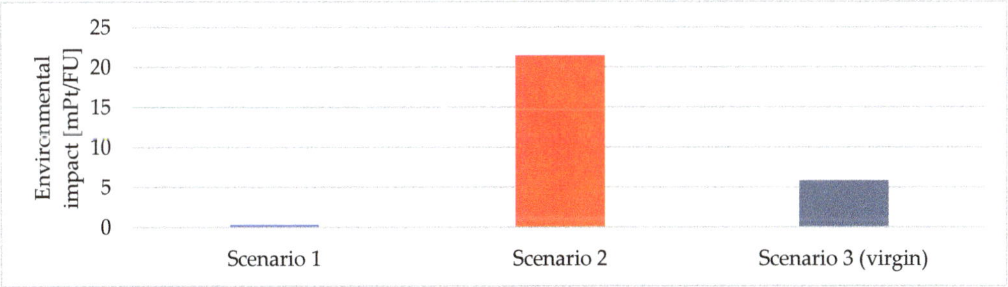

Figure 5. The potential environmental impact (single score) for the three scenarios [mPt/FU].

Table 5. The environmental impact per functional unit-characterized results for Scenarios 1–3.

Impact Category	Unit	Scenario 1	Scenario 2	Scenario 3 (Virgin)
Acidification	mol H^+ eq	1.42×10^{-2}	3.06	4.50×10^{-1}
Climate change	kg CO_2 eq	5.12	8.54×10^{1}	8.16×10^{1}
Ecotoxicity, freshwater	CTUe	5.30×10^{1}	2.43×10^{2}	1.83×10^{2}
Particulate matter	disease inc.	1.39×10^{-7}	9.17×10^{-6}	6.14×10^{-6}
Eutrophication, marine	kg N eq	2.80×10^{-3}	2.01	6.77×10^{-2}
Eutrophication, freshwater	kg P eq	1.00×10^{-4}	9.60×10^{-3}	3.20×10^{-3}
Eutrophication, terrestrial	mol N eq	3.08×10^{-2}	1.14×10^{1}	7.40×10^{-1}
Human toxicity, cancer	CTUh	6.48×10^{-10}	2.94×10^{-8}	1.53×10^{-8}
Human toxicity, non-cancer	CTUh	1.89×10^{-8}	1.01×10^{-6}	5.30×10^{-7}
Ionizing radiation	kBq U-235 eq	1.13×10^{-1}	1.65	2.41
Land use	Pt	9.19	1.50×10^{2}	2.20×10^{2}
Ozone depletion	kg CFC11 eq	8.30×10^{-8}	5.81×10^{-7}	4.15×10^{-7}
Photochemical ozone formation	kg NMVOC eq	1.35×10^{-2}	5.86	2.32×10^{-1}
Resource use, fossils	MJ	6.86×10^{1}	1.06×10^{3}	9.76×10^{2}
Resource use, minerals and metals	kg Sb eq	3.37×10^{-7}	3.03×10^{-6}	4.72×10^{-6}
Water use	m^3 depriv.	1.44	2.15×10^{1}	1.40×10^{1}

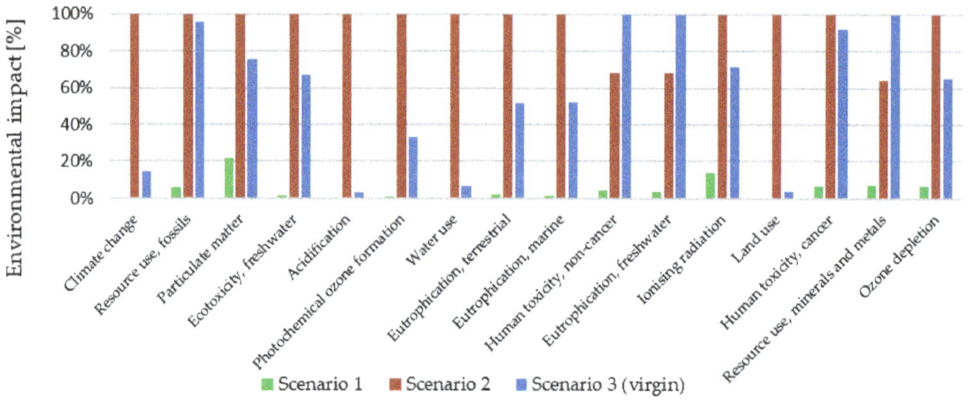

Figure 6. The environmental impact (characterized) for three Scenarios under analysis. Source: SimaPro software.

Based on the data from Table 4 and Figure 6, the following key conclusions can be drawn. Scenario 1 clearly has the lowest environmental impact across all analyzed impact categories, including CO_2 emissions (5.12 kg CO_2 eq), acidification (0.0142 mol H^+ eq), and fossil resource use (68.6 MJ). This makes it the most environmentally friendly option among the three considered. Scenario 2 generates the highest environmental burdens in most categories, particularly in terms of climate change (85.4 kg CO_2 eq), eutrophication (freshwater, marine, and terrestrial), and land use (150 Pt). Scenario 3 ranks in the middle, falling between Scenarios 1 and 2. Although it has a lower impact than Scenario 2 in most categories, it is still significantly more environmentally burdensome than Scenario 1, especially in terms of CO_2 emissions (81.6 kg CO_2 eq) and fossil fuel consumption (976 MJ). Figure 7 shows flow diagrams of the three scenarios in the impact category Climate Change.

For Scenario 1, the most influential factor is Ethylene glycol, which accounts for over 33% of the result. In Scenario 2, electricity has the greatest impact on the results, representing over 80% of the total impact. In Scenario 3, electricity also has the largest environmental impact, but it accounts for 50% of the result (209 MJ).

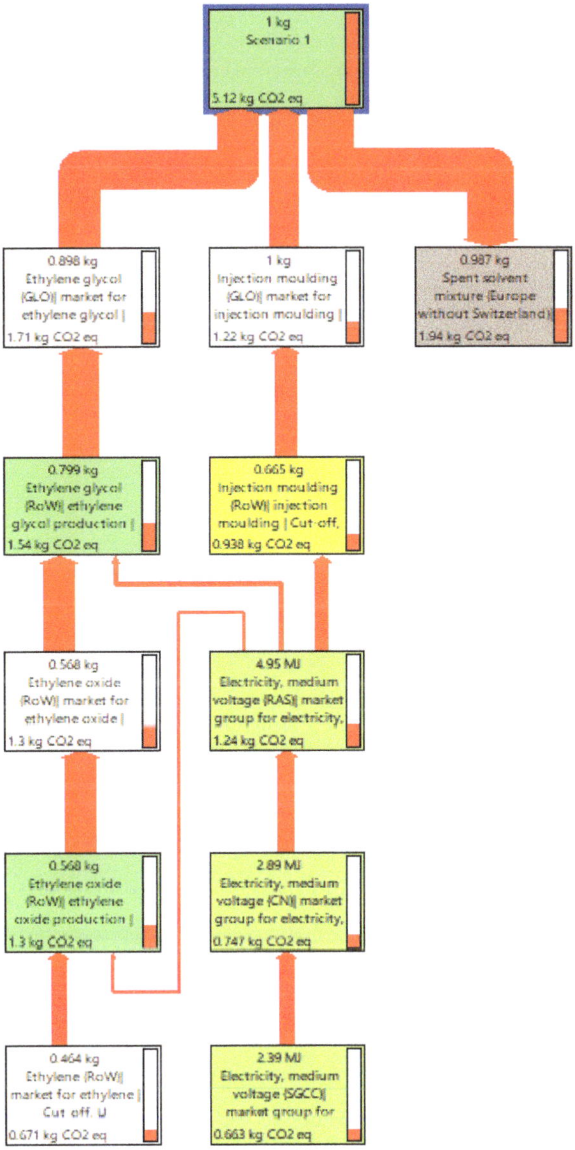

(a)

Figure 7. *Cont.*

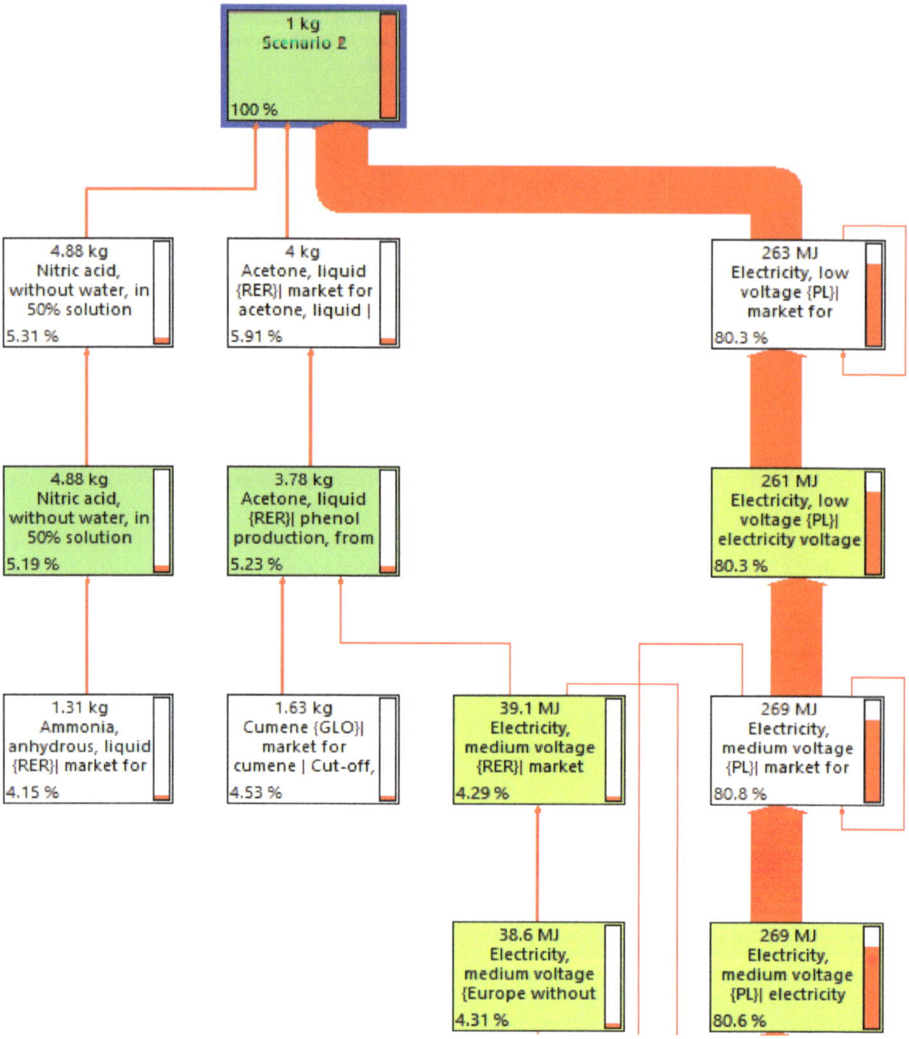

(**b**)

Figure 7. *Cont.*

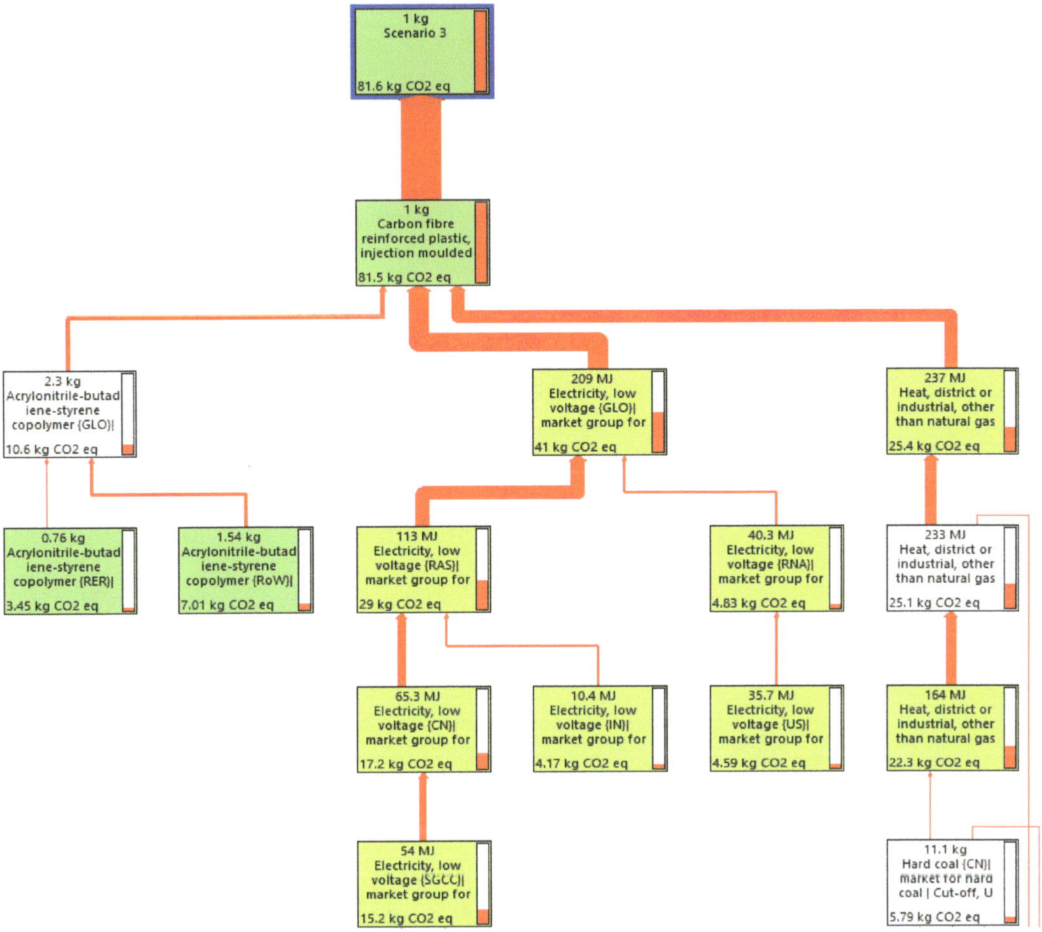

(**c**)

Figure 7. Flow diagrams reflecting the processes contributing mostly to the Climate Change (kg CO_2 eq/FU) (**a**) Scenario 1; (**b**) Scenario 2; (**c**) Scenario 3. Source: SimaPro software.

4. Socio-Economic Analysis

Recycling significantly influences the economy through job creation, increased competitiveness, cost savings, and overall improvements in societal well-being. It generates employment across various sectors, including waste collection, material processing, and manufacturing of new products from recycled materials, with industries such as wind energy and automotive manufacturing directly benefiting from secondary raw materials. This process not only reduces dependence on imported resources but also lowers production costs, enhancing the economy's resilience. Furthermore, innovation in recycling technologies fosters the development of new industrial sectors, driving long-term growth and sustainability.

Despite the costs of establishing waste collection systems and processing facilities, recycling brings financial benefits by reducing landfill usage and lowering expenses tied to raw material extraction. Beyond economics, recycling boosts environmental awareness through education and public campaigns, encouraging more sustainable consumption habits. Health benefits also arise from decreased pollution, leading to better air, water, and soil quality, and lowering pollution-related disease rates. In terms of quality of life, recycling supports cleaner surroundings, reduces illegal dumping, and promotes the circular economy by increasing access to valuable secondary materials.

However, challenges persist, such as high infrastructure costs, uneven access to recycling facilities, and low public motivation due to insufficient information systems. Addressing these barriers is essential to maximizing the socio-economic potential of recycling and building a more sustainable, circular economy. Table 6 shows the possible impact of recycling on the economy and the community.

Table 6. Possible impact of recycling on the economy and the community.

Category	Impact/Description
Job creation	Recycling generates employment in sectors such as collection, processing, and the production of new products from secondary raw materials.
Increase in economic competitiveness	The use of secondary raw materials reduces production costs and makes the economy less dependent on the import of primary raw materials.
	Innovations in recycling technology drive the development of new industrial sectors.
Costs and savings related to recycling	Costs of selective waste collection systems and processing technologies.
	Savings resulting from a reduced need for waste landfilling and lower costs of extracting natural resources.
Social environmental awareness	Environmental education and changing consumption habits.
	Social campaigns and local government initiatives promoting waste segregation.
Health benefits	Reduced environmental pollution improves air, water, and soil quality.
	Lower emissions of harmful substances decrease the incidence of pollution-related diseases.
Impact on quality of life	Recycling promotes a cleaner environment and reduces the number of illegal dumpsites.
	Increased availability of secondary raw materials supports the development of a circular economy.
Environmental awareness	Increasing environmental awareness influences the formation of more sustainable consumption habits, promotes waste segregation, and encourages responsible resource management. Environmental education and social campaigns inspire people to take everyday actions for environmental protection, such as reducing plastic use, reusing products, or choosing local and eco-friendly goods. Greater awareness makes society more likely to demand systemic change, support pro-environmental initiatives, and pressure companies and institutions to adopt more sustainable strategies. In the long run, this leads to improved quality of life, reduced pollution, and healthier ecosystems, benefiting both people and the planet.

Carbon Fiber Market

The carbon fiber market has been growing dynamically in recent years, driven mainly by demand in the aerospace, automotive, and wind energy sectors. In 2019, the market was valued at USD 4.7 billion, with forecasts suggesting it could grow to USD 13.3 billion by 2029, representing a compound annual growth rate (CAGR) of 11% [44].

Figure 8 shows the size of the global carbon fiber market from 2020 to 2030, broken down by raw material: PAN (polyacrylonitrile) and pitch, with values in USD billion. The data indicate steady market growth, with the total size increasing from USD 3.7 billion in 2020 to USD 10.5 billion in 2030. PAN remains the dominant raw material, consistently accounting for the majority of the market share, while pitch contributes a smaller but gradually growing portion. The market experiences consistent year-over-year growth, with a notable compound annual growth rate (CAGR) of 10.9% from 2025 to 2030. This expansion reflects rising demand for carbon fiber across industries such as aerospace, automotive, wind energy, and sporting goods.

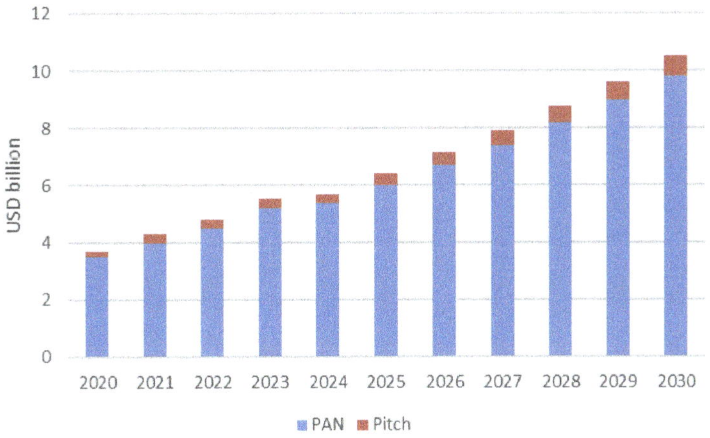

Figure 8. Carbon fiber market.

In terms of value, the global carbon fiber market is expected to reach approximately USD 8.5 billion by 2030 [45]. Earlier estimates from 2019 suggested the market could hit USD 8 billion by 2026, with a CAGR of 10.8%.

Geographically, the market spans key regions, including Europe, Asia-Pacific, North America, the Middle East and Africa, and Latin America. While the demand is rising, the market faces challenges, including the high cost of carbon fiber and the lack of standardized manufacturing technologies, which may slow down wider adoption.

The competitive landscape highlights Toray Industries Inc. as a leading player, with various other companies positioned across different levels of market share and product footprint. The ecosystem involves raw material providers like Minglang and Lanhai, carbon fiber manufacturers such as Toray and SGL Carbon, and end users including major global brands like Teijin, Mercedes, and Audi.

Overall, the data suggests that while the carbon fiber market holds immense growth potential, its expansion depends on addressing cost barriers and streamlining production processes to meet rising global demand.

The key drivers of this growth are the push to reduce vehicle and structure weight to improve energy efficiency and lower emissions. Carbon fibers, thanks to their lightweight and high-strength properties, are widely used in aerospace, automotive manufacturing, wind energy, and sports and recreational equipment production [4–7].

5. Discussion

This study was conducted to compare two carbon fiber recycling methods with virgin production using environmental gate-to-gate life cycle assessment (LCA) methodologies. The results reveal that not all recycling processes have a lower environmental impact than virgin carbon fiber production. However, the research identified that the most environmentally favorable method is a developed chemical recycling technology, referred to as Scenario 1. This process involves solvolysis performed in ambient-pressure batch reactors, equipped with a reflux condenser and an inert gas supply, using a solution of ethylene glycol and potassium hydroxide.

Scenario 2 is still under development, and therefore it is possible that it could also demonstrate a lower environmental impact than virgin carbon fiber production. Sensitivity analysis plays an important role in all scenarios, helping to identify and reduce environmental hotspots. In Scenario 1, ethylene glycol emerges as the most impactful factor, contributing over 33% to the total environmental burden. In Scenario 2, electricity has the highest impact, accounting for over 80% of the total footprint. Similarly, in Scenario 3, electricity remains the most significant factor, representing 50% of the impact (equivalent to 209 MJ).

If energy is the hotspot, it can potentially be replaced with renewable or waste-derived energy sources. If the hotspot is a chemical substance, exploring potential substitutes could be a viable solution. These considerations could serve as the foundation for future research directions.

The article also discusses other recycling processes, as carbon fiber recycling is increasingly relevant due to growing market demand. The expanding use of carbon fiber across industries—including aerospace, automotive, wind energy, and sports equipment—is expected to substantially increase the volume of waste generated both during production and at the end of product life. Therefore, there is a pressing need to develop and deploy innovative, sustainable strategies to manage this waste stream.

The average annual growth in demand for carbon fiber-reinforced composites (CFRCs) is estimated at approximately 11%.

One key strategy is to explore and refine carbon fiber recycling methods. However, not all recycling processes are inherently environmentally beneficial, which is why conducting a life cycle assessment (LCA) is crucial. LCA enables a comprehensive evaluation of environmental impacts across the entire process—including resource use, energy consumption, emissions, and waste generation. Through such analysis, it becomes possible to assess whether a recycling method truly offers environmental advantages compared to traditional disposal or the use of virgin materials.

A socio-economic analysis highlights a gap in the market that could be filled if the carbon fiber recycling industry expands. Potential benefits include the creation of new jobs, increased market competitiveness, and reduced pollution levels. Given that the global carbon fiber market is projected to reach approximately USD 8.5 billion by 2025, future studies should explore the exact number of new jobs that could be generated and assess the health impact on workers from implementing cleaner recycling technologies.

It is especially important to evaluate a specific recycling process in direct comparison with virgin carbon fiber production. The global production of virgin carbon fiber is dominated by several major manufacturers known for their advanced materials and extensive production capacities. Leading companies in this sector include Toray Industries (Tokyo, Japan), one of the largest and most established producers [46]; Zoltek (a subsidiary of Toray, St. Louis, MI, USA), specializing in industrial-grade carbon fibers; Mitsubishi Chemical (Tokyo, Japan), known for high-performance fibers used in aerospace and automotive industries [47]; Hexcel Corporation (Stamford, CT, USA), a key supplier for aerospace composites [48]; and SGL Carbon (Wiesbaden, Germany), which serves diverse markets including automotive and wind energy [49]. These producers contribute significantly to the global supply chain, driving innovation and scaling production to meet growing demand. However, for the purposes of life cycle assessment and environmental impact calculations in this study, secondary data from the Ecoinvent database was employed to ensure consistency and comparability. This is particularly relevant in light of the new ISO 14075:2024 standard published at the end of 2024, which reinforces the growing trend in economic thinking—not solely focused on profit, but increasingly on environmental sustainability and human well-being, including worker health.

Nevertheless, due to the high cost of virgin carbon fiber, there is a growing need to develop new and more cost-effective solutions. Recycling carbon fiber composites plays a crucial role in addressing this challenge. Although the concept of recycling is not entirely new, several companies have already established industrial-scale recycling operations—for example, ELG Carbon Fibre Ltd., Coseley, UK [50]. Their efforts not only demonstrate the feasibility of recycling but also highlight the ongoing need for improved, economically viable recycling technologies to meet increasing demand sustainably.

While the composites industry must thoroughly evaluate the economic and environmental aspects of carbon fiber composite recycling, it is equally important to acknowledge the critical role of the resin component. The type of resin used significantly influences the choice and effectiveness of the recycling method. Although this factor is not addressed in the current article, future research may place strong emphasis on resin analysis, as it represents a key element in developing efficient and sustainable recycling strategies.

In conclusion, while carbon fiber recycling is a promising path forward, only those solutions that demonstrate environmental benefits or are cleaner and safer for human health (as validated through tools like LCA) should be prioritized for large-scale implementation.

6. Conclusions

The recycling of carbon fiber-based components aligns with the principles of a circular economy and sustainable development. One method for recovering carbon fibers from components is solvolysis. However, research indicates that not all recycling processes necessarily benefit the environment—in some cases, producing material from virgin resources may have a lower environmental impact than recycling. Nevertheless, proper process optimization and hotspot identification can improve LCA results. Studies confirm the importance of environmental analyses to select processes that truly reduce environmental impact.

The results show that, at the current stage of research, among the two analyzed recycling (solvolysis) processes, Scenario 2 has the highest negative environmental impact compared to carbon fiber production from virgin materials (Scenario 3). Additionally, Scenario 2 has a specific hotspot (electricity consumption by water bathes used to generate heat), which could be optimized, for example, by utilizing waste heat.

Scenario 2 is still in the development stage, and its potential is continuously improving, offering opportunities for better performance and environmental outcomes through ongoing optimization and waste minimization. On the other hand, Scenario 1 provides environmental benefits compared to virgin fiber production.

The presented research constitutes a preliminary approach to the topic of chemical recycling of composite waste and demonstrates the potential of the studied process for their disposal and the recovery of valuable oligomers and fibers (in this example, carbon fibers). Our further research will focus on increasing the number of cycles in which the liquid solvolysis products are recirculated into the process, aiming to concentrate the oligomers released from the resin matrix. In the next stage, we will develop a method for purifying these oligomers and using them for the synthesis of new resin. The results have been presented at TRL 1–3, and future scale-up will enable us to address challenges related to energy recovery.

Author Contributions: Conceptualization, N.P. and K.P.; methodology, N.P., K.P., N.G.-C., M.M., J.K. and A.L.; software, N.P., K.P., N.G.-C., M.M., J.K. and A.L.; validation, N.P., K.P., N.G.-C., M.M., J.K., D.M., E.F., E.A., A.L., M.S., S.W. and S.S.; formal analysis, N.P., K.P., J.K., A.L., M.S. and S.W.; investigation, N.P., N.G.-C. and M.M.; resources, N.P., K.P., D.M., E.F., E.A., M.S., S.W. and S.S.; data curation, N.P. and K.P.; writing—original draft preparation, N.P., N.G.-C. and M.M.; writing—review and editing, N.P., N.G.-C., M.M., A.L., M.S. and S.W.; visualization, N.P., K.P., J.K. and A.L.; supervision, N.P., K.P., J.K. and A.L.; project administration, N.P. and K.P.; funding acquisition, N.P. and K.P. All authors have read and agreed to the published version of the manuscript.

Funding: The Silesian University of Technology funded this research: 08/030/BK_25/0151. This research was partially funded by the Horizon Europe project "European recycling and circularity in large composite structures" (EuReComp), under Grant Agreement no. 101058089.

Institutional Review Board Statement: Not applicable.

Informed Consent Statement: Not applicable.

Data Availability Statement: The original contributions presented in this study are included in the article. Further inquiries can be directed to the corresponding author.

Conflicts of Interest: The authors declare no conflicts of interest.

References

1. Alwaeli, M.; Gołaszewski, J.; Pizoń, J.; Andrzejewska, A.; Szwan, P.; Ceglarz, K.; Buchta, W.; Jachnik, T.; Złotoś, J. Weryfikacja przydatności kruszywa produkowanego z odpadów hutniczych do produkcji betonu wraz z analizą LCA (ang. Verification of the suitability of aggregate produced from metallurgical waste for concrete production, including LCA analysis). *PM News* **2019**, *19*, 28. (In Polish)
2. Towards the Circular Economy Vol. 1: An Economic and Business Rationale for an Accelerated Transition. The Ellen MacArthur Foundation's Report. 2013, pp. 1–98. Available online: https://www.ellenmacarthurfoundation.org/towards-the-circular-economy-vol-1-an-economic-and-business-rationale-for-an (accessed on 12 November 2024).
3. Completing the Picture: How the Circular Economy Tackles Climate Change. The Ellen MacArthur Foundation. Available online: https://www.ellenmacarthurfoundation.org/completing-the-picture (accessed on 19 November 2024).
4. Carbon Fiber Market Projected to Grow Significantly as Increasing Demand for Growing Number of Industries Rises. GlobeNewswire. Available online: https://www.globenewswire.com/news-release/2025/01/14/3009365/0/en/Carbon-Fiber-Market-Projected-to-Grow-Significantly-as-Increasing-Demand-for-Growing-Number-of-Industries-Rises.html (accessed on 7 February 2025).
5. Carbon Fiber Market Worth $23.2 Billion by 2033—Exclusive Report. Marketsand Markets. Available online: https://www.prnewswire.com/news-releases/carbon-fiber-market-worth-23-2-billion-by-2033--exclusive-report-by-marketsandmarkets-302036673.html (accessed on 7 February 2025).
6. Carbon Fiber Market Size & Forecast [Latest]. Marketsand Markets. Available online: https://www.marketsandmarkets.com/Market-Reports/carbon-fiber-396.html (accessed on 7 February 2025).
7. Carbon Fiber Market Size & Trends. Grandviewresearch. Available online: https://www.grandviewresearch.com/industry-analysis/carbon-fiber-market-analysis (accessed on 7 February 2025).
8. Tamez, M.B.A.; Taha, I. A Review of Additive Manufacturing Technologies and Markets for Thermosetting Resins and Their Potential for Carbon Fiber Integration. *Addit. Manuf.* **2021**, *37*, 101748. [CrossRef]
9. Zhu, X.; Li, W. Takagi–Sugeno Fuzzy Model Based Shaft Torque Estimation for Integrated Motor–Transmission System. *ISA Trans.* **2019**, *93*, 14–22. [CrossRef] [PubMed]
10. Zhao, Z.; He, L.; Yang, Y.; Wu, C.; Li, X.; Hedrick, J.K. Estimation of Torque Transmitted by Clutch during Shifting Process for Dry Dual Clutch Transmission. *Mech. Syst. Signal Process.* **2016**, *75*, 413–433. [CrossRef]
11. Ursache, Ş.; Cerbu, C.; Hadăr, A. Characteristics of Carbon and Kevlar Fibres, Their Composites and Structural Applications in Civil Engineering—A Review. *Polymers* **2024**, *16*, 127. [CrossRef]

12. Agarwal, N.; Rangamani, A.; Bhavsar, K.; Virnodkar, S.S.; Fernandes, A.A.A.; Chadha, U.; Srivastava, D.; Patterson, A.E.; Rajasekharan, V. An Overview of Carbon-Carbon Composite Materials and Their Applications. *Front. Mater.* **2024**, *11*, 1374034. [CrossRef]
13. Wang, B.; Ma, S.; Yan, S.; Zhu, J. Readily recyclable carbon fiber reinforced composites based on degradable thermosets: A review. *Green Chem.* **2019**, *21*, 5781–5796. [CrossRef]
14. Borjan, D.; Knez, Ž.; Knez, M. Recycling of Carbon Fiber-Reinforced Composites—Difficulties and Future Perspectives. *Materials* **2021**, *14*, 4191. [CrossRef]
15. Rani, M.; Choudhary, P.; Krishnan, V.; Zafar, S. A review on recycling and reuse methods for carbon fiber/glass fiber composites waste from wind turbine blades. *Compos. Part B-Eng.* **2021**, *215*, 108768. [CrossRef]
16. Oliveux, G.; Dandy, L.O.; Leeke, G.A. Current status of recycling of fibre reinforced polymers: Review of technologies, reuse and resulting properties. *Prog. Mater. Sci.* **2015**, *72*, 61–99. [CrossRef]
17. Jiang, G.; Pickering, S.J.; Walker, G.S. Softening behaviour of epoxy polymers used in carbon fibre reinforced composites during pyrolysis. *Polym. Degrad. Stab.* **2008**, *93*, 800–810. [CrossRef]
18. Witik, R.A.; Teuscher, R.; Michaud, V.; Ludwig, C.; Månson, J.-A.E. Carbon fibre reinforced composite waste: An environmental assessment of recycling, energy recovery and landfilling. *Compos. Part A Appl. Sci. Manuf.* **2013**, *49*, 89–99. [CrossRef]
19. Carbon Fiber Market Analysis: Industry Market Size, Plant Capacity, Production, Operating Efficiency, Demand & Supply, End-User Industries, Sales Channel, Regional Demand, Foreign Trade, Company Share, Manufacturing Process, 2015–2035. Chemanalyst. Available online: https://www.chemanalyst.com/industry-report/carbon-fiber-market-676 (accessed on 24 May 2025).
20. Global Carbon Fibers & Carbon Fiber Reinforced Plastics (CFRP) Market—Products, Applications and Production Capacities. Industry-Experts. Available online: https://industry-experts.com/verticals/chemicals-and-materials/carbon-fibers-and-carbon-fiber-reinforced-plastics-cfrp-a-global-market-overview (accessed on 24 May 2025).
21. Meng, F.; Cui, Y.; Pickering, S.; Mckechnie, J. From aviation to aviation: Environmental and financial viability of closed-loop recycling of carbon fibre composite. *Compos. Part B Eng.* **2020**, *200*, 108362. [CrossRef]
22. Mishnaevsky, L. Sustainable End-of-Life Management of Wind Turbine Blades: Overview of Current and Coming Solutions. *Materials* **2021**, *14*, 1124. [CrossRef]
23. Mishnaevsky, J.L. Recycling of wind turbine blades: Recent developments. *Curr. Opin. Green Sustain. Chem.* **2023**, *39*, 100746. [CrossRef]
24. Wajda, A.; Brociek, R.; Pleszczyński, M. Optimization of Energy Recovery from Hazardous Waste in a Waste Incineration Plant with the Use of an Application. *Processes* **2022**, *10*, 462. [CrossRef]
25. Xu, M.; Ji, H.; Wu, Y.; Di, J.; Meng, X.; Jiang, H.; Lu, Q. The Pyrolysis of End-of-Life Wind Turbine Blades under Different Atmospheres and Their Effects on the Recovered Glass Fibers. *Compos. Part B Eng.* **2023**, *251*, 110493. [CrossRef]
26. Ginder, R.; Ozcan, S. Recycling of Commercial E-glass Reinforced Thermoset Composites via Two Temperature Step Pyrolysis to Improve Recovered Fiber Tensile Strength and Failure Strain. *Recycling* **2019**, *4*, 24. [CrossRef]
27. Muzyka, R.; Mumtaz, H.H.; Sobek, S.; Werle, S.; Adamek, J.; Semitekolos, D.; Charitidis, C.A.; Tiriakidou, T.; Sajdak, M. Solvolysis and oxidative liquefaction of the end-of-life composite wastes as an element of the circular economy assumptions. *J. Clean. Prod.* **2024**, *478*, 143916. [CrossRef]
28. Branfoot, C.; Folkvord, H.; Keith, M.; Leeke, G.A. Recovery of Chemical Recyclates from Fibre-Reinforced Composites: A Review of Progress. *Polym. Degrad. Stab.* **2023**, *215*, 110447. [CrossRef]
29. Czajkowski, A.; Wajda, A.; Poranek, N.; Bhadoria, S.; Remiorz, L. Prediction of the Market of End-of-Life Photovoltaic Panels in the Context of Common EU Management System. *Energies* **2023**, *16*, 284. [CrossRef]
30. Bogacka, M.; Iluk, T.; Pikoń, K.; Sciążko, M.; Stec, M.; Czaplicki, A.; Wajda, A. Life Cycle Assessment (LCA) of the Use of Metallurgical Process Gas for Heat and Electricity, Combined with Salt Removal from Discarded Water. *Sustainability* **2022**, *14*, 1205. [CrossRef]
31. Repp, L.; Hekkert, M.; Kirchherr, J. Circular Economy-Induced Global Employment Shifts in Apparel Value Chains: Job Reduction in Apparel Production Activities, Job Growth in Reuse and Recycling Activities. *Resour. Conserv. Recycl.* **2021**, *171*, 105621. [CrossRef]
32. Lawal, S.O. The Economics of Recycling: A Review Compiled with Tax and Subsidiary, Implication for Government, Decision-Makers, Enterprises, Community, and Analysis Cost/Benefit and Market. *ASEAN J. Sci. Eng. Mater.* **2024**, *3*, 141–164.
33. Lindhal, M.; Ekermann, L. Structure for Categorization of EcoDesign Methods and Tools. In *Re-Engineering Manufacturing for Sustainability, Proceedings of the 20th CIRP International Conference on Life Cycle Engineering, Singapore, 17–19 April 2013*; Nee, A.Y.C., Song, B., Ong, S.K., Eds.; Springer: Singapore, 2013; pp. 117–122. [CrossRef]
34. Fiksel, J. *Design for Environment. A Guide to Sustainable Product Development*, 2nd ed.; McGraw-Hill: New York, NY, USA, 2012; p. 410.
35. Joachimiak-Lechman, K. LCA in Environmental Management Systems—Results of Individual Interviews with Selected Enterprises from Poland and Sweden. *J. Environ. Prot.* **2013**, *4*, 694–700. [CrossRef]

36. *ISO 14040*; Environmental Management—Life Cycle Assessment—Principles and Framework. International Organization for Standardization: Brussels, Belgium, 2006.
37. *ISO 14044*; Environmental Management—Life Cycle Assessment—Requirements and Guidelines. International Organization for Standardization: Brussels, Belgium, 2006.
38. Guinée, J.B.; Gorrée, M.; Heijungs, R.; Huppes, G.; Kleijn, R.; Koning, A.; van de Oers, L.; Sleeswijk, A.W.; Suh, S.; Udo de Haes, H.A.; et al. *Handbook on Life Cycle Assessment. Operational Guide to the ISO Standards. III: Scientific Background*, 1st ed.; Kluwer Academic Publishers: Dordrech, The Netherlands, 2002; pp. 1–314.
39. European Commission, Joint Research Centre, Institute for Environment and Sustainability. *ILCD International Reference Life Cycle Data System (ILCD) Handbook—General Guide for Life Cycle Assessment—Detailed Guidance*, 1st ed.; Office of the European Union: Luxembourg, 2010; pp. 1–417. Available online: https://eplca.jrc.ec.europa.eu/uploads/ILCD-Handbook-General-guide-for-LCA-DETAILED-GUIDANCE-12March2010-ISBN-fin-v1.0-EN.pdf (accessed on 13 March 2025).
40. Marinis, D.; Markatos, D.; Farsari, E.; Amanatides, E.; Mataras, D.; Pantelakis, S. A Novel Plasma-Enhanced Solvolysis as Alternative for Recycling Composites. *Polymers* **2024**, *16*, 2836. [CrossRef]
41. Ekvall, T.; Björklund, A.; Sandin, G.; Jelse, K. Modeling recycling in life cycle Assessment. In *Technical Report: Towards Sustainable Waste Management*; The Swedish Life Cycle Centre: Gothenburg, Sweden, 2020; pp. 1–138. Available online: https://www.lifecyclecenter.se/wp-content/uploads/2020_05_Modeling-recyling-in-life-cycle-assessment-1.pdf (accessed on 6 December 2023).
42. Semitekolos, D.; Terzopoulou, S.; Amanatides, E.; Sajdak, M.; Zecchi, S.; Marinis, D.; Farsari, E.; Sobek, S.; Werle, S.; Tagliaferro, A.; et al. Performance Restoration of Chemically Recycled Carbon Fibres Through Surface Modification with Sizing. *Polymers* **2025**, *17*, 33. [CrossRef]
43. Recommendation 2021/2279/EU Commission Recommendation (EU) of 15 December 2021 on the Use of the Environmental Footprint Methods to Measure and Communicate the Life Cycle Environmental Performance of Products and Organizations. Available online: https://eur-lex.europa.eu/legal-content/EN/TXT/PDF/?uri=CELEX:32021H2279 (accessed on 13 March 2025).
44. Zhang, Z. Recovery and Reuse of Thermosetting Carbon Fibre Reinforced Composites. In Proceedings of the 4th International Conference on Materials Chemistry and Environmental Engineering, London, UK, 13–19 January 2024; Available online: https://www.ewadirect.com/proceedings/ace/article/view/11998/pdf (accessed on 26 May 2025).
45. Global Carbon Fiber Capacity to Double by 2030 | Carbon Fibers & Carbon Fiber Reinforced Plastics Market Products, Applications and Production Capacities Report 2025. GlobeNewswire. Available online: https://www.globenewswire.com/news-release/2025/05/02/3073125/28124/en/Global-Carbon-Fiber-Capacity-to-Double-by-2030-Carbon-Fibers-Carbon-Fiber-Reinforced-Plastics-Market-Products-Applications-and-Production-Capacities-Report-2025.html (accessed on 26 May 2025).
46. Carbon Fiber Composite Materials. Toray. Available online: https://www.toray.com/global/products/carbon-fiber/ (accessed on 24 May 2025).
47. Carbon Fiber & Composite Materials. Mitsubishi Chemical Group. Available online: https://www.mcgc.com/english/ (accessed on 24 May 2025).
48. Carbon Fiber. Hexcel. Available online: https://www.hexcel.com/Products/Carbon-Fiber/ (accessed on 24 May 2025).
49. Overview of our Materials. Sglcarbon. Available online: https://www.sglcarbon.com/en/markets-solutions/material/material-overview/ (accessed on 24 May 2025).
50. Inston, D. ELG Carbon Fibre Open House Showcases Recycling Technology. Composites World 2017. Available online: https://www.gen2carbon.com/wp-content/uploads/2021/07/Composites-World-June-2017.pdf (accessed on 26 May 2025).

materials

MDPI

Article

Valorizing Organic Waste: Selenium Sulfide Production Mediated by Sulfate-Reducing Bacteria

Shahrzad Safinazlou [1], Ahmad Yaman Abdin [1], Eduard Tiganescu [1], Rainer Lilischkis [2], Karl-Herbert Schäfer [3], Claudia Fink-Straube [4], Muhammad Jawad Nasim [1,*] and Claus Jacob [1,*]

[1] Division of Bioorganic Chemistry, School of Pharmacy, Saarland University, 66123 Saarbruecken, Germany; shsa00006@uni-saarland.de (S.S.); yaman.abdin@uni-saarland.de (A.Y.A.); s9edtiga@stud.uni-saarland.de (E.T.)

[2] Working Group Informatic and Microsystems Technology Department, University of Applied Sciences Kaiserslautern, Amerikastrasse 1, 66482 Kaiserslautern, Germany

[3] Working Group Enteric Nervous Systems (AGENS), University of Applied Sciences Kaiserslautern, Amerikastrasse 1, 66482 Kaiserslautern, Germany; karlherbert.schaefer@hs-kl.de

[4] INM—Leibniz Institute for New Materials, 66123 Saarbrücken, Germany; claudia.fink-straube@leibniz-inm.de

* Correspondence: jawad.nasim@uni-saarland.de (M.J.N.); c.jacob@mx.uni-saarland.de (C.J.); Tel.: +49-681-302-57335 (M.J.N.); +49-681-302-3129 (C.J.)

Abstract: Selenium sulfide, the active ingredient of traditional antidandruff shampoos, is industrially produced from selenium dioxide (SeO_2) and hydrogen sulfide (H_2S) under acidic conditions. This reaction can also be carried out with natural H_2S and H_2S generated by sulfate-reducing bacteria (SRB). These bacteria are robust and, by relying on their conventional growth medium, also thrive in "waste" materials, such as a mixture of cabbage juice and compost on the one side, and a mixture of spoiled milk and mineral water on the other. In these mixtures, SRB are able to utilize the DL-lactate and sulfate (SO_4^{2-}) present naturally and produce up to 4.1 mM concentrations of H_2S in the gas phase above a standard culture medium. This gas subsequently escapes the fermentation vessel and can be collected and reacted with SeO_2 in a separate compartment, where it yields, for instance, pure selenium sulfide, therefore avoiding the need for any cumbersome workup or purification procedures. Thus "harvesting" H_2S and similar (bio-)gases produced by the fermentation of organic waste materials by suitable microorganisms provides an elegant avenue to turn dirty waste into valuable clean chemical products of considerable industrial and pharmaceutical interest.

Keywords: cabbage juice; compost; hydrogen sulfide; milk; mineral water; selenium sulfide; silage; sulfate-reducing bacteria; waste-to-value

1. Introduction

Recent years have witnessed an increased interest in developing sustainable green approaches to produce molecules relevant for pharmaceutical and cosmetic industries [1–4]. Whilst in many cases traditional methods remain unavoidable to meet strict production and quality standards, they tend to be resource and energy intensive. Alternative, more sustainable and green approaches promise to reduce this demand for resources, the environmental impact and costs associated with the production of such molecules and the notion of turning "waste products" into valuable compounds, possibly with the aid of low-energy fermentation processes, are especially attractive for several of those reasons [5–7].

We have recently turned our attention to the production of selenium sulfide. The conventional synthesis involves the reduction of Se^{4+} from selenium dioxide (SeO_2) by sulfide ions (S^{2-}) from sodium sulfide (Na_2S), leading to the formation of elemental selenium (Se^0), which subsequently reacts with additional S^{2-} to form polymeric selenium sulfide (Se_nS_{8-n}). The reaction proceeds via nucleophilic attack and redox coupling, resulting in a precipitate of Se–S compounds. The reaction takes place in the presence of glacial acetic acid, which supports the precipitation of selenium sulfide. Once the reaction is completed, the product is separated by centrifugation, washed with water and dried to obtain the desired compound [8–11]. Whereas SeO_2 is not easily available outside the mining and chemical industry, H_2S is an abundant natural product found, for instance, in volcanic and mineral waters, and may therefore be harvested from a sustainable, green source [12–15]. And indeed, it has been possible to perform the synthesis of selenium sulfide with natural spring water from the spring of Bad Nenndorf in Lower Saxony, Germany, as reported by us in this journal [16,17]. Although this approach, based on a natural, inorganic H_2S source is very promising, it also has its limitations. The geographical distribution of and access to these wells is often limited, and this may require complex logistics and the transport of waters across longer distances. Furthermore, the concentrations of H_2S in these waters are modest, and this, in turn, requires the handling of considerable amounts/volumes of water and extensive filtration to harvest the selenium sulfide produced [16].

We have therefore followed up this initial approach and turned our attention to other, more readily available organic sources of H_2S. Inspired by the fact that H_2S often occurs in high concentrations in cesspits, the microbial fermentation of organic materials and especially organic waste may provide an alternative, more convenient access to this gas [18,19]. Indeed, sulfate-reducing bacteria (SRB) reduce sulfate (SO_4^{2-}) to H_2S through the dissimilatory sulfate reduction (DSR) pathway where SO_4^{2-} serves as a terminal electron acceptor [20,21]. Reduction of SO_4^{2-} to H_2S involves several intermediates and the process of DSR requires the transfer of eight electrons in total [22,23]. In contrast to the assimilatory sulfate reduction pathway, which incorporates sulfate into sulfur-containing amino acids and proteins, the dissimilatory pathway produces H_2S as a stable product which can be readily collected for synthetic applications, offering a more amenable approach [24,25]. Notably, several of these SRB rely on DL-lactate as carbon source, and since lactate is found naturally in many types of agricultural and food wastes, such as silage and spoiled milk, this strategy indeed has attracted our immediate attention [26–29].

Here, we report our first investigations of the selenium sulfide synthesis with H_2S produced from organic sources and naturally occurring mineral SO_4^{2-} sources. Among the organic sources are common, highly abundant, readily available and low-cost waste materials, including the fermentation juice of vegetables, spoiled milk and compost, whereas mineral water from gypsum-rich soils may serve as natural source for SO_4^{2-}.

It should be emphasized from the beginning that to turn these–often dirty–waste materials into the value of a clean chemical product is a considerable challenge, especially if the methods employed need to be "green", (cost) effective, low energy, and suitable for large-scale processing avoiding extensive purification. Thus, a special and sophisticated sequence of biological and chemical processes had to be devised in order to meet these various challenges, and although the ensuing strategy presented here may look a little "redneck" at first, its design is far from trivial.

It should also be noted that this is an initial feasibility study and that optimization of the various processes is, of course, always possible yet needs to be the topic of a further follow-on manuscript.

2. Materials and Methods

2.1. Materials

Chemicals were purchased from Merck (Darmstadt, Germany) unless stated otherwise. *Desulfovibrio desulfuricans* (DSM 642) was purchased from the Leibniz Institute DSMZ–German Collection of Microorganisms and Cell Cultures GmbH (Braunschweig, Germany) [30]. Amongst various compost compositions available commercially, "Bio Hochbeet Kompost 2 torffrei" was selected and purchased from a local retail store, Bauhaus (Saarbruecken, Germany), at a cost of less than three Euros per 10 kg. Fermented cabbage juice (Alnatura Sauerkraut Saft), kefir (Kalinka Kefir (low fat)) and low-fat milk are available in shops and thus were purchased from local supermarkets such as Netto and Müller (Saarbruecken, Germany). The same products can be collected for free once they have met their expiration date. Ensinger Schiller Quelle mineral water, used as a mineral water source, was purchased from the Alldrink local beverage store in Saarbruecken, Germany. Plastic fermentation tubes (airlock for trapping H_2S) were purchased from Amazon. Grass silage was collected from the Bannsteinhof farm in Zweibruecken, Rhineland-Palatinate, Germany (49°14′56.4″ N, 07°21′53.6″ E) in August 2024. Red clover was harvested from Saarland University, Saarbruecken, Germany (49°15′32.0″ N, 07°02′25.4″ E in October 2024 and identified as *Trifolium pratense* using the PlantApp application (Scale Up, Izmir, Turkey) [31]. Mifloran intense containing *Bifidobacterium* and *Lactobacillus* species used for silage fermentation were purchased from the local Reformhaus store in Saarbruecken, Germany.

2.2. Design of Fermentation Equipment

The design of the fermentation equipment is shown in Figure 1. It consists of a standard fermentation vessel enabling the anaerobic fermentation with SRB and the collection of the gasses produced in a U-shaped airlock, which at the same time serves as a reaction vessel for the reaction of H_2S with SeO_2.

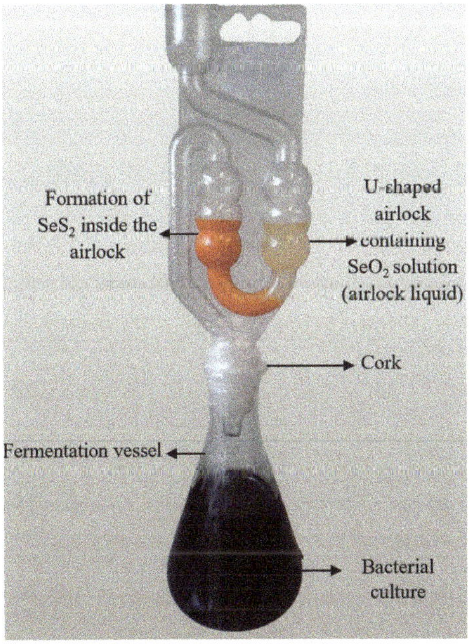

Figure 1. SRB were incubated under anaerobic conditions in specially designed fermentation equipment, including a fermentation vessel coupled to fermentation tube (airlock). This photo of actual SRB fermentation equipment is provided by the first author (SS).

2.3. Cultivation of SRB

D. desulfuricans is a SO_4^{2-} reducing anaerobic bacterium which was cultured according to the special instructions provided by the Leibniz Institute DSMZ–German Collection of Microorganisms and Cell Cultures GmbH (Braunschweig, Germany) for handling the anaerobic bacteria [30]. In brief, the standard cultivation medium comprises three different solutions, i.e., solutions A, B and C. Solution A consisted of K_2HPO_4 (0.5 g), NH_4Cl (1.0 g), Na_2SO_4 (1.0 g), $CaCl_2•2H_2O$ (0.1 g), $MgSO_4•7H_2O$ (2.0 g), sodium DL-lactate (2.0 g), yeast extract (1.0 g) and sodium resazurin (0.1% w/v, 0.5 mL), dissolved in 980 mL of distilled water. Solution B consisted of a solution of $FeSO_4•7H_2O$ (0.5 g) dissolved in 10 mL of distilled water and solution C contained sodium thioglycolate (0.1 g) and ascorbic acid (0.1 g) dissolved in 10 mL of distilled water. Solution A was boiled for 5 min and subsequently cooled to room temperature under a continuous flow of nitrogen gas to maintain anaerobic conditions. This step was followed by the addition of solutions B and C. Lastly, the pH of the solution was adjusted to 7.0 using a few drops of NaOH and maintained without further dynamic adjustment during incubation. After autoclaving, the medium was carefully transferred to the fermenting equipment under anaerobic conditions, followed by the inoculation of the bacteria. The cultures were incubated at 30 °C in the Environmental Shaker–Incubator ES20 toprovide, at stable temperature and agitation throughout the fermentation process (Grant Instruments, Royston, UK).

The composition of the medium was later optimized to find the essential ingredients required for the cultivation of bacteria and the production of H_2S as shown in the Supplementary Section, Table S1. Then, the essential ingredients of the medium were gradually replaced with more common "household" ingredients and eventually waste products, i.e., silage, cabbage juice and spoiled milk were used as a substitute for DL-lactate while compost and mineral water were added as a source of SO_4^{2-} and minerals (summarized in Table S2 in Supplementary Materials).

2.4. Quantification of DL-Lactic Acid

DL-lactic acid, a key component of the standard medium, serves as a carbon source for bacterial growth and the synthesis of essential cellular structures. Thus, the initial step during replacement with silage, spoiled milk, etc., involved quantifying the concentration of DL-lactic acid in commercially available fermented cabbage juice and kefir obtained from local supermarkets, as well as monitoring its concentration over time in homemade red clover silage and spoiled milk. The quantification of DL-lactic acid was carried out using the protocol described in the literature [32]. Briefly, the standard curve was prepared employing different concentrations of DL-lactic acid ranging from 0 to 300 mg L^{-1}. Solutions containing 100 μL of DL-lactic acid were diluted with 400 μL of distilled water followed by the addition of 3 mL H_2SO_4 (95%). The reaction mixtures were heated at 95–100 °C for 10 min and then cooled to room temperature, followed by the addition of $CuSO_4•5H_2O$ solution (50 μL, 4% w/v) and *p*-phenylphenol solution (100 μL, 1.5% w/v in 95% ethanol). After mixing, the resultant acetaldehyde-*p*-hydroxyphenyl complex was incubated at room temperature for 30 min and the absorbance was recorded at 570 nm using a BioTek Epoch 2 Microplate Spectrophotometer (BioTek Instruments GmbH, Bad Friedrichshall, Germany). A similar procedure was performed for milk and red clover samples. Briefly, samples were prepared as follows: 300 mL milk samples in tightly closed containers were used for measuring their DL-lactic acid content. Red clover samples were collected and chopped into small pieces, divided and stored in small tightly sealed glass containers, each containing approximately 16 g of red clover. Subsequently, the samples were inoculated with 100 μL culture containing *Bifidobacterium* and *Lactobacillus* species to promote fermentation at room temperature. Experiments were conducted in triplicate, with three subsamples obtained from each replicate ($n = 9$). Results are presented as arithmetic means.

2.5. Quantification of H_2S

Following the preparation and inoculation of the culture media, the SRB initiated H_2S production, which accumulated in the headspace of the fermentation vessel and dissolved in the solution within the airlock. H_2S was collected in the headspace (40 mL) of an Erlenmeyer flask (115 mL) containing bacterial culture (75 mL). The headspace-to-liquid ratio of approximately 0.5 was maintained to ensure sufficient space for gas accumulation without excessive pressurization. The quantification of H_2S in the headspace was performed employing the methylene blue (MB) assay as described in the literature, with a few modifications [33]. The standard curve was prepared using different concentrations of anhydrous sodium sulfide (Na_2S) ranging from 0 to 0.6 mM. An amount of 100 µL from the gas phase (headspace) was removed with a syringe and mixed with 1 mL $FeCl_3 \bullet 6H_2O$ (30 mM) dissolved in HCl (1.2 M) and 1 mL *N*,*N*-dimethyl-*p*-phenylenediamine dihydrochloride (DMPD, 20 mM) dissolved in HCl (7.2 M). The reaction mixture was incubated at room temperature for 15 min, and the absorbance was recorded at 670 nm using a microplate spectrophotometer. Experiments were in triplicate, with three subsamples from each replicate ($n = 9$). Results are presented as arithmetic means.

2.6. Production of Selenium Sulfide

The reaction of H_2S with SeO_2 solution (100 mM) was carried out in the airlock directly attached to the incubation assembly as shown above in Figure 1. The samples of selenium sulfide were collected from the airlock after 310 h of incubation as the production of H_2S was completed by then and stabilized at baseline level. The selenium sulfide suspension was centrifuged at 3000 rpm for 15 min. The supernatant was discarded subsequently, and the pellet was washed with distilled water (50 mL) three times to eliminate unreacted SeO_2. The collected selenium sulfide was dried and stored in the dark at room temperature until further use.

2.7. Characterization of Selenium Sulfide

The selenium sulfide obtained was characterized using analytical techniques, such as CHNS analysis, Scanning Electron Microscopy with Energy Dispersive X-ray Spectroscopy (SEM-EDX), Raman spectroscopy and Optical Emission Spectrometry combined with Inductively Coupled Plasma (ICP-OES). CHNS analysis was carried out employing a Vario MICRO cube CHN-elemental analyzer (Elementar GmbH, Langenselbold, Germany). SEM-EDX analysis was conducted with a ZEISS Supra 40 field emitter microscope (Carl Zeiss NTS GmbH, Oberkochen, Germany) coupled to a Bruker Quantax EDX system (Bruker Nano GmbH, Berlin, Germany). Raman spectroscopy was performed on a Renishaw InVia microscope (Wotton-under-Edge, Gloucestershire, UK) coupled with an excitation laser adjusted to a 532 nm wavelength. ICP-OES analysis was carried out using an Ultima 2 tool (Horiba Jobin-Yvon, Longjumenau, France) coupled with a Czerny-Turner-type monochromator with a focal length of 1 m. Commercially available selenium sulfide (Merck, Darmstadt, Germany) served as reference material.

3. Results

Together, our investigations have shown that SRB are able to reduce SO_4^{2-} to H_2S in the presence of a suitable carbon source, and that this source may indeed also be organic waste material, such as fermented cabbage juice and spoiled milk. Importantly, the H_2S produced by these bacteria escapes the fermentation culture in millimolar concentrations for up to 310 h and can be collected from the gas phase and reacted directly, for instance, with SeO_2 to synthesize selenium sulfide. Unlike the fermentation mixture itself, the product formed is highly pure and therefore does not require extensive purification as may be necessary if the reaction was to be performed in the cultivation medium directly.

3.1. Fermentation Under Standard Conditions

3.1.1. Production and Quantification of H_2S

In the first step, the ability of SRB to convert SO_4^{2-} to H_2S in the fermentation equipment was evaluated under standard conditions using the growth medium recommended by the supplier (see DSMZ medium 63: desulfovibrio (postgate) medium). As presented in Figure 2, the fermentation proceeds at 30 °C for almost 13 days, around 310 h, during which time the growth medium transitions in color from yellowish-grey to black.

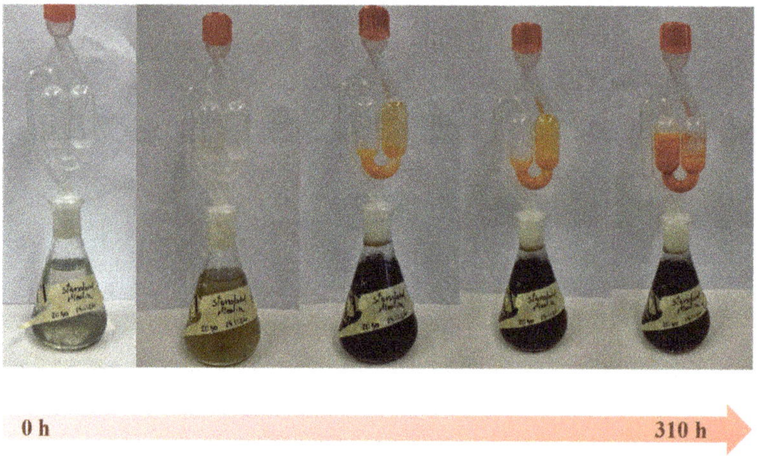

Figure 2. The SRB were cultured in standard media to generate H_2S which then reacted with SeO_2 present in the airlock to produce selenium sulfide. A time-dependent increase in the concentration of H_2S as well as selenium sulfide was observed for up to 310 h.

The relevant H_2S concentration in the gas phase was quantified by the MB assay as described in Section 2.5. A maximum H_2S concentration of 4.1 mM was observed in the gas phase, gradually increasing from zero over 45 h, and subsequently decreasing over the following 265 h (Figure 3). The concentration of H_2S in the growth medium itself could not be quantified due to interference of the medium constituents with the MB assay (see Section 4).

3.1.2. Selenium Sulfide Production

The airlock, filled with 15 mL of 100 mM SeO_2 solution (pH = 2.1), showed the formation of an orange solid starting after around 20 h of incubation. This solid (average weight of 32.7 mg after 310 h, obtained from 75 mL bacterial culture) was collected, washed, and identified as selenium sulfide using a combination of analytical techniques, as explained in Section 2.7. This selenium sulfide produced from biologically generated H_2S is referred to as "biologically produced selenium sulfide" throughout the manuscript. An overview of the relevant analytical data is provided in Table 1. The yield of selenium sulfide produced in the airlock liquid was calculated to be 15.9% based on the amount of SO_4^{2-} present in the culture medium and a formal composition of Se_5S_3 for the product (see below).

Table 1. Characterization of the selenium sulfide produced using standard medium.

Characterization Method	Biologically Produced Selenium Sulfide (wt%)	Commercial Selenium Sulfide (wt%)
CHNS	S: 35.02	S: 47.98
ICP-OES	S: 21.50 Se: 78.50	S: 43.21 Se: 56.79
EDX	S: 24.05 Se: 75.95	S: 45.02 Se: 54.98

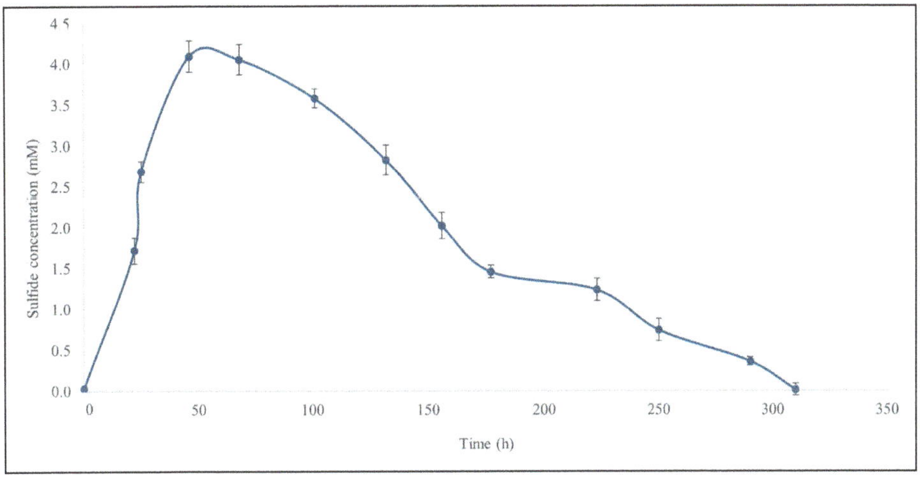

Figure 3. The concentration of H_2S produced under standard conditions increases rapidly in the gas phase for the first 45 h, reaching 4.1 mM, and then gradually declines over the following 265 h. The experiments were performed in triplicate and on three different occasions ($n = 9$). Results are presented as mean ± SD. See text for further details.

Notably, the elemental selenium-to-sulfur ratio tends to vary slightly from batch to batch, as may be expected for a mixture of 29 different compounds theoretically possible based on eight-membered sulfur–selenium rings with the general formula Se_nS_{8-n} [16,34]. The ICP-OES analysis revealed mass percentage values corresponding to a molar ratio of 4.8:3.2 (Se:S), aligning with a formula of Se_5S_3. EDX analysis provided a ratio of 4.5:3.5 (Se:S), further confirming the predominant average Se_5S_3 composition. Essentially, these results suggest an average formal Se to S stoichiometry of 5:3 yet do not allow a firm assignment of exact chemical structures and their respective abundances (Table 1). This ratio is comparably high for selenium, probably due to the excess of SeO_2 used in the airlock which was deemed necessary to capture the H_2S gas under these experimental and design conditions. No carbon, hydrogen or nitrogen impurities were detected in the sample according to CHNS analysis, confirming the high purity of the selenium sulfide samples obtained. SEM-EDX was employed to quantify the elemental composition and physical characteristics of the compound and in essence confirmed the presence of selenium and sulfur in the sample (Figure 4). The presence of oxygen was observed by EDX in commercially available selenium sulfide from Merck as well as biologically produced selenium sulfide. Multiple attempts to obtain solid state ^{77}Se Nuclear Magnetic Resonance (^{77}Se NMR) spectra were unsuccessful, perhaps due to the highly amorphous nature of the sample.

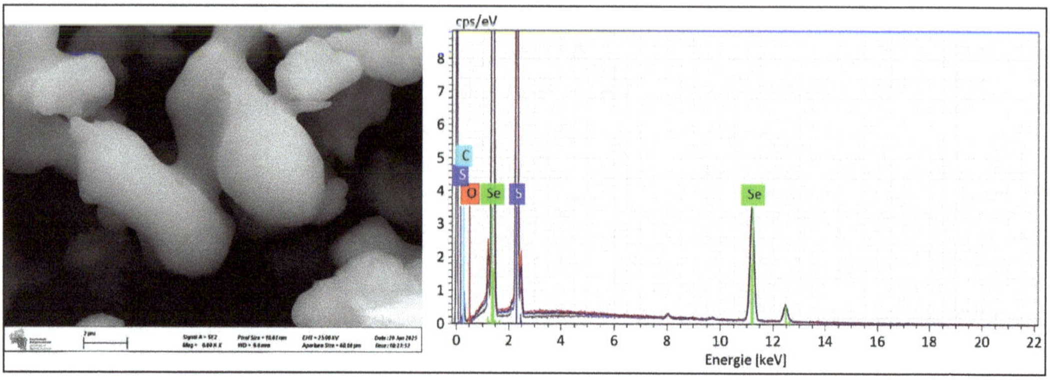

Figure 4. The typical sample of biologically produced selenium sulfide was analyzed using SEM which revealed the presence of agglomerated selenium sulfide particles (**left**). EDX data affirmed the presence of sulfur and selenium in the sample (**right**).

The produced selenium sulfide was also analyzed using Raman spectroscopy to investigate the vibrational modes and to establish the structural fingerprints. Commercially available selenium sulfide was analyzed first as reference (Figure 5, spectrum A) followed by biologically generated selenium sulfide (Figure 5, spectrum B). The key Raman vibrational modes of selenium sulfide include Se–Se stretching vibrations occurring around 250 cm^{-1}, S–Se stretching vibrations in the range of 300–400 cm^{-1} and S–S stretching vibrations between 400–500 cm^{-1} while peaks with the frequency of lower than 200 cm^{-1} are related to the non-stretching vibration area [35]. A comparative analysis of the Raman spectra of commercial and biologically produced selenium sulfide confirms the presence of all these vibrational modes in our sample, indicating its structural similarity to commercial selenium sulfide (Figure 5).

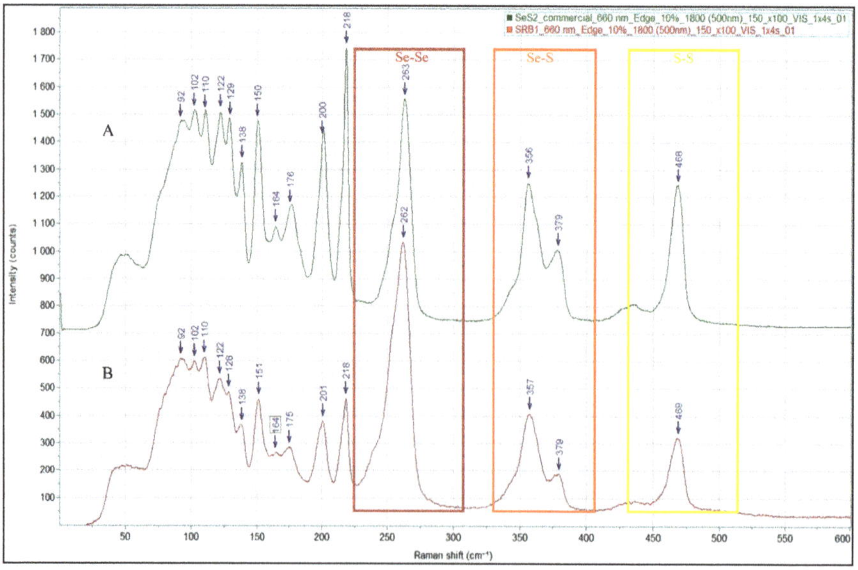

Figure 5. The vibrational modes of biologically produced selenium sulfide (B) provide similar structural fingerprints as that of commercial selenium sulfide (A). Se–S, Se–Se and S–S vibrational stretches are marked in the figure.

In addition to the comparison between the Raman spectra of the biosynthesized and commercial selenium sulfide, a spectral comparison between selenium sulfide and the elemental forms of sulfur and selenium was performed. While the Se–S vibrational fingerprint appeared exclusively in the selenium sulfide samples, slight shifts in the Se–Se and S–S vibrational modes were observed compared to the elemental references (Figure S1 in the Supplementary Section).

3.2. Variations on the Theme

In the second step, various components present in the growth medium were either omitted completely or replaced by more conventional materials. These replacements followed a specific traffic-light strategy which went from fine chemicals (red) to less expensive commercial household products (yellow) and eventually waste materials (green). Thus, the replacement strategy in the first step included lactic acid from the drug store, (by-)products of food manufacture and agriculture, and, eventually, organic waste materials. Figure 6 sums up this "replacement" strategy aiming at readily available, ecologically and economically viable substitutes for quality and refined chemicals from industry.

It should be emphasized that the rationale behind this strategy has been to sequentially minimize the number of components and to focus, if possible, on waste materials available in bulk quantities. Thus, not every theoretically possible combination of materials was tried, since several combinations have been deemed worth investigating further, whereas others were not considered as such. Furthermore, this section of the study also did not yet aim at optimization of the various processes. The selection of waste materials was primarily based on the presence of key ingredients required for the bacterial growth. Fermented cabbage juice, kefir, red clover silage, grass silage and spoiled milk are rich in DL-lactic acid whilst compost, ash and mineral water serve as source of essential minerals and SO_4^{2-}.

Eventually, it was possible to minimize the number and amount of fine chemicals and conduct the fermentation successfully with either a combination of commercial fermented cabbage juice and compost (see Section 3.2.1), or with spoiled milk and mineral water (see Section 3.2.3).

3.2.1. The Sauerkraut Connection

The first replacement avenue takes advantage of the fact that SRB require DL-lactic acid as a carbon source. The juice of fermented cabbage, naturally containing between approximately 0.61 g L^{-1} and 11.60 g L^{-1} DL-lactic acid, mostly thanks to *Lactobacillus* bacteria converting lactose to DL-lactic acid, could be used as an alternative carbon source [36]. Although rich in DL-lactic acid and minerals, fermented cabbage juice does not contain sulfate to support H_2S production. The need for an additional sulfate source could be met by adding compost, which is rich in carbon, nitrogen, phosphorous and sulfur and thus seems to be sufficient to supply the missing components that are not naturally present in fermented cabbage (Sauerkraut) juice, such as SO_4^{2-} ions. The composition of compost was analyzed using CHNS analysis alongside the data provided by the manufacturer (Figure 6).

The "minimal Kraut medium" (comprising of Sauerkraut juice and compost) was able to support the reduction of SO_4^{2-} to H_2S by the SRB culture as shown in Figure 7. Compared to the conventional growth medium, the yields of H_2S in the gas phase and thus selenium sulfide seemed to be lower (an average of 5 mg selenium sulfide was collected from 75 mL culture after 170 h), yet this is hardly surprising as the fermentation was not fully optimized during this initial feasibility study. The amount of selenium sulfide obtained was insufficient to perform the complete physical characterization analysis.

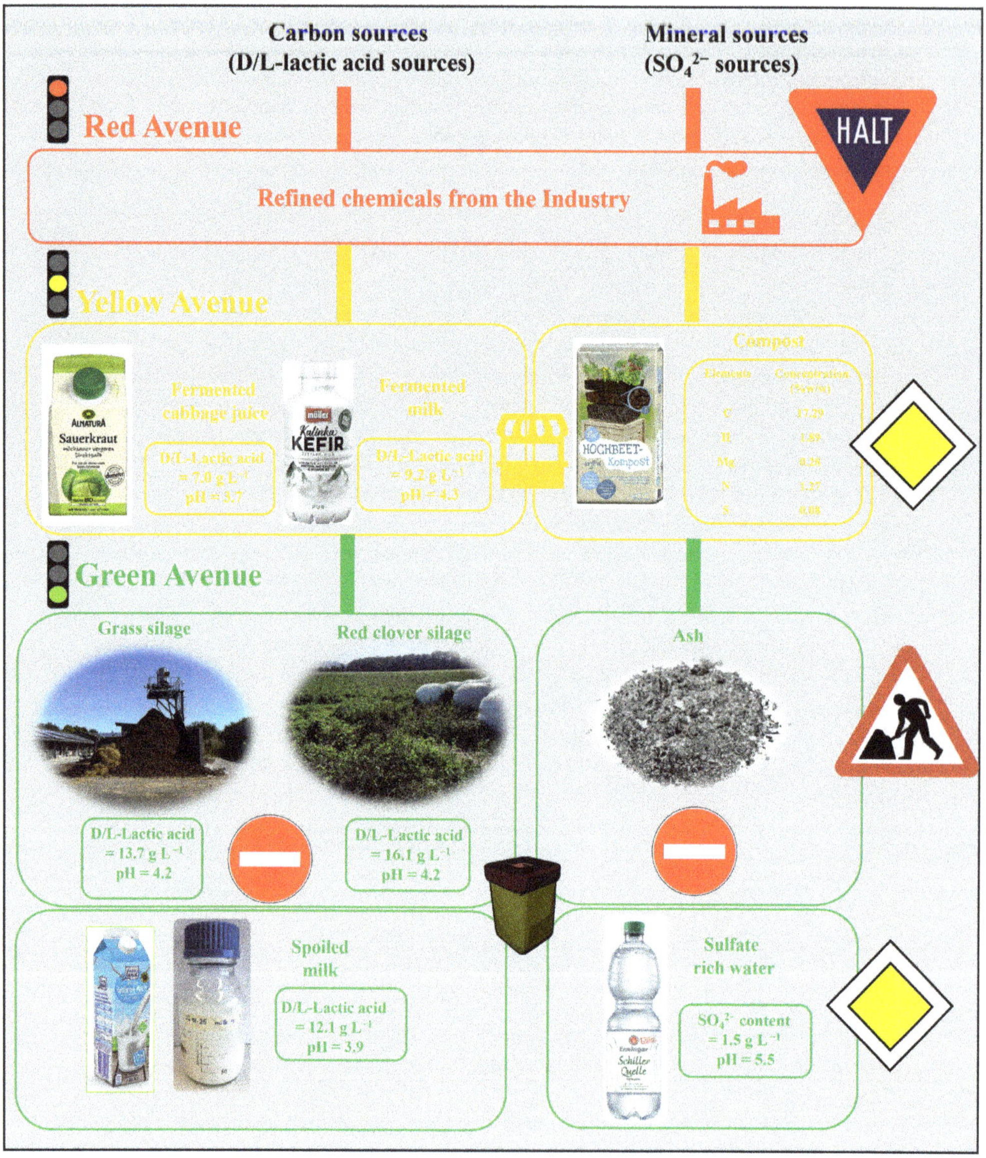

Figure 6. The two main components of standard medium, i.e., DL-lactic acid and minerals were replaced with cheaper and low-grade alternatives for the sake of upcycling organic wastes. Silages (grass and red clover), fermented cabbage juice, fermented milk (kefir) and spoiled milk serve as sources of DL-lactic acid whilst compost, ash and mineral water serve as sources of minerals and especially SO_4^{2-}. The traffic signs indicate which avenues were most promising, closed or are still under construction. Photos are provided by the authors.

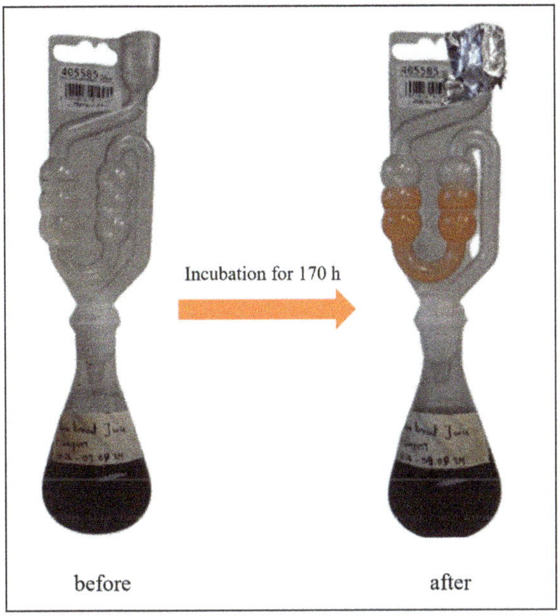

Figure 7. A mixture of fermented cabbage juice and compost was utilized as growth medium to culture SRB. The pH was adjusted to 7.0 using a few drops of ammonium hydroxide (NH_4OH). The resultant H_2S produced by the bacteria can be exploited to produce selenium sulfide.

3.2.2. The Silage Saga

Fermented cabbage–and its juice–is not necessarily a waste material, and thus this avenue was not investigated further. A less valuable alternative was therefore considered and found in the juice produced during the silage process which indeed mirrors the Sauerkraut fermentation with Lactic Acid Bacteria (LAB) yet yields a worthless liquid. Notably, the starting materials for silage, such as red clover, are not heat sterilized or salted, thus the silage juice is not only rich in DL-lactic acid (16.1 g L^{-1}) and fairly acidic (pH = 4.2); it also contains a wide range and variety of other compounds and components and therefore is often considered as an outright nuisance by farmers (Figure 8). Regardless if "real life" grass silage from the Bannsteinhof farm, or homemade red clover silage was used in place of cabbage juice, the SRB did no longer flourish or produce selenium sulfide. This is somewhat disappointing since the homemade silage did contain up to 16.1 g L^{-1} DL-lactic acid and with a pH of 4.2 in these important aspects was not that dissimilar to the commercial cabbage juice employed successfully in Section 3.2.1.

3.2.3. The Milky Way

In addition to fermented cabbage juice, DL-lactic acid is also found in fermented milk products such as yogurt and kefir, both of which contain up to 12 g L^{-1} and 10 g L^{-1} DL-lactic acid, respectively [37,38]. In these food products, bacteria such as *Lactobacillus* species convert lactose to DL-lactic acid. DL-lactic acid also often forms when milk expires and simply "goes off" and the resulting spoiled milk may also contain up to 10% DL-lactic acid [37]. In fact, the bacteria responsible for this conversion of lactose to DL-lactic acid are omnipresent, infecting many natural products, from red clover to milk. Moreover, the non-growing starter cultures also pose a severe challenge to the food industry due to their prolonged metabolic activity during storage leading to post-production acidification. In stored yogurts, for instance, slow but continuous lactic acid generation causes not only shorter shelf life but also considerable alterations in flavor as well as acidity [39–42].

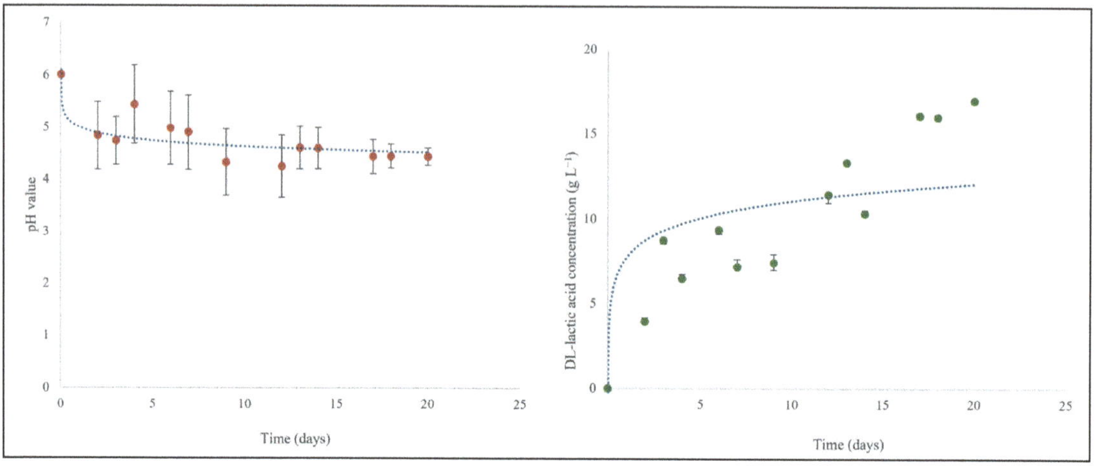

Figure 8. Red clover silage was produced in the laboratory from the clover collected at Saarland University and showed a time-dependent decrease in pH from 6.0 to 4.2 (**left**) and increase in the DL-lactic acid concentration from 0 to 16.1 g L^{-1} (**right**) over a period of 20 days. The experiments were performed in triplicate and on three different occasions (n = 9). Results are presented as mean ± SD. See the text for further details.

Commercial kefir containing 9.2 g L^{-1} D/L-lactic acid (pH 4.3) was therefore investigated and turned out to be a rather good source of D/L-lactic acid for SRB fermentation in combination with Na_2SO_4. In order to replace commercial kefir with spoiled milk, 1.5% homogenized low-fat milk was obtained from the local Netto supermarket and simply allowed to "rot" at room temperature for up to 40 days. During this time, the concentration of DL-lactic acid in the sample increased from 0 to 12.1 g L^{-1}, whilst the pH value continuously dropped from pH 6.7 to pH 3.9 as shown in Figure 9. After 40 days, this lactose-to-DL-lactic acid "fermentation" was complete as indicated by a more or less steady pH value of 3.9 and a DL-lactic acid concentration of 12.1 g L^{-1}.

Spoiled milk can be used in combination with a SO_4^{2-} source, in this case Na_2SO_4, to provide essential nutrients for the growth of SRB and the production of H_2S. Inoculation of this culture with SRB resulted in the generation of up to 1 mM H_2S in the headspace (Figure 10). The H_2S was further reacted with SeO_2 to provide selenium sulfide which was subsequently characterized using the above-mentioned techniques. The results are presented in Supplementary Section Table S3 and Figures S2 and S3.

Alternatively, SO_4^{2-} can be replaced with sulfate found in natural sources such as mineral waters. One of these natural mineral water sources is the SO_4^{2-} rich spa water originating from the natural gypsum-rich underground sedimentary rock formations around Ensingen in Baden-Wuerttemberg (Germany, 48°58′ N, 8°57′ E) and reportedly containing 1.5 g L^{-1} of SO_4^{2-} [43]. Following the successful inoculation of SRB in a 1:2 mixture of spoiled milk and mineral water neutralized by a few drops of NH_4OH, H_2S production in the headspace after 45 h exceeded an astonishing 4.3 mM before gradually declining over the next 190 h (Figure 11). Thus, the "redneck" fermentation with spoiled milk and Ensinger mineral water was as efficient as the one following the standard protocol and employing various expensive fine chemicals (see Figure 11).

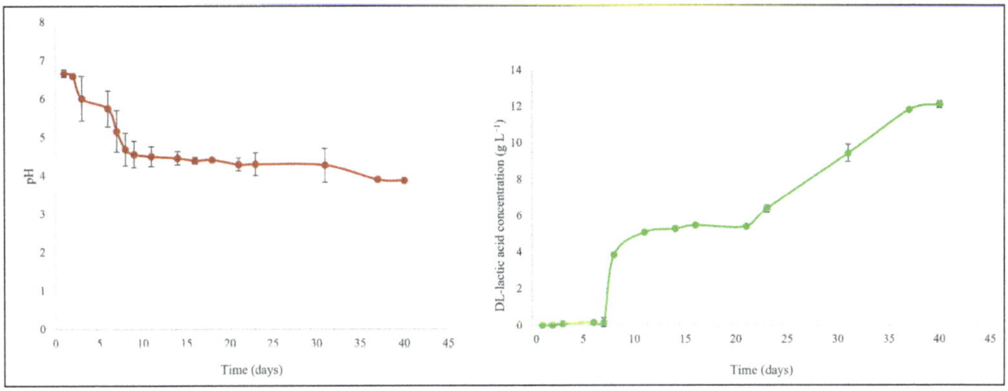

Figure 9. A time-dependent decrease in pH from 6.7 to 3.9 (**left**) and an increase in the DL-lactic acid concentration from 0 to 12.1 g L^{-1} (**right**) was observed for milk which was allowed to rot at room temperature for 40 days. The experiments were performed in triplicate and on three different occasions (n = 9). Results are presented as mean ± SD. See text for further details.

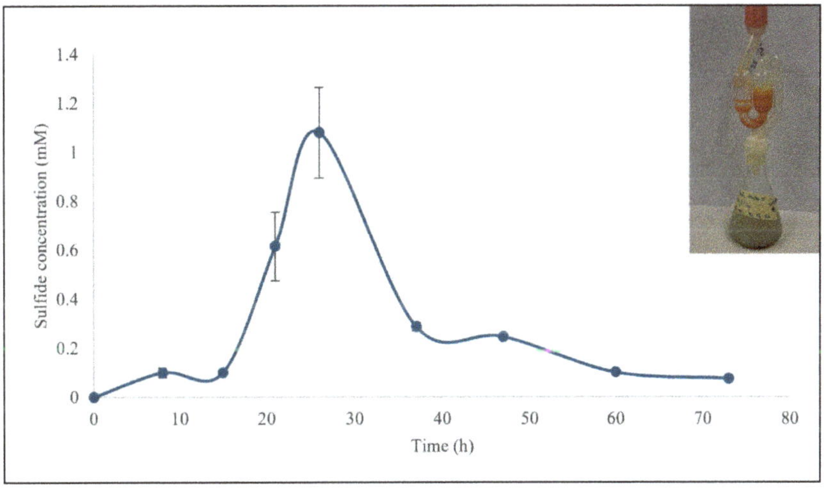

Figure 10. SRB were able to produce H_2S when a mixture of spoiled milk and Na_2SO_4 was used as culture medium. H_2S was produced over 60 h in the headspace, reaching a maximum concentration of just above 1 mM after 25 h. The experiments were performed in triplicate and on three different occasions (n = 9). Results are presented as mean ± SD. See text for further details.

After 235 h, an average of 48.3 mg of selenium sulfide was obtained. The selenium sulfide was then thoroughly washed (see Section 2.6), dried and subsequently analyzed. CHNS analysis, ICP-OES, and EDX confirmed the presence of selenium and sulfur in the sample. The elemental composition is more similar to the commercial selenium sulfide compared to the standard medium, with the Se to S ratio in ICP-OES corresponding to 3:5 and in EDX corresponding to 2.8:5.2, resulting in the overall stoichiometric ratio of approximately 1:2 (Supplementary Section, Table S4, Figure S4). CHNS analysis also affirmed the absence of any carbon, hydrogen and nitrogen in the sample. The Raman spectroscopy confirmed similar structural fingerprints of selenium sulfide as those of standard selenium sulfide from Merck (Supplementary Section, Figure S5).

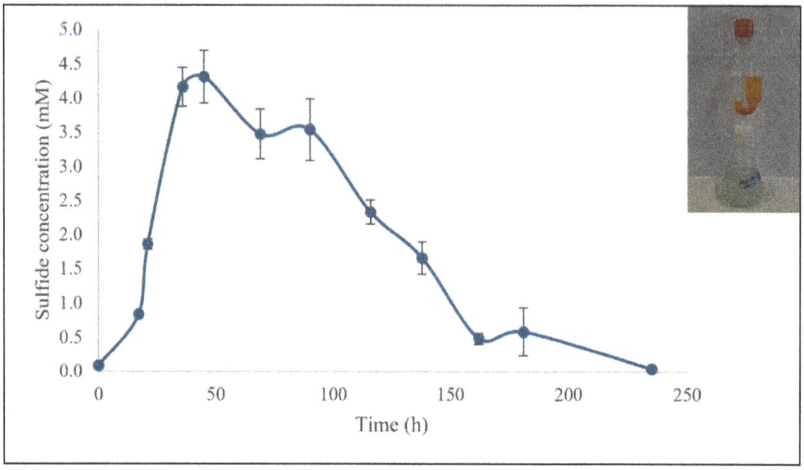

Figure 11. SRB were able to produce H_2S when a mixture of spoiled milk and mineral water was used as culture medium. A time-dependent increase in the concentration of H_2S was observed in the headspace, reaching more than 4.3 mM, followed by a gradual decrease in the subsequent hours. The experiments were performed in triplicate and on three different occasions ($n = 9$). Results are presented as mean ± SD. See text for further details.

4. Discussion

In essence, the successful production of selenium sulfide from SeO_2 and H_2S produced via anaerobic SRB fermentation of a mixture of spoiled milk and Ensinger mineral water confirms the feasibility of a "redneck" chemistry using "dirty" waste materials as one of the relevant components in place of commercially available fine chemicals. Although the processes underlying these fermentations still could be refined and optimized in order to achieve even higher yields and thus fulfil economic prospects, this feasibility study already demonstrates and underlines a few important points.

First and foremost, microbial fermentation seems to be a valuable avenue to generate H_2S under mild conditions, at room temperature and from readily available organic (by-) products such as spoiled milk and natural spa waters as well as a combination of fermented cabbage juice and compost. This differs from ccommercial H_2S which is either recovered from gas mixtures or produced chemically. Both of these means are energy intensive as well as environmentally harmful. Compared to our previous studies utilizing H_2S-rich natural waters, H_2S produced by fermentation is more accessible and also less tedious to collect and to process. The concentrations of 4.1–4.3 mM in the headspace of the airlock, for instance, compare favorably with the mineral waters used in our previous study, which contained only up to 2.4 mM H_2S. The escape of H_2S gas from such sulfide-rich springs reduces its available sulfur content. Therefore, sulfate-rich mineral water such as Ensinger Schiller Quelle serves as a better and more reliable sulfur source with a SO_4^{2-} concentration of around 15.6 mM, providing more than six times the sulfur content of the H_2S-rich Bad Nenndorf spring water used previously and without the risk of SO_4^{2-} escaping in the gas phase.

The collection of the selenium sulfide produced from the water from Bad Nenndorf was also tedious as it required extensive filtration. Using SRB, H_2S can be produced from SO_4^{2-} rather easily in higher concentrations and also, literally, escapes the dirty fermentation mixture and can be captured continuously in the airlock. This strategy avoids the need for subsequent filtration or purification steps and also enables an easy harvest of considerable amounts of product.

Indeed, besides simple low-energy fermentation, purification is the second most relevant cornerstone of this fermentation process. There are reports in the literature describing the production of selenium sulfide using yeast cultured in the presence of selenite (SeO_3^{2-}) and sulfite (SO_3^{2-}) [44]. Considering the simplicity of selenium sulfide production in such a culture, this process at first seems to be highly attractive. Yet, despite the fact that some selenium sulfide may indeed have formed under these conditions, purification of the material is tedious, cumbersome and, in our opinion and experience, neither economical nor especially ecological. It is also not possible to determine or use H_2S concentrations in such "dirty" media directly, employing, for instance, the MB assay, as mentioned above. This also applies to the reaction of H_2S with SeO_2 in the medium, as a product formed in the medium is difficult to purify.

Thus, separating the fermentation process on the one side from the product formation and purification process on the other has considerable attraction. In this context, H_2S is therefore of special interest, as it escapes the culture in a more or less pure form together with other gases such as CO_2 and subsequently can be captured and used as a starting material for further products, in our case selenium sulfide. Other H_2S-based compounds, such as thiazolidine-2-one derivatives, are, of course also in the pipeline/airlock as attractive molecules using a similar setup for their synthesis [45]. In addition to organic and inorganic compounds, one may also employ H_2S directly, for instance in fuel cells [46].

The setup itself has been designed as a simple, cost-effective and robust system, and indeed, the use of the airlock as a combination of traditional airlock for anaerobic fermentation and simultaneously a reaction vessel has proved an elegant solution to the anaerobic SRB fermentation, at least on the laboratory scale. Upscaling and refining this process is, of course, possible, and this may indeed include the collection of H_2S and its storage or transport to a separate reaction chamber. It is important to mention that the volume of headspace may also affect the production of H_2S and this could be fine-tuned together with other relevant variables, such a volume of media and temperature [47]. This is an issue of technical (bio-) chemistry that may be addressed in subsequent studies considering the optimization of this process.

Optimization itself may include a finer adjustment of individual components or eventually the use of mutant strains of SRB able to produce H_2S in higher yields. The choice of fermented cabbage juice as a replacement for commercial DL-lactic acid has been a good starting point, since this juice is more sustainable and thus considerably cheaper than industrially produced DL-lactic acid. Cabbage waste is a significant issue in the agricultural sector. In the cabbage market, the global production of cabbage was estimated at 70 million tons in 2018 and is projected to reach 88 million tons in 2025 with an average wastage rate of 14%, according to data from Gauteng, South Africa [48–50]. In theory, this accounts for approximately 16 million tons of discarded cabbage for 2025, which could be recovered and repurposed as a sustainable carbon source, along with expired food-grade cabbage juice [50–52]. Of course, fermented cabbage–and its juice–is not necessarily a waste material. In Germany, large crowds of Krauts simply love their Sauerkraut and consume its juice like table water *(personal communication by the corresponding author)*.

Although initial attempts to move on from cabbage juice and to utilize silage or silage juices have failed due to their apparent toxicity, this avenue is indeed still very promising and may need to be revisited again in the hope that some SRB or other lactate-consuming bacteria may be more resistant and thus able to thrive in the presence of this liquid. Considering that silage juice is rich in many nutrients, it may also be possible, for instance, to pretreat it before using it in SRB culture or to switch from red clover to maize or another green waste material just as it was possible to eventually switch from cabbage juice to spoiled milk. In any case, the silage saga must go on as silage juice is a major side-product of agricultural fermentation and so far more of a nuisance to farmers and the environment than a potential mine for sustainable processes—and therefore fulfils many of the conditions for turning this inherent waste into considerable value—easily and on a very large scale indeed.

Then again, milk is among the most discarded food items today. In the UK, for instance, it comes second on the list of most wasted food items right after wasted bread and before wasted bananas. Each year, around 500,000 tons of milk literally go down the drain in the UK alone, and this milk may be used in fermentation, together with the tons of dairy products, yogurt, kefir, etc., also dumped each year [53–55]. This concept becomes particularly appealing as the amount of selenium sulfide produced using spoiled milk and the spa water from Ensinger Schiller Quelle exceeds the amount obtained with standard media. Unlike fermented cabbage juice, which is not universally considered a waste material, spoiled milk is clearly a waste product. For these reasons, spoiled milk and mineral water from rocks rich in gypsum emerge as the most promising alternatives to the standard medium in this experiment.

Ideally, a sustainable process not only requires "worthless" or even "nuisance" educts—it should also minimize its own waste. As for the waste produced by this microbial fermentation, neither the compost, the spoiled milk nor the SRB pose any environmental risk and thus may be discarded after use, for instance for further downstream fermentation in biogas production. SRB are harmless soil bacteria found, for instance, in soils, freshwater and saltwater ecosystems, sediments, mud and even the rhizosphere of plants and, in any case, are sensitive to dioxygen, and thus unlikely to pose any danger to plants, animals or humans [56,57]. A mixed culture of SRB can, therefore, be exploited in the future for further fine-tuning of this robust system of H_2S production. Since the amount of Fe^{2+} ions also impacts the production of H_2S, one could sprinkle a pinch of iron and steel slag or red mud on the growth media to further enhance the efficiency of the set up, and could also consider gypsum waste from building sites in place of mineral waters [47,58].

Eventually, the notion of using cheap organic by-products or waste as carbon sources to fuel the microbial fermentation to produce valuable (raw) chemicals may be expanded further. Considering that such fermentations are indeed "robust" or, as one may call it, "dirty", conventional extractions and purifications may be difficult and expensive, thus focussing on "phase separations", for instance, by involving gasses, may be the most economical way forward. Here, H_2S stands out, also thanks to its low solubility in aqueous media (117 mmol L^{-1} at 20 °C), yet it is not alone, and other biologically generated gasses may also be produced by waste fermentation. In addition to the more usual suspects, such as methane (CH_4) and hydrogen (H_2), they also include chemically interesting gases such as ammonia (NH_3) from nitrogen-rich waste materials, including urine.

5. Conclusions

In order to make chemical and pharmaceutical processes sustainable, one could consider a number of strategies, from recycling to replacement. Turning organic waste into value is one of these promising avenues since it utilizes readily available yet hitherto worthless or even nuisance materials. Although there is no free lunch, there are plenty of by-products and leftovers on the table which can be used, from cabbage juice and silage on one the side to spoiled milk, natural mineral water and compost on the other. These materials, together with inorganic wastes, such as gypsum, are omnipresent in modern society, they can be found in considerable quantities and, thanks to a continued supply, are unlikely to run out in the near future.

The real challenge is to find or produce valuable materials from this waste without excessive processing or purification which eventually may negate the economical or ecological gains. Indeed, stimulating chemistry in a waste environment is far from trivial, and purification may be like looking for the literal needle in the haystack—or gold in a French soccer star's manure—a cumbersome and unpleasant experience of little economical value [59].

In this context, fermentation by robust (soil) bacteria may resolve some of these issues and involving a gas such as H_2S or NH_3 certainly provides an elegant solution to many of these problems. In the future, these processes need to be studied, optimized and upscaled further in order to gain ecological and economic value. Based on our initial studies presented here, this seems to be possible. To put is poetically, it is not a sham—the right design may really turn poo into shampoo!

Supplementary Materials: The following supporting information can be downloaded at https://www.mdpi.com/article/10.3390/ma18122784/s1, Table S1: The gradual elimination of ingredients of standard medium to determine the most essential ingredients required for the growth of *D. desulfuricans*; Table S2: Substitution of ingredients of standard medium with more common/waste ingredients; Table S3: The elemental composition of selenium sulfide harvested from the SRB cultured in spoiled milk and Na_2SO_4. The Se:S ratio slightly differs from that obtained from the standard medium, as confirmed by CHNS, ICP-OES and EDX analysis; Table S4: The elemental composition of selenium sulfide harvested from the SRB cultured in a 1:2 mixture of spoiled milk and mineral water analyzed by CHNS, ICP-OES and EDX analysis; Figure S1: Raman spectra of elemental selenium, elemental sulfur, commercial selenium sulfide and biosynthesized selenium sulfide show that the Se–S vibrational band appears only in the selenium sulfide samples, while slight shifts in the Se–Se and S–S modes are observed compared to the elemental references; Figure S2: The spoiled milk and Na_2SO_4 mediated selenium sulfide powder was analyzed using SEM which revealed the presence of agglomerated selenium sulfide particles (left). The presence of sulfur and selenium at 34.56 wt% and 65.44 wt% was affirmed by EDX (right); Figure S3: The structural fingerprints of selenium sulfide produced using a mixture of spoiled milk and Na_2SO_4. The sample selenium sulfide (A) and commercial selenium sulfide (B). Similar vibrational modes could be observed between biologically produced selenium sulfide and commercial selenium sulfide; Figure S4: Selenium sulfide produced using the spoiled milk and mineral water was analyzed using SEM (left). EDX analysis confirmed the presence of sulfur and selenium at 39.29 wt% and 60.71 wt%, respectively (right); Figure S5: The comparison of the structural fingerprints of selenium sulfide produced using a mixture of spoiled milk and mineral water (A) and commercial selenium sulfide (B) as well as the standard medium mediated selenium sulfide (C).

Author Contributions: Conceptualization, C.J. and M.J.N.; methodology, S.S., A.Y.A. and E.T.; validation, C.J., M.J.N., R.L., K.-H.S. and C.F.-S.; formal analysis, C.J.; investigation, S.S., A.Y.A. and E.T.; data curation, M.J.N.; writing—original draft preparation, C.J.; writing—review and editing, M.J.N., R.L., K.-H.S., C.F.-S., S.S., A.Y.A. and E.T.; supervision, C.J., M.J.N., R.L., K. H.S. and C.F. S. All authors have read and agreed to the published version of the manuscript.

Funding: The authors would like to acknowledge the financial support provided by the University of Saarland, Saarbruecken, Germany.

Institutional Review Board Statement: Not applicable.

Informed Consent Statement: Not applicable.

Data Availability Statement: The original contributions presented in this study are included in the article and Supplementary Materials. Further inquiries can be directed to the corresponding authors.

Acknowledgments: The authors would like to thank Sodomir Popojuk and Ken Rory for advice, discussions and proof-reading of the manuscript. The authors would like to express their gratitude to Anna Elizabeth Schmitz for her assistance in collecting the silage, as well as to Tobon Ostonaeva and Selma Benfares for their valuable support in preparing the reagents. Special thanks go to many other colleagues from the Academiacs International network (www.academiacs.eu) (accessed on 2 April 2025) and the "Pharmasophy" team for their helpful discussions and advice.

Conflicts of Interest: The authors declare no conflicts of interest.

References

1. Heath, R.S.; Ruscoe, R.E.; Turner, N.J. The Beauty of Biocatalysis: Sustainable Synthesis of Ingredients in Cosmetics. *Nat. Prod. Rep.* **2022**, *39*, 335–388. [CrossRef] [PubMed]
2. Ganesh, K.N.; Zhang, D.; Miller, S.J.; Rossen, K.; Chirik, P.J.; Kozlowski, M.C.; Zimmerman, J.B.; Brooks, B.W.; Savage, P.E.; Allen, D.T.; et al. Green Chemistry: A Framework for a Sustainable Future. *Environ. Sci. Technol. Lett.* **2021**, *8*, 487–491. [CrossRef]
3. Martinengo, B.; Diamanti, E.; Uliassi, E.; Bolognesi, M.L. Harnessing the 12 Green Chemistry Principles for Sustainable Antiparasitic Drugs: Toward the One Health Approach. *ACS Infect. Dis.* **2024**, *10*, 1856–1870. [CrossRef] [PubMed]
4. Wynendaele, E.; Furman, C.; Wielgomas, B.; Larsson, P.; Hak, E.; Block, T.; Van Calenbergh, S.; Willand, N.; Markuszewski, M.; Odell, L.R.; et al. Sustainability in Drug Discovery. *Med. Drug Discov.* **2021**, *12*, 100107. [CrossRef]
5. Razouk, A.; Tiganescu, E.; von Glahn, A.J.; Abdin, A.Y.; Nasim, M.J.; Jacob, C. The Future in the Litter Bin—Bioconversion of Food Waste as Driver of a Circular Bioeconomy. *Front. Nutr.* **2024**, *11*, 1325190. [CrossRef]
6. Wang, Q.; Li, H.; Feng, K.; Liu, J. Oriented Fermentation of Food Waste towards High-Value Products: A Review. *Energies* **2020**, *13*, 5638. [CrossRef]
7. Sabater, C.; Ruiz, L.; Delgado, S.; Ruas-Madiedo, P.; Margolles, A. Valorization of Vegetable Food Waste and By-Products Through Fermentation Processes. *Front. Microbiol.* **2020**, *11*, 581997. [CrossRef]
8. Zhu, S.H.; Zhu, S.M.; Zhu, L. Production Method of Selenium Sulfide. CN 201010571362, 25 November 2010.
9. Steudel, R.; Laitinen, R. Cyclic Selenium Sulfides. In *Inorganic Ring Systems*; Springer: Berlin/Heidelberg, Germany, 1982; pp. 177–197.
10. Laitinen, R.S. Selenium Sulfide Ring Molecules. *Acta Chem. Scand.* **1987**, *41*, 361–376. [CrossRef]
11. Laitinen, R.S.; Pakkanen, T.A. Selenium-77 NMR Spectroscopic Characterization of Selenium Sulfide Ring Molecules Se_nS_{8-n}. *Inorg. Chem.* **1987**, *26*, 2598–2603. [CrossRef]
12. Ausma, T.; De Kok, L.J. Atmospheric H_2S: Impact on Plant Functioning. *Front. Plant Sci.* **2019**, *10*, 743. [CrossRef]
13. Aiuppa, A.; Moussallam, Y. Hydrogen and hydrogen sulphide in volcanic gases: Abundance, processes, and atmospheric fluxes. *Comptes Rendus. Géoscience* **2024**, *356*, 85–108. [CrossRef]
14. Carbajo, J.M.; Maraver, F. Sulphurous Mineral Waters: New Applications for Health. *Evid.-Based Complement. Altern. Med.* **2017**, *2017*, 8034084. [CrossRef] [PubMed]
15. Malone Rubright, S.L.; Pearce, L.L.; Peterson, J. Environmental Toxicology of Hydrogen Sulfide. *Nitric Oxide* **2017**, *71*, 1–13. [CrossRef]
16. Tiganescu, E.; Safinazlou, S.; Abdin, A.Y.; Lilischkis, R.; Schäfer, K.-H.; Fink-Straube, C.; Nasim, M.J.; Jacob, C. Selenium Disulfide from Sustainable Resources: An Example of "Redneck" Chemistry with a Pinch of Salt. *Materials* **2024**, *17*, 5733. [CrossRef]
17. Quellen. Available online: https://www.badnenndorf.de/gesundheit-wellness/quellen (accessed on 3 April 2025).
18. Tian, G.; Yeung, M.; Xi, J. H_2S Emission and Microbial Community of Chicken Manure and Vegetable Waste in Anaerobic Digestion: A Comparative Study. *Fermentation* **2023**, *9*, 169. [CrossRef]
19. Austigard, Å.D.; Svendsen, K.; Heldal, K.K. Hydrogen Sulphide Exposure in Waste Water Treatment. *J. Occup. Med. Toxicol.* **2018**, *13*, 10. [CrossRef]
20. Kushkevych, I.; Vítězová, M.; Vítěz, T.; Bartoš, M. Production of Biogas: Relationship between Methanogenic and Sulfate-Reducing Microorganisms. *Open Life Sci.* **2017**, *12*, 82–91. [CrossRef]
21. Kushkevych, I.; Vítězová, M.; Vítěz, T.; Kováč, J.; Kaucká, P.; Jesionek, W.; Bartoš, M.; Barton, L. A New Combination of Substrates: Biogas Production and Diversity of the Methanogenic Microorganisms. *Open Life Sci.* **2018**, *13*, 119–128. [CrossRef]
22. Kushkevych, I.; Cejnar, J.; Treml, J.; Dordević, D.; Kollar, P.; Vítězová, M. Recent Advances in Metabolic Pathways of Sulfate Reduction in Intestinal Bacteria. *Cells* **2020**, *9*, 698. [CrossRef]
23. Kushkevych, I.V. Dissimilatory Sulfate Reduction in the Intestinal Sulfate-Reducing Bacteria. *Біологічні студії/Stud. Biol.* **2017**, *10*, 197–228. [CrossRef]
24. Bagheri Novair, S.; Biglari, Z.; Asgari Lajayer, B.; Shu, W.; Price, G.W. The Role of Sulphate-Reducing Bacteria (SRB) in Bioremediation of Sulphate-Rich Wastewater: Focus on the Source of Electron Donors. *Process Saf. Environ. Prot.* **2024**, *184*, 190–207. [CrossRef]
25. Gopi Kiran, M.; Pakshirajan, K.; Das, G. An Overview of Sulfidogenic Biological Reactors for the Simultaneous Treatment of Sulfate and Heavy Metal Rich Wastewater. *Chem. Eng. Sci.* **2017**, *158*, 606–620. [CrossRef]

26. Li, Z.; Yang, J.; Guo, H.; Kumseranee, S.; Punpruk, S.; Mohamed, M.E.; Saleh, M.A.; Gu, T. Carbon Source Starvation of a Sulfate-Reducing Bacterium–Elevated MIC Deterioration of Tensile Strength and Strain of X80 Pipeline Steel. *Front. Mater.* **2021**, *8*, 794051. [CrossRef]
27. Kung, L.; Shaver, R.D.; Grant, R.J.; Schmidt, R.J. *Silage Review:* Interpretation of Chemical, Microbial, and Organoleptic Components of Silages. *J. Dairy Sci.* **2018**, *101*, 4020–4033. [CrossRef]
28. Kung, L.; Shaver, R. Interpretation and Use of Silage Fermentation Analysis Reports. *Focus Forage* **2001**, *3*, 1–5.
29. Bouteille, R.; Gaudet, M.; Lecanu, B.; This, H. Monitoring Lactic Acid Production during Milk Fermentation by in Situ Quantitative Proton Nuclear Magnetic Resonance Spectroscopy. *J. Dairy Sci.* **2013**, *96*, 2071–2080. [CrossRef]
30. Leibniz Institute. DSMZ: Video Tutorials. Available online: https://www.dsmz.de/shop-1/faqs/video-tutorials (accessed on 3 April 2025).
31. PlantApp—Botanist in Your Pocket. Available online: https://plantapp.app/ (accessed on 3 April 2025).
32. Taylor, K.A.C.C. A Simple Colorimetric Assay for Muramic Acid and Lactic Acid. *Appl. Biochem. Biotechnol.* **1996**, *56*, 49–58. [CrossRef]
33. Li, Z.-G. Chapter Six—Quantification of Hydrogen Sulfide Concentration Using Methylene Blue and 5,5′-Dithiobis(2-Nitrobenzoic Acid) Methods in Plants. In *Methods in Enzymology*; Cadenas, E., Packer, L., Eds.; Hydrogen Sulfide in Redox Biology, Part A; Academic Press: Cambridge, MA, USA, 2015; Volume 554, pp. 101–110.
34. Tiganescu, E.; Abdin, A.Y.; Razouk, A.; Nasim, M.J.; Jacob, C. The Redox Riddle of Selenium Sulfide. *Curr. Opin. Chem. Biol.* **2023**, *76*, 102365. [CrossRef]
35. Han, X.; Tong, Y. Vibrational Dynamics of SxSey Ring Clusterson AB Initio Ccalculation. *Chalcogenide Lett.* **2018**, *15*, 459–466.
36. Yoon, K.Y.; Woodams, E.E.; Hang, Y.D. Production of Probiotic Cabbage Juice by Lactic Acid Bacteria. *Bioresour. Technol.* **2006**, *97*, 1427–1430. [CrossRef]
37. Alm, L. Effect of Fermentation on L(+) and D(−) Lactic Acid in Milk. *J. Dairy Sci.* **1982**, *65*, 515–520. [CrossRef]
38. Farag, M.A.; Jomaa, S.A.; Abd El-Wahed, A.; R. El-Seedi, H. The Many Faces of Kefir Fermented Dairy Products: Quality Characteristics, Flavour Chemistry, Nutritional Value, Health Benefits, and Safety. *Nutrients* **2020**, *12*, 346. [CrossRef] [PubMed]
39. Shah, N.P.; Lankaputhra, W.E.V.; Britz, M.L.; Kyle, W.S.A. Survival of *Lactobacillus acidophilus* and *Bifidobacterium bifidum* in Commercial Yoghurt during Refrigerated Storage. *Int. Dairy J.* **1995**, *5*, 515–521. [CrossRef]
40. Settachaimongkon, S.; van Valenberg, H.J.F.; Gazi, I.; Nout, M.J.R.; van Hooijdonk, T.C.M.; Zwietering, M.H.; Smid, E.J. Influence of *Lactobacillus plantarum* WCFS1 on Post-Acidification, Metabolite Formation and Survival of Starter Bacteria in Set-Yoghurt. *Food Microbiol.* **2016**, *59*, 14–22. [CrossRef]
41. Nugroho, A.D.W.; Kleerebezem, M.; Bachmann, H. A Novel Method for Long-Term Analysis of Lactic Acid and Ammonium Production in Non-Growing Lactococcus Lactis Reveals Pre-Culture and Strain Dependence. *Front. Bioeng. Biotechnol.* **2020**, *8*, 580090. [CrossRef]
42. Shangguan, Y.; Yang, D.; Zhao, L.; Rao, L.; Liao, X. High-Pressure-Induced Viable but Non-Culturable Lactic Acid Bacteria Inhibit Its Post-Acidification. *Bioresour. Technol.* **2025**, *422*, 132221. [CrossRef] [PubMed]
43. Ensinger Schiller Quelle—Ensinger. Available online: https://www.ensinger.de/produkte/ensinger-schiller-quelle (accessed on 3 April 2025).
44. Asghari-Paskiabi, F.; Imani, M.; Rafii-Tabar, H.; Razzaghi-Abyaneh, M. Physicochemical Properties, Antifungal Activity and Cytotoxicity of Selenium Sulfide Nanoparticles Green Synthesized by Saccharomyces Cerevisiae. *Biochem. Biophys. Res. Commun.* **2019**, *516*, 1078–1084. [CrossRef]
45. Wu, J.; He, R.; Cheng, S.; Han, L.; Hong, H.; Zhu, N. Simultaneous Immobilization of CO_2 and H_2S by Propargyl Amines under Mild Conditions: Efficient Synthesis of Thiazolidine-2-Ones. *ACS Sustain. Chem. Eng.* **2022**, *10*, 1214–1219. [CrossRef]
46. Aguilar, L.; Zha, S.; Cheng, Z.; Winnick, J.; Liu, M. A Solid Oxide Fuel Cell Operating on Hydrogen Sulfide (H_2S) and Sulfur-Containing Fuels. *J. Power Sources* **2004**, *135*, 17–24. [CrossRef]
47. Jia, R.; Tan, J.L.; Jin, P.; Blackwood, D.J.; Xu, D.; Gu, T. Effects of Biogenic H_2S on the Microbiologically Influenced Corrosion of C1018 Carbon Steel by Sulfate Reducing *Desulfovibrio vulgaris* Biofilm. *Corros. Sci.* **2018**, *130*, 1–11. [CrossRef]
48. Kowalski, A.; Agati, G.; Grzegorzewska, M.; Kosson, R.; Kusznierewicz, B.; Chmiel, T.; Bartoszek, A.; Tuccio, L.; Grifoni, D.; Vågen, I.M.; et al. Valorization of Waste Cabbage Leaves by Postharvest Photochemical Treatments Monitored with a Non-Destructive Fluorescence-Based Sensor. *J. Photochem. Photobiol. B Biol.* **2021**, *222*, 112263. [CrossRef] [PubMed]
49. IndexBox. Global Cabbage Market to Reach 80M Tonnes by 2025. Available online: https://www.globaltrademag.com/global-cabbage-market-to-reach-80m-tonnes-by-2025/ (accessed on 4 April 2025).
50. Le Roux, B.; Van der Laan, M.; Vahrmeijer, T.; Annandale, J.G.; Bristow, K.L. Water Footprints of Vegetable Crop Wastage along the Supply Chain in Gauteng, South Africa. *Water* **2018**, *10*, 539. [CrossRef]
51. World Cabbage Production by Country. Available online: https://www.atlasbig.com/en-gb/countries-by-cabbage-production (accessed on 3 April 2025).
52. Juicernet. *Average Juice Yields for Fruits & Vegetables*; Juicernet: Jupiter, FL, USA, 2021.

53. Holden, P. *The Woe of Wasted Milk*; Sustainable Food Trust: Bristol, UK, 2023.
54. Campbell, C.G.; Feldpausch, G.L. *Invited Review:* The Consumer and Dairy Food Waste: An Individual plus Policy, Systems, and Environmental Perspective. *J. Dairy Sci.* **2022**, *105*, 3736–3745. [CrossRef] [PubMed]
55. The Estimated Amount, Value, and Calories of Postharvest Food Losses at the Retail and Consumer Levels in the United States | Economic Research Service. Available online: https://www.ers.usda.gov/publications/pub-details?pubid=43836 (accessed on 3 April 2025).
56. Wu, Y.-F.; Huang, H.; Zhang, J.; Hu, G.; Wang, J.; Peng, C.; Kappler, A.; Zhao, F.-J. Sulfate-Mediated Fe(III) Mineral Reduction Accelerates Arsenic Mobilization by a *Desulfovibrio* Strain Isolated from Paddy Soil. *Sci. Total Environ.* **2024**, *954*, 176529. [CrossRef]
57. Kushkevych, I.; Hýžová, B.; Vítězová, M.; Rittmann, S.K.-M. Microscopic Methods for Identification of Sulfate-Reducing Bacteria from Various Habitats. *Int. J. Mol. Sci.* **2021**, *22*, 4007. [CrossRef]
58. Jia, R.; Wang, D.; Jin, P.; Unsal, T.; Yang, D.; Yang, J.; Xu, D.; Gu, T. Effects of Ferrous Ion Concentration on Microbiologically Influenced Corrosion of Carbon Steel by Sulfate Reducing Bacterium *Desulfovibrio vulgaris*. *Corros. Sci.* **2019**, *153*, 127–137. [CrossRef]
59. Hodgetts, R. Gold-Coated Steak Backlash Earns Franck Ribery 'Heavy Fine'. Available online: https://www.cnn.com/2019/01/06/sport/franck-ribery-gold-steak-dubai-bayern-munich-fine-spt-intl/index.html (accessed on 3 April 2025).

materials

MDPI

Article

Durability Considerations in Replacing Blast Furnace Slag with Low-Grade Calcined Clay and Natural Pozzolan in Quaternary Cements

Juan Manuel Etcheverry [1,*], Laurent Detemmerman [1,2], Krist Degezelle [2], Vadim Grigorjev [1], Laurena De Brabandere [1] and Nele De Belie [1,*]

1 Magnel-Vandepitte Laboratory for Structural Engineering and Building Materials, Ghent University, 9052 Ghent, Belgium
2 DEVAGRO, Beton and Recyclage Harelbeke, 8530 Harelbeke, Belgium
* Correspondence: juanmanuel.etcheverry@ugent.be (J.M.E.); nele.debelie@ugent.be (N.D.B.)

Abstract: Belgium and the EU-27 face a shortage of suitable supplementary cementitious materials (SCMs) capable of supporting high levels of Portland cement substitution. To reduce CO_2 emissions from the cement industry, blended cements incorporating low-grade calcined clay, limestone, and lava (a natural pozzolan) are investigated. Calcined clay is combined with limestone to produce a limestone–calcined clay cement (LC3). The reactivity of these new blends is assessed using isothermal calorimetry and compared to a reference blend with ground-granulated blast-furnace slag (GGBFS). Results show that mixtures with calcined clay develop slightly lower 28-day strength than those with GGBFS, while blends with lava exhibit strength gains only at later ages due to delayed pozzolanic activity. Overall, concrete made with low-grade calcined clay and lava achieves comparable compressive strength to the reference (CEM III/A), but with higher capillary porosity, leading to increased water absorption, drying shrinkage, and reduced freeze–thaw resistance. Despite these durability limitations, the sustainability assessment reveals that the LC3 mix with low-grade clay and lava has a lower global warming potential per unit strength at 28 days than CEM III/A and is competitive with CEM III/B.

Keywords: calcined clay; lava; natural pozzolan; reactivity assessment; concrete durability

1. Introduction

The production of Portland cement (PC) is responsible for a large part of global carbon dioxide emissions, estimated between 6% and 10% [1,2]. Given the expected rise in demand for concrete, there is a considerable need for alternative supplementary cementitious materials (SCMs) with local availability. The replacement of PC with SCMs is an important part of the CEMBUREAU plan to decarbonise the cement industry by 2050 [3,4]. SCMs reduce emissions by partially replacing PC, as they are typically not produced at the same high temperature of PC (some of them do not even require thermal activation) and because there is no limestone decomposition, the main process during which CO_2 is released. A well-established and frequently used SCM is ground-granulated blast-furnace slag (GGBFS). However, the supply of GGBFS itself is expected to be significantly reduced in the near future due to changes in the steel production process; therefore, it is unable to meet the large demand of the cement industry. The amount of slag available is currently 5–10% of the amount of cement produced worldwide [5]. GGBFS is a valuable SCM (with a comparable price to PC) widely regarded as environmentally friendly in the cement industry, largely due to its favourable allocation methods (i.e., environmental burdens are mostly attached to steel as the main product because of the classification of the slag as an industrial by-product). Therefore, local cement and concrete producers are constantly in search of alternative SCMs that can replace GGBFS without significantly increasing the PC content in concrete mixes (maintaining a similar SCM content as in CEM III cements, while keeping strength development and durability performance).

An existing alternative is limestone–calcined clay cement (LC3). Previous research found that by combining limestone (LL) with calcined clay (CC), up to 50% of PC can be replaced without compromising performance [5]. In addition, there is enough proof that the development of a microstructure is not limited by the availability of portlandite [6]. Usually, clays with a kaolinite content in the range 40–60% are required as a raw resource which are then calcined around 650–800 °C to produce reactive calcined clays [7]. The properties of the CC depend mostly on the kaolinite content of the raw clay, with the fineness being the second most important parameter [8]. During the calcination process, kaolinite transforms into metakaolin. This metakaolin is mainly responsible for reactivity and in combination with PC and limestone, it is able to react within the first week, allowing the LC3 mixture to reach a similar strength to PC mixes. Non-kaolinitic clay phases such as illite or montmorillonite are also reactive after calcination but to a lesser degree compared to kaolinite and require broader calcination temperatures to reach dihydroxylation (500–900 °C) [5,9,10].

A different SCM that has recently received attention is recycled concrete fines. The crushing of concrete debris is often conducted with the aim of re-using the coarse aggregates in recycled aggregate concrete (RAC) and the recycled concrete powder is generally the last fraction to be repurposed [11]. Natural pozzolans, such as lava (LV), can also be used as an SCM and have the advantage of being locally available, similarly to clay, and thus able to partially sustain the large demand for cement [12]. Secondly, no additional heat is required for the activation of the material. However, natural pozzolans tend to possess only a limited reactivity and the pozzolanic activity is often only noticeable after an extended curing period [3].

The replacement of GGBFS with less reactive SCMs (usually containing less calcium) does not only affect the mechanical strength of concrete but also the durability characteristics. The reactivity of the SCM influences the development of the pore structure. When comparing blended cements containing GGBFS with traditional PC mixes, a lower connectivity of the pore network tends to be pointed out as beneficial for the durability characteristics of concrete. The ingress of water (carrying harmful substances) is part of almost every deleterious process in concrete structures [13]. In addition, water transport is key for freeze–thaw resistance. Concrete may deteriorate by the freezing of water inside saturated pores. Especially for clays with a kaolinite content above 40%, a significant pore refinement (in comparison to PC) has been reported, which is expected to contribute to freeze–thaw resistance. Previous studies showed that concrete mixes made with calcined clay containing sufficient kaolinite result in an improved freeze–thaw resistance comparable to that of concrete with GGBFS [14,15]. In contrast, clays containing large fractions of illite and montmorillonite have been reported to increase capillary porosity [9,13], consistent with lower reactivity, rather than kaolinite-type clays. Therefore, the kaolinite content of the calcined clay is not only important for the mechanical strength but also for the durability of concrete [16].

This study investigates the reactivity of an LC3 blend using a low-grade CC in combination with minor additions of the SCMs available in Benelux as potential replacement for CEM III/A. Lava is incorporated with the aim of enhancing the long-term performance of the blended cement containing low-grade calcined clays. Furthermore, naturally carbonated recycled fines (from stockpiles) are used, which are unlikely to have any chemical reactivity [17]. However, they can reduce the extraction of raw limestone, since they contain a high fraction of $CaCO_3$. A small decrease in workability (in comparison to natural limestone filler) can be acceptable if it allows for the re-utilization of waste material. Isothermal calorimetry is used to determine the reactivity of the blended cements within the first days of hydration and to optimize the sulphate content of ternary and quaternary cement blends. Compressive strength tests on mortar samples, along with drying and autogenous shrinkage assessments, are used to identify the most promising mix for scaling up to concrete and conducting durability tests. Quaternary blends containing PC, LL, calcined clay, and lava are assessed for porosity [18], pore connectivity, shrinkage, and freeze–thaw resistance with de-icing salts. The results are compared to those of a reference mix containing GGBFS. In summary, this study aims to replace CEM III/A while maintaining strength development and to assess properties that could potentially compromise durability in non-structural concrete field applications.

2. Materials and Methods

2.1. Materials

2.1.1. Cementitious Materials

The reference cement blend (CEM III/A according to EN 197-1) is prepared with 50% GGBFS and 50% PC (CEM I 52.5 N). The limestone is a commercial high-purity $CaCO_3$ filler. The calcined clay is sourced from a local quarry and calcined in the laboratory at 750 °C for 4 h. Thermogravimetric analysis (TGA) reveals a kaolinite content around 20% (Figure S1 in the Supporting Information), which classifies CC as a low-grade calcined clay. The lava is a volcanic material originating from the Eifel region. The lava is commercially available as a 0–4 mm fraction. Recycled concrete fines are sourced from the waste fraction of the concrete debris crushing process. The fine fraction (0–2 mm) is obtained as a secondary product of the process. Spraying with a ~1 wt.% phenolphthalein-ethanol solution reveals that the recycled fines are carbonated, consistent with long-term exposure (from several months to years) to the environment in stockpiles.

The CC, lava (LV), and recycled fines (F) are milled in the laboratory to reach a similar fineness to PC (see particle size distribution (PSD), D_{v10}, D_{v50}, and D_{v90} in Table 1). The particle size distribution (PSD) is determined using an Occhio FC200 M-HR optical particle analyser, Occhio s.a., Angleur, Belgium, which determines particle sizes in the range of 200 nm to 800 µm. The chemical composition is determined with X-ray fluorescence spectrometry (XRF), Malvern Panalytical, Malvern, UK. The loss on ignition (LOI) obtained for LL is utilized to convert the CaO in $CaCO_3$. The particle density is determined using the Le Chatelier flask according to ASTM C188. The fractiles of the PSD ($D_{v,10}$, $D_{v,50}$, and $D_{v,90}$), the oxide composition, and the particle density of the materials are reported in Table 1. The full particle size distribution curves can be observed in the Supporting Information File (Figure S2).

Table 1. Physical properties and chemical composition of the binders.

	PC	CC	LV	F	LL	GGBFS
$D_{v,10}$ (µm)	9.50	7.01	6.39	4.95	6.42	5.86
$D_{v,50}$ (µm)	21.47	27.93	16.40	14.38	13.16	13.32
$D_{v,90}$ (µm)	37.81	49.32	38.24	33.78	34.29	24.21
Density (g/cm^3)	3.15	2.30	2.75	2.71	2.69	2.89

Table 1. *Cont.*

	PC	CC	LV	F	LL	GGBFS
Chemical composition (%)						
CaO	73.20	10.78	17.39	46.22	-	40.25
Fe_2O_3	3.96	12.16	26.93	3.25	0.09	0.43
SiO_2	12.69	36.23	28.99	36.30	0.39	30.48
Al_2O_3	3.79	9.44	7.01	5.91	0.13	12.18
MgO	1.45	2.28	3.71	2.17	0.50	8.13
SO_3	2.17	0.16	0.16	2.01	-	0.08
K_2O	0.24	1.52	2.80	0.50	-	0.45
Na_2O	0.87	0.56	1.18	1.35	-	0.58
TiO_2	1.56	1.15	2.73	2.21	-	-
MnO	0.06	0.02	0.20	0.06	-	0.25
$CaCO_3$	-	-	-	-	98	-
LOI (%)					41.3	

2.1.2. Blended Cements

Six different blended cements are initially prepared to investigate the effectiveness of the SCMs in the mixes. The blended cements are made using different amounts of CC, lava, recycled fines, and limestone, as well as gypsum. Binder formulations are reported in Table 2. An in-house-made CEM III/A with 50% GGBFS is used as the main reference. A mix incorporating 45% GGBFS, 10% limestone filler, and 45% PC is also included to clarify the effect of partially substituting GGBFS with LL in paste and mortar experiments. Paste specimens are prepared as described in Section 2.2.2. Mortar prisms with the six different blended cements (Table 2) are prepared according to EN 196-1, using a sand-to-binder ratio of 3:1 and a w/b ratio of 0.50.

Table 2. Binder formulations for the studied pastes and mortar systems (in wt%).

Mixture ID	PC	CC	LL	F	LV	GGBFS	Gypsum	Superplasticizer (mL/ 100 g PC)
LC3-65	65	23.5	11.5	-	-	-	3.50	0.20
20CC-5LL-10F	65	20	5	10	-	-	3.50	0.20
20CC-9LL-6F-10LV	55	20	9	6	10	-	3.50	0.20
25CC-10LL-15LV	50	25	10	-	15	-	3.50	0.20
45GGBFS-10LL	45	-	10	-	-	45	2.75	
CEM III/A (Ref. mix)	50	-	-	-	-	50	2.75	

2.1.3. Concrete Mixes

Two batches of concrete samples are made, one with the CEM III/A and another with the 25CC-10LL-15LV. The latter quaternary blended cement (concrete) is designated as a CEM II/C-M according to EN 197-5. The CEM III/A concrete with 50% GGBFS is used as a reference. The mix design of both concrete types is given in Table 3. The binder content is the same in both concrete mixtures, 340 kg per m^3, and the concrete is made with 100% coarse recycled aggregates (RAs) with BENOR certification. The water absorption after 24 h of the RAs is 4.9%. To compensate for the absorption, additional water is added to the mixture. The flow and the air content of the fresh concrete is measured according to EN 12350-5 and EN 12350-7, resulting in a similar flow and air content for both mixtures.

Table 3. Mix design of the two concrete types. PC, CC, LL, LV, and GGBFS are considered as binders for the calculation of the w/b ratio.

Material (kg/m^3)	CEM II/C-M	CEM III/A
Limestone 0/4	346	349
Recycled aggregate 4/20	889	897
Sea sand 0/2	508	513
CEM I 52.5 N	173	173
Calcined clay	87	-
Limestone powder	35	-
Lava	51	-
GGBFS	-	173
SIKA ViscoCrete 1090	2.5	1.5
$Water_{free}$	173	173
$Water_{add}$	44	44
w/b ratio	0.5	0.5
Flow (EN 12350-5) (mm)	440	440
Air content (EN 12350-7)	1.7%	1.7%

2.2. Methods

2.2.1. Water Demand

A standard consistency paste (SCP) test is performed to determine the influence of calcined clay and lava on the water demand of the mixtures. The necessary w/b ratio for a standard consistency paste is determined according to ASTM C187. Three mixtures (Table 4) are tested, all including 65% PC and 10% limestone. The remaining 25% is either quartz (as reference), CC, or lava. These ratios are chosen to have a similar SCM replacement ratio to the LC3 system in Table 2.

Table 4. Mixes for water demand determination on calcined clays and lava.

ID \ Mix	PC	LL	Quartz	CC	LV
PC + LL + Q	65	10	25	-	-
PC + LL + CC	65	10	-	25	-
PC + LL + LV	65	10	-	-	25

The SCP test shows a noticeable difference in the mixing water necessary to reach a standard consistency paste. The mixtures with quartz, lava, and calcined clay need a w/b ratio of 0.27, 0.24, and 0.35, respectively. Thus, calcined clay increases the water demand considerably, whereas lava keeps the water demand constant or even slightly lower compared to the mix containing quartz. This means that lava may increase workability. The reduced workability of mixtures with CC must be compensated with either additional water or a superplasticizer addition. In this study, the water-to-binder (w/b) ratio is always kept constant and differences in workability between the different mixtures are compensated by superplasticizer addition.

2.2.2. Isothermal Calorimetry

Isothermal calorimetry is conducted at 40 °C using a TAM Air calorimeter, TA Instruments, New Castle, DE, USA. All dry materials and mixing water are accurately weighed and preconditioned in an oven at 40 °C for 24 h before testing. The paste mixes are prepared by adding the demineralised water (w/b = 0.40) and superplasticizer to the mixtures with CC to achieve a sufficient flow. No superplasticizer is needed for the mixes containing GGBFS (Table 2). The paste is mixed at 1600 rpm for 2 min. Then, 14 g of paste is taken from the mixing container (70 g) using a plastic pipette and transferred to the glass vial. The lid is placed, and the container is lowered into the calorimeter, where the heat released is recorded for 4 days.

In addition to isothermal calorimetry on the novel binder mixes, a modified version of the R3 test method is used to determine the reactivity of the novel SCMs. The modified version of the R3 test [19] uses a potassium hydroxide solution (KOH) and calcium hydroxide (CH) (differing from the KOH, CH, and a potassium sulphate solution in ASTM C1897-20) and test temperature of 50 °C instead of 40 °C

2.2.3. Mechanical Strength Tests

The mortar prisms described in Section 2.1 are demoulded after 24 h and cured in water (20 °C) until the required test age (2, 7, 28, 56, and 90 days) for the determination of the compressive strength.

The compressive strength of the concrete mixes (Table 3) is determined by casting 150 mm cubes according to EN 12390-2/3. The samples are cured in water (20 °C) until the required test age (2, 7, 28, 56, and 90 days) for the determination of the compressive strength.

2.2.4. Durability Assessment

All concrete specimens are demoulded after 24 h and stored in a curing room (20 °C and >95% RH). The exceptions are the specimens used to measure the shrinkage, which, immediately after demoulding, are stored in a climate-controlled chamber at 20 °C and 60% RH.

The samples for porosity and the capillary imbibition tests (10 cm diameter and 5 cm height) are sawn from larger cylinders (10 cm diameter and 20 cm height). Top and bottom surfaces (2.5 cm height) are discarded. Curing time for all concrete samples for durability is 60 days in total.

2.2.5. Capillary Imbibition

Three cylindrical samples (5 cm in height and 10 cm in diameter) are tested for capillary imbibition for each concrete type. Samples are cured at 95% RH and 20 °C for 60 days and preconditioned by placing them in an oven at 40 °C until the difference in the mass over 24 h is lower than 0.1%. Immediately after, the samples are placed in plastic bags for three days to enable the samples to reach a homogeneous moisture distribution (IRAM 1871:2021). The sides of the specimens are wrapped with aluminum-butyl tape (waterproof) to ensure unidirectional water flow during capillary imbibition. The level of water in contact with the specimens is always in between 1 and 3 mm. The mass of the samples is recorded after 15 min, 30 min, 1 h, 2 h, 3 h, 4 h, and 6 h. Next, the specimens are weighed every 24 h for one week and weekly up to 56 days.

2.2.6. Porosity

First, three cylindrical samples (5 cm in height and 10 cm in diameter) are put into a vacuum chamber for 2 h when the water is allowed to enter the container while maintaining the vacuum. Next, the samples are left inside the water-filled tank for 24 h. Then, the samples are weighed in open air (M_a) and under water (M_W). The capillary porosity (Equation (1)) and open porosity (Equation (2)) are determined by drying the samples at 40 °C (M_{d40}) or 105 °C (M_{d105}), until a mass difference over a period of 24 h of less than 0.1%.

$$CP\ (\%) = \frac{M_a - M_{d40}}{M_a - M_w} \cdot 100 \quad (1)$$

$$OP\ (\%) = \frac{M_a - M_{d105}}{M_a - M_w} \cdot 100 \quad (2)$$

2.2.7. Mercury Intrusion Porosimetry (MIP)

The porosity (MIP) is tested on concrete samples made with CEM III/A and CEM II/C-M. The hydration of the samples was stopped by using 2 cycles of solvent exchange (isopropanol) followed by 1 day of drying under light vacuum [20]. MIP was conducted using a Pascal 140/440 system (Thermo Fisher, Waltham, MA, USA).

Two grams of sample was placed in the dilatometer. To ensure complete filling of the intergranular voids, a vacuum was initially applied before allowing mercury to enter. A preliminary pressure of 200 kPa was then exerted in the low-pressure chamber to facilitate pre-intrusion. The dilatometer was subsequently transferred to the high-pressure unit, where the pressure was progressively increased up to 400 MPa while recording the intrusion curve. For pore entry size calculations, a mercury contact angle of 142° and a surface tension value of 0.48 N/m were used.

2.2.8. Shrinkage

Shrinkage is measured on mortar samples (Table 2) and concrete samples (Table 3). The size of the testing specimens is $4 \times 4 \times 16$ cm^3 and $10 \times 10 \times 40$ cm^3, respectively. The initial shrinkage measurements are taken immediately after demoulding the prisms, (24 h after casting). Samples for autogenous shrinkage are wrapped in aluminum tape immediately after demoulding. Autogenous and total shrinkage are measured in accordance with EN 12390-16, with the exception that the mortar prisms are cast in moulds containing metallic pins on the extremes which allows us to monitor the change in length using a digital comparator (accuracy 1 µm) instead of DEMEC gauge points. The change in length for concrete prisms is measured using DEMEC gauge points glued on the surfaces of the prisms, as defined by EN 12390-16.

2.2.9. Freeze–Thaw Resistance

The freeze–thaw resistance of concrete is tested using cylindrical concrete samples (10 cm in diameter and 5 cm in height). The test and the preparations are performed in accordance with EN 12390-9 after a curing age of 60 days, as previously described. A 3 mm layer of 3% NaCl solution is put on the top surface (sawn samples) to include the influence of de-icing salts. The scaling damage is weighed after 7, 14, 28, 42, and 56 freeze–thaw cycles. Each cycle consists of a temperature variation from 22 °C to −15 °C (measured via a thermocouple on the surface of a dummy specimen with a layer of water and de-icing salts inside the freeze–thaw chamber.

2.2.10. Environmental Impact Calculations

The ISO 14040/44 2006 methodology is utilized for the life cycle assessment (LCA) of the concrete mixes. Although replacing PC with SCMs reduces the global warming potential (GWP) of the mixes, the reactivity of the SCMs as well as their production process should be considered in the calculations. An environmental performance metric is utilized to assess the environmental impact of the various formulations, taking into account their mechanical efficiency. This metric evaluates the CO_2 emissions per m^3 of concrete to deliver 1 MPa of strength.

The CO_2 indicator is calculated for a cubic metre of concrete (cradle-to-gate), which involves all required intermediate treatments for (recycled) aggregates, calcined clays, limestone, PC, lava, and GGBFS. The impacts are regionalized using the current electricity and fuel mixes in Belgium, as detailed by [21]. The RA is considered waste material, with no allocation of environmental impacts to its generation. CO_2 burdens for primary crushing operations, classifying operations, and a secondary crushing are assigned [22]. A transportation distance of 100 km is considered (processing and distribution). Although the treatment and CO_2 emission of the SCMs may vary with location, availability of raw materials, and production processes, this study adopted the following: calcined clay and limestone production values from Pillai et al. [23] and the PC and GGBFS from [24], as listed in Table 5. The blast furnace slag is a by-product of steel production, and the basic intermediate treatment to produce the GGBFS includes granulation, drying, and grinding operations. An economic allocation procedure based on a previous work of some of the authors is applied [24], in which 0.84% of the impacts related to the steelmaking process are allocated to GGBFS production [25]. The burdens for extraction, primary crushing, classifying, and milling operations are allocated, with a 200 km transport distance considered for lava. Transportation is assumed to occur by truck (0.015 kg CO_2 eq/kg lava), aligning with the fact that 80% of cement transport within the EU-27 is carried out by truck [26].

Table 5. CO_2 burdens for the materials used in this study.

Material	Granular Skeleton + Mixing	PC	Clay	Lava	Limestone	GGBFS
kg CO_2 equivalent/kg	0.0206	0.7166	0.1494	0.0500	0.0026	0.2443

3. Results and Discussion

3.1. Isothermal Calorimetry

3.1.1. Modified R3 Test on SCMs

The results of the modified R3 test on the anhydrous calcined clay, lava, and inert quartz are reported in Table 6, providing an initial indication of the reactivity of the CC and lava.

Table 6. Modified R3 test results. Heat of dissolution at 3, 7, and 10 days for calcined clay, lava, and reference with inert quartz, and portlandite consumption of lava and calcined clays at the end of the experiment (10 days).

Material	Cumulative Heat Release (J/g SCM)			CH Consumption (g/100 g Solids)
	3 Days	7 Days	10 Days	
Lava	61.6	97.5	137.0	4.36
Calcined clay	220.3	273.4	297.8	13.34
Quartz	19.5	31.0	44.8	-

According to the classification (modified R3 test) proposed by [19], the CC is classified as pozzolanic with mid reactivity. The lava is situated on the boundary between inert and low reactivity materials. Compared to CC, lava has a higher heat release between 7 and 10 days which relates to its higher late degree of pozzolanic activity. In terms of pozzolanic activity, CH consumption for lava is much lower than for CC.

In summary, the CC not only reacts much faster, but it is also more reactive than the lava. CC is unable to achieve the reactivity of GGBFS, which is approximately 450 J/g SCM after 10 days at 50 °C [19], this seems to be associated with low kaolinite content, which is about 20%. Table 6 shows the cumulative heat release (after 3, 7, and 10 days) and CH consumption following the modified R3 test. The CH consumption is measured via thermogravimetric analysis and the consumption of portlandite from quartz mixes is used as a reference to calculate the extra consumption of CH due to the pozzolanic activity of CC and lava at 10 days.

3.1.2. Blended Cements

Figure 1a–c display the heat flow and total heat curves for the different binders obtained by isothermal calorimetry (40 °C) and are normalized per gram of PC. An initial induction period, followed by the alite dissolution peak, is observed in Figure 1a. The alite dissolution peak is also often indicated as a C(-A)-S-H and CH precipitation peak [27]. This peak is usually followed by a peak or hump associated with the formation of secondary ettringite [27], also often referred to as an aluminate peak. Figure 1a shows that the hydration of alite is significantly enhanced for all blended systems (compared to PC) because of the filler effect provided by the SCMs. Please note that, in this study, all SCMs (except the coarsest fraction of the calcined clays) are finer than PC (Table 1). Figure 1b shows that the addition of GGBFS appears to have a minor (but still measurable) effect in the duration of the main peak. In any case, the finesses of the SCMs and the clinker contents are much more important factors, as explained by [27].

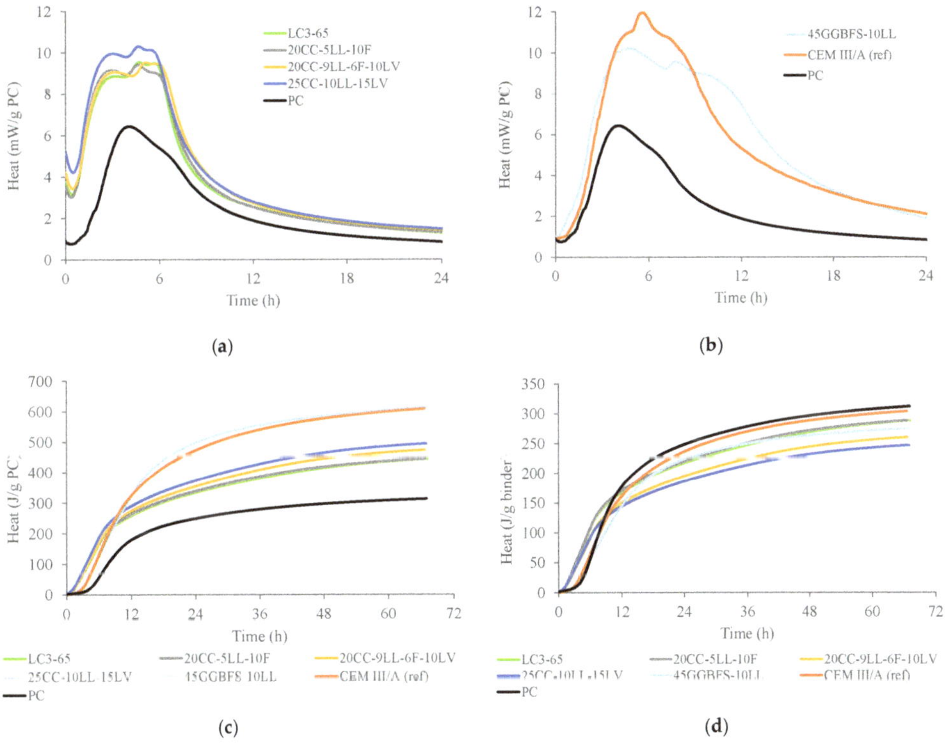

Figure 1. (**a**,**b**) Heat flow (mW/g PC), (**c**) cumulative heat (J/g PC), and (**d**) cumulative heat (J/g binder) for the PC and blended systems from Table 2.

Concerning the aluminates peak, mixes containing calcined clays (Figure 1a) have limited differences in the sulphate depletion point (valley in between alite and secondary ettringite peak). A third peak or hump is identified in Figure 1a,b, which are accelerated in comparison to previous studies [27,28] due to testing conditions at 40 °C instead of 20 °C, as previously explained by some authors [24,29]. For the 45GGBFS-10LL system (Figure 1b), the addition of limestone (in comparison to CEM III/A), resulted in a slightly delayed AFm peak (as well as the sulphate depletion point) due to limestone being slightly coarser than GGBFS. Note that the same amount of sulphate was used for both mixes.

Figure 1c shows the cumulative heat during the first 3 days of hydration (approximately equivalent to 7 days at 20 °C) for all the systems, in which the lowest heat/g of PC is for the PC system, followed by the calcined clay containing mixes, and then GGBFS mixes. These values are in agreement with the reactivity observed in the R3 test, which revealed that GGBFS is more reactive at early ages than both the low-grade calcined clays and the lava.

When similar data are plotted in J/g binder (Figure 1d), it can be observed that LC3-65 and CEM III/A mixes can achieve a similar heat release during the first days of hydration simply because of the 15% extra PC added in LC3-65 versus CEM III/A. Despite a limited contribution of the lava which seems to be observed at early ages (expected according to the late pozzolanic reactivity observed via the R3 test), a contribution of the lava within the first 28 days is expected, which cannot be measured via calorimetry so is clarified via compressive strength measurements in next section.

Figure 1d shows the cumulative heat (J/g binder) release during the first days of hydration measured via isothermal calorimetry, providing a clear indication that the early development of microstructures can be slightly slower for the calcined clay-containing systems at a similar clinker content than for the CEM III/A reference. It is worth mentioning that these differences (50 J/g binder during the first week of hydration) can be considered rather minor when the final goal is to produce concrete mixes with a similar concrete strength class at 28 days [28].

3.2. *Mechanical Strength*

3.2.1. Mortars Compressive Strength

Figure 2 displays the compressive strength development for standardized mortar samples made with the binders in Table 2. The development of the compressive strength (2–28 days of hydration) of mortar mixes with low-grade calcined clay and 65% PC (LC3-65 and 20CC-5LL-10F) is similar to the reference mix CEM III/A and 45GGBFS-10LL. However, to achieve similar strength development in low-grade CC-containing systems, the proportion of PC in these mixes needs to be higher than in CEM III/A, with a required PC content of 65% compared to 50%. The two mixtures with low-grade calcined clay and lava (20CC-9LL-6F-10LV and 25CC-10LL-15LV) have the lowest compressive strength up to 56 days, when all the systems appear to reach a strength in the order of 55–60 MPa (except for CEM III/A which reaches a higher strength), consistent with the low reactivity of the lava at early ages (see Table 6).

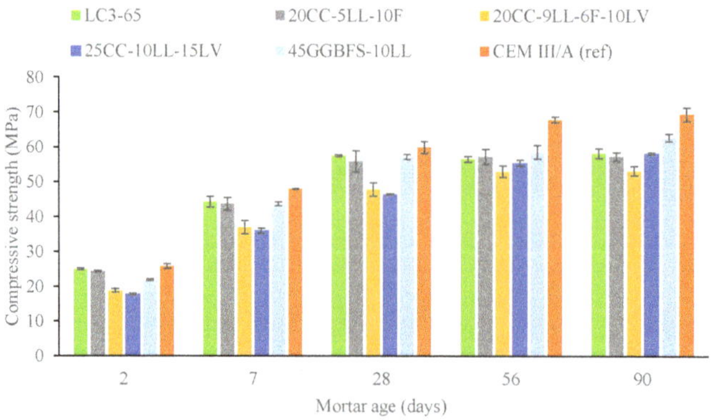

Figure 2. Compressive strength evolution of standardized mortar mixes made with the binders in Table 2.

In this study, the limestone–calcined clay cement blends were only compared to a CEM III/A in terms of strength because that is the material that we expect to replace. In any case, CEM III/A tends to have a similar early age heat release (at 40 °C) to PC (Figure 1d). The two mixtures, LC3-65 and 20CC-5LL-10F, were found to be able to reach a similar strength development as a CEM III/A within the first 28 days of hydration. Nevertheless, they reached a plateau from 28 to 90 days The most likely explanation for this plateau is the low kaolinite content of the raw clay locally available; 40 to 60% MK content is the optimum for LC3 technology. The mix containing increasing contents of lava and clay (25CC-10LL-15LV) was able to increase the compressive strength beyond 28 days, reaching the same strength as the LC3-65 and 45GGBFS-10LL mixes. This indicates that, despite the metakaolin content of the local clays being lower than the optimum for LC3 production, low-grade clays are still a valuable source for fast-reacting SCMs [14,30] and it may be feasible to use a low-grade limestone–calcined clay 'doped' with late reacting pozzolanic material (in this case lava) as an alternative binder.

The 20CC-5LL-10F mixture with recycled fines yields a similar performance to its respective counterpart mix (LC3-65) without fines, similarly to the isothermal calorimetry results discussed in the previous section. Thus, the results suggest that it is possible to (partially) substitute raw limestone with naturally carbonated recycled fines as an alternative source of calcium carbonate, which is mostly interesting from a circularity perspective as it decreases the need for natural resources and it avoids the downcycling of naturally carbonated recycled fines (currently mostly used for soil stabilization). A recent study from some of the authors resulted in a similar outcome when adding recycled fines in concrete mixes with a high PC content [17]. Phase assemblage and microstructural evaluations would be necessary to clarify the underlaying mechanisms that resulted in the similar strength development observed among several systems in this section.

Overall, the early compressive strength of the mortars is in line with the isothermal calorimetry results. The correlation between the cumulative heat release at 24 h and the mortar compressive strength at 2 days is plotted in Figure 3 (orange) as well as between the cumulative heat release at 48 h and the mortar compressive strength at 7 days (blue). The slope of both trend lines is similar, linking the heat release to the strength development process.

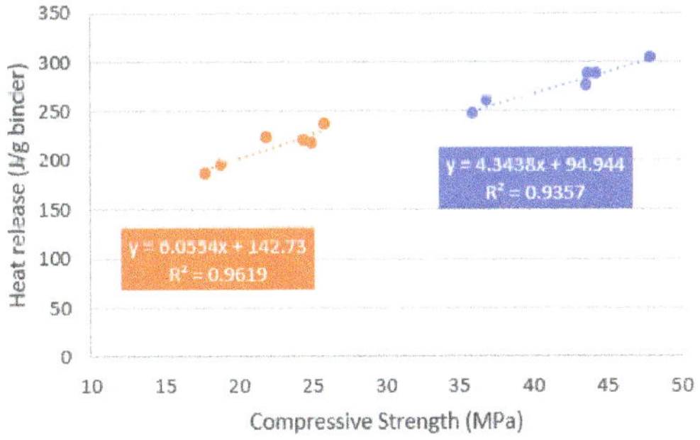

Figure 3. Correlation between heat release (paste scale) and compressive strength (mortar mixes).

The decent mechanical performance of 25CC-10LL-15LV (high enough to allow for demoulding at early ages), supported by the early chemical reactivity of the CC shows the potential use as a CEM III/A replacement without a significant drop in compressive strength at 90 days. Thus, this blended cement is further tested at concrete scale and further designated as CEM II/C-M. A mix with CEM III/A is utilized as the reference for testing the performance at the concrete scale.

3.2.2. Concrete Compressive Strength

The results of the concretes' compressive strength are given in Figure 4. The early age strength of the CEM II/C-M is slightly lower than the CEM III/A reference, which is in agreement with previous discussions at the paste and mortar scale. The compressive strength of the concrete mixes (where the binder is ~10% in volumetric terms) is not only determined by the binder, but also by the granular skeleton, which has an important effect. This aspect explains that the differences in strength between CEM II/C-M and CEM III/A are less marked than for standardized mortar samples (binder represents ~17% in volumetric terms).

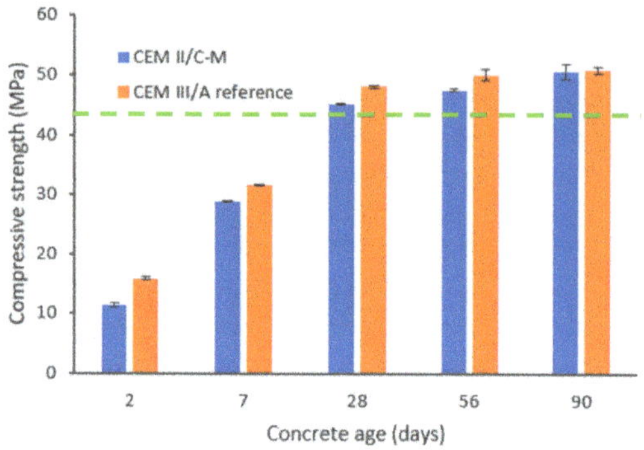

Figure 4. Compressive strength evolution of concrete mixes from Table 3. The 42.5 MPa cement strength class is indicated by the green dashed line.

Both concrete types achieved the design's compressive strength (28 days) of 42.5 MPa at concrete scale and a compressive strength at 90 days of 50 MPa. This shows that CC and likely LV continued forming hydration products beyond 28 days, whereas for GGBFS, the development of strength after 28 days is less pronounced.

3.3. Durability

3.3.1. Porosity

The porosity measurements are presented in Figure 5. The capillary and open porosity of CEM II/C-M increase by 22% and 19%, respectively, compared to CEM III/A. MIP results confirm this trend, indicating a more porous cement matrix. Several factors may explain this increase:

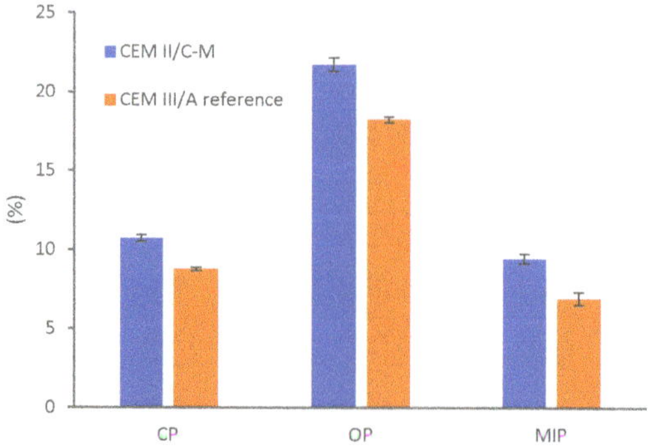

Figure 5. Capillary, open, and MIP porosity of concrete made with CEM III/A and CEM II/C-M.

(i) SCM reactivity and phase assemblage: the reactivity of the incorporated SCMs and the resulting phase assemblage are influenced by the reduction in calcium content. Compared to GGBFS, CC and LV contribute less calcium, affecting matrix densification [31].

(ii) The addition of up to 15% LL has been reported to enhance hydration in low-SCM systems, where calcium availability is not a limiting factor [6]. However, in PC blends with only 50% PC, replacing a reactive SCM (e.g., GGBFS) with LL can lead to a dilution effect, increasing porosity relative to the reference mix [5,13]. Note that in Figure 2, the 90 days compressive strength is affected by less than 10% (compare CEM III/A and 45GGBFS-10LL mixes).

(iii) Cement volume differences: due to the lower particle density of CC, a higher binder volume (per m^3) could be obtained when dosed by mass (kg/m^3), potentially increasing porosity. As discussed in the next section, this effect is expected to be limited.

3.3.2. Capillary Imbibition and Water Absorption

The capillary imbibition of the concrete specimens is plotted in Figure 6 as a function of the fourth root of time [32]. It is immediately noticeable that the capillary imbibition rate of the CEM II/C-M is higher than the reference mix. This is a logical consequence of the higher capillary porosity of this concrete type. Nevertheless, the increase in the primary capillary imbibition rate is 59%, which is remarkably higher than the increase in capillary porosity (22%).

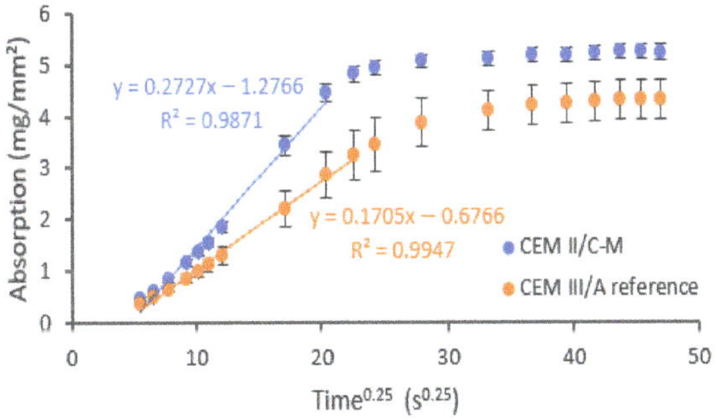

Figure 6. Capillary water uptake of the CEM II/C-M specimens and the CEM III/A references.

The capillary imbibition rate mainly depends on the connectivity of the pores and the amount of capillary pores [33]. The addition of 'low-reactive' pozzolans tend to show a higher absorption and capillary porosity [9]. LC3 mixes made with illite and montmorillonite were also reported to increase sorptivity compared to LC3 with kaolinite clays. Since the transport properties of concrete mostly depend on the capillary water ingress, this increase in capillary imbibition rate might negatively influence the durability characteristics of the CEM II/C-M concrete in comparison to CEM III/A.

The cement volume differences mentioned in Section 3.3.1 could also explain the increased capillary imbibition. Due to the fact that the CC has a low particle density, a larger paste volume is added when using dosages expressed in weight (kg/m^3), leading to a larger volume of binder per m^3 of concrete. If there is a higher volume of paste in the CEM II/C-M than in CEM III/A, then a higher capillary imbibition rate as well as a higher water absorption might be expected [34,35]. The difference in cement volume is calculated in Figure 7. These volumes are obtained using the density of the binders from Table 1. When comparing both cement volumes, the CEM II/C-M has a 0.5% higher volume of cement than the CEM III/A. Thus, this increase in binder content in volumetric terms can (only) partially explain the higher imbibition rate of the CEM II/C-M, but the higher connectivity of the capillary porosity is still deemed to be the dominant factor.

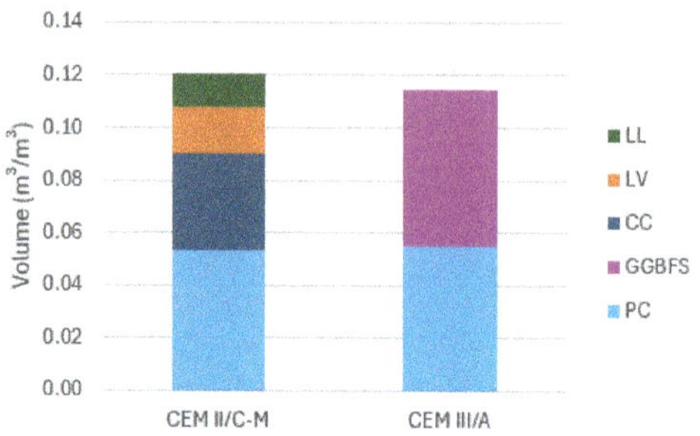

Figure 7. Binder volume for CEM II/C-M and CEM III/A calculated using the densities from Table 1.

3.3.3. Shrinkage

The incorporation of less reactive SCMs than GGBFS may result in a slower strength development, which can potentially lead to issues such as shrinkage cracking (tensile strength lower than tensile stress). The high water demand of calcined clays may exacerbate shrinkage-related cracking. This phenomenon is closely linked to the early age behaviour of the material, where increased evaporation combined with delayed hydration can cause shrinkage cracking or contribute to autogenous shrinkage under self-desiccation conditions [36]. In addition, the lower reactivity of these SCMs was reported to lead to a more open pore structure, particularly at early ages. A more porous matrix might result in higher drying rates and increased moisture loss, potentially accelerating drying shrinkage.

Shrinkage measurements were performed on both mortar prisms and concrete prisms. The results of the concrete shrinkage strain given in Figure 8 show a higher drying shrinkage in the first 14 days for the CEM II/C-M. The drying shrinkage rate slows down for this concrete after 28 days compared to the CEM III/A. Autogenous shrinkage is comparable for both concrete types, especially within the first 28 days. Afterwards, a slightly higher autogenous shrinkage is noticed for the CEM II/C-M, consistent with the greater pozzolanic activity of clay and lava. This is in line with the findings at the mortar scale (Figure 9), where specimens with CC and limestone also showed a higher drying shrinkage and a moderately higher autogenous shrinkage compared to the reference mixtures with PC + GGBFS.

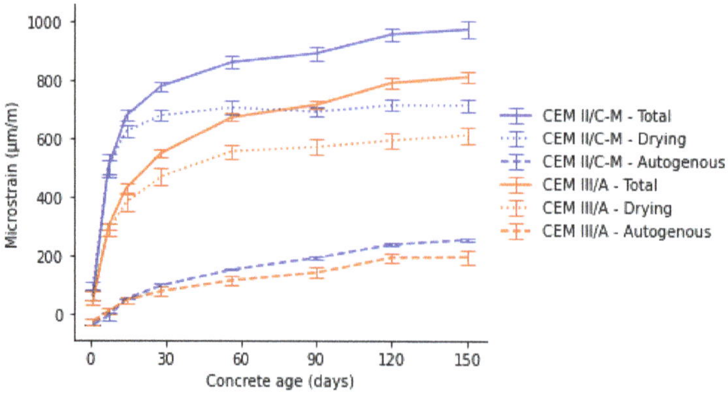

Figure 8. Total, drying, and autogenous shrinkage strain of concrete after 150 days.

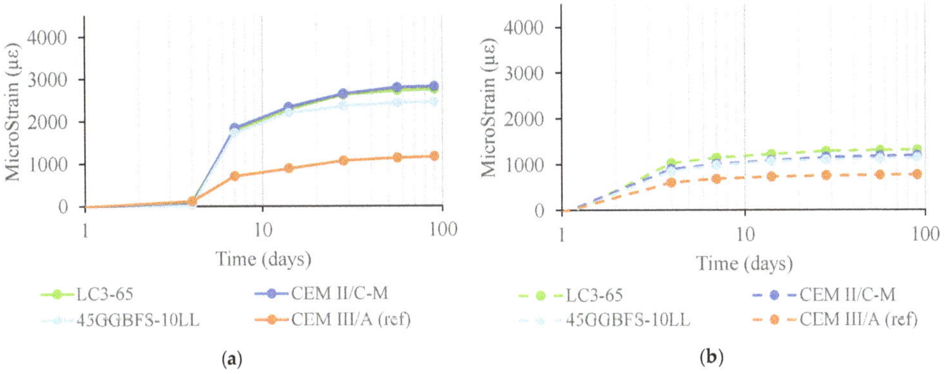

Figure 9. (**a**) Drying and (**b**) autogenous shrinkage strain of selected mortar mixes from Table 2.

Results at the mortar scale (Figure 9a) confirmed that although calcined clay cement mixes shrink more than CEM III/A, such an effect is not necessarily linked to the differences in reactivity between the low-grade clay and GGBFS. Instead, it is the limestone addition that boosts the drying shrinkage. This is clear when the results of mixes CEM III/A and 45GGBFS-10LL are compared. There, the addition of 10% limestone yields shrinkage values in the same order as for all limestone–calcined clay-containing mixes in this study. In addition, the comparison of drying shrinkage between mixes LC3-65 and CEM II/C-M also confirms that the reduction in the PC content of the mixes does not necessarily yield higher shrinkage values (within the boundaries of these mixes; 65% PC in LC3-65 and 50% PC in CEM II/C-M). Figure 9b shows the autogenous shrinkage for the mortar mixes described and indicates that the effect of adding limestone is mostly seen within the first week, with all the mixes showing parallel trends after this timeframe.

The higher drying shrinkage strain, resulting in a larger total shrinkage, could also be explained by the higher accessible porosity of the CEM II/C-M. An increased connectivity of the pore network facilitates a more rapid evaporation of moisture [37]. The combination of higher early age shrinkage and lower early age strength might be seen as a greater risk of shrinkage cracking in the CEM II/C-M. However, comparing the micro-strain values of the concrete mixtures (Figure 8) with those of the mortar mixtures (Figure 9) reveals that shrinkage is not exclusively dependent on the binder mix design. In fact, the type of aggregates and the granular skeleton play a fundamental role—one that is even more significant than the binder itself—in shrinkage behaviour.

None of the materials tested in this study have shown signs of (micro)cracking due to shrinkage (visual inspection). Specific environmental conditions such as high temperature, excessive wind (both favouring greater drying rates), and/or very low temperature during casting (delaying hardening) must be considered as relevant parameter for on-site applications of either CEM III/A or CEM II/C-M mixes.

3.3.4. Freeze–Thaw Resistance

The high quantity of non-kaolinitic phases in the locally available calcined clay is the main motivation for addressing the possible increase in scaling damage for which the failure mechanism is extensively discussed. This study does not aim to discuss the effect of salt (NaCl) concentration on the process. That is because of three reasons, (i) two concrete types are tested using the same type and concentration of salt, (ii) NaCl crystallization itself (from a brine) cannot take place above −22 °C [38] (testing setup reaches −15 °C), and (iii) it is very well known that salt crystallization has a pessimum effect at 3% concentration, with greater or lower values than 3% resulting in a much lower amount of descaled material.

The freeze–thaw resistance of the two concrete types was tested with the inclusion of de-icing salts. The cumulative scaling damage after a certain number of freeze–thaw cycles is shown in Figure 10. The lower freeze–thaw resistance of the CEM II/C-M compared to the reference CEM III/A is evident. A constant trend is observed where the scaling damage of the CEM II/C-M is nearly four times higher than the CEM III/A. The linear trend observed for both concrete samples rules out the possible damage related to macrocracking of one of the binder types, which would lead to scaling damage with an exponential trendline [38]. To provide the reader with an order of magnitude, a maximum value of 1.0 kg/m^2 is recommended by NBN EN 1339 2003 after 28 cycles of freeze–thaw with de-icing salts for paving blocks.

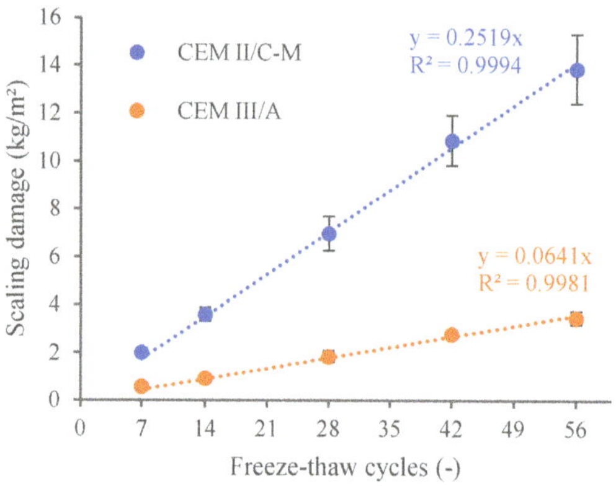

Figure 10. Cumulative scaling damage from 7 to 56 freeze–thaw cycles.

The freeze–thaw damage mechanisms (with de-icing salts) have been extensively studied for construction and building materials. The size of the pores appears to play a relevant role on the failure mechanism. Hydraulic pressure, crystallization pressure, and glue-spall mechanisms have been reported as possible failure mechanisms in buildings materials [39]. For hydraulic pressure, freeze–thaw resistance mainly depends on three aspects: degree of saturation (higher than ~90% [38]), the porosity, and pore connectivity. The damage occurs when pores are saturated with water and the water freezes. However, the freezing process occurs at different temperatures depending on the pore size, with greater pore diameters (>100 μm) freezing at temperatures (ΔT) near to 0 °C and nanopores (<10 nm) at temperatures in the range of −16 °C. These temperature values (ΔT) are calculated using the Gibbs–Thomson equation (Equation (3)) deducted in [39], assuming an ice front in an existing air void (with radius r) propagating into a cylindrical pore with radius r_p. This expression assumes constant entropy for the liquid and the crystal within that temperature range [$\Delta S_{fv} = 1.12 \times 10^6$ J/ (°C. m^3)], a curvature $K_{cl} = 2/r$ for the ice bulging in the pore, and $r = r_p - \delta$, with δ (thickness of the unfrozen layer of water against the pore wall) equal to ~1.0 nm for cement paste [39].

$$(r_p - \delta) \cong -\frac{2\gamma_{CL}}{\Delta S_{fv}\Delta T} \quad (3)$$

Based on the previous equation, it can be noted that pores with a diameter lower than 6 nm (mainly gel pores), could not have a significant influence in the freeze–thaw damage using the aforementioned testing setup (for which the lowest temperature was −15 °C). This is consistent with the literature, indicating that water inside gel pores rarely freezes under these testing conditions [40].

According to hydraulic pressure, when the material starts to freeze (for instance from the outer part of the sample), small capillary pores act as an internal restriction for the water in bigger pores that, thermodynamically speaking, has to freeze first. This pressure is the result of the ice front pushing the water deeper in the sample (note that there is an increase in volume when water freezes and water has some viscosity (η), so that it will experience some resistance to this movement from the pore walls). There might be a local failure when this pressure created on the pore wall exceeds the tensile strength of the material. If the material is not close to its saturation degree, the presence of air will (partially) allow this pressure to be released.

Equation (4) expresses the flow rate (dv/dt) as a function of the gradient pressure (ΔP) on the cylindric pore (with radius R and length L).

$$\frac{dv}{dt} = \Delta P \frac{\pi R^4}{8L} \tag{4}$$

This equation can be rearranged by considering dv/dt equal to the area of cylindrical pore multiplied by dx'/dt, where dx' is to increase the movement of the waterfront deeper into the pore. Then, dx' is equal to the advance of the ice front (dx) multiplied by $(1 - \rho_{Ice}/\rho_W)$, with ρ being the density of ice and water, respectively. Therefore, Equation (5) can be obtained.

$$\Delta P = 8\eta \frac{L}{R^2}\left(1 - \frac{\rho_{ice}}{\rho_{water}}\right)\frac{dx}{dt} \tag{5}$$

By plotting Equation (5) for different pore diameters, Figure 11 is obtained. This figure reveals that the hydraulic pressure per unit length exponentially decreases with the increase in pore diameter.

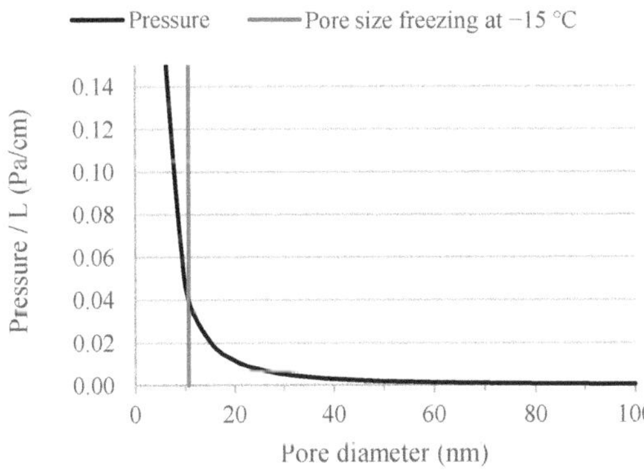

Figure 11. Pressure per unit of length as a function of pore diameter calculated using Equation (5).

An air void analysis could have been performed to identify if there are significant differences in the L (usually called air void spacing in aired intruded concrete) between the two mixes. However, this parameter was only reported significant with the addition of an air entraining agent that creates hydrophilic pores acting as pressure sinks [41]. Figure S3 in the Supporting Information shows the freeze–thaw resistance of samples saturated under vacuum before the NaCl solution was added, aiming to fill any hydrophilic pores potentially created by the superplasticizer addition. Only very limited variations in freeze–thaw damage are observed. This indirectly suggests that differences in the spacing factor (L) between air voids are not the main reason for the fourfold increase in freeze–thaw damage associated with CEM II/C-M.

Naturally, to estimate the influence of the net pressure exerted in the different mixes, the volumes of pores within specific pore ranges should be accounted for. From MIP analyses (and taking into account the possible limitations of the pore model used for MIP), it can be observed that concrete with CEM II/C-M is more porous (9.5%) than CEM III/A (6.9%) within the pore range of 100,000 to 4 nm. However, the porosity in the range 100 to 10 nm amounts for CEM II/C-M to ~2.2% while for CEM III/A, ~3.0% (Figure 12). If the same analysis is made in the range 30 to 10 nm (the most critical range of pores observed in Figure 11), the porosity of CEM II/C-M is ~1.2% while for CEM III/A it is ~1.5%.

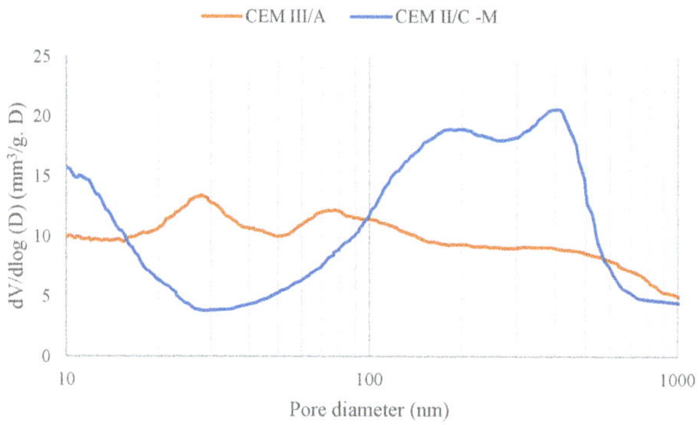

Figure 12. Pore size distribution of CEM III/A and CEM II/C-M obtained from MIP measurements.

This seems inconsistent with CEM II/C-M having a freeze–thaw resistance four times worse than CEM III/A, while both materials have similar mechanical properties. The same occurs when the threshold pore size is analyzed (in the order of 900 nm for concrete with CEM III/A and 700 nm for concrete with CEM II/C-M). To confirm this behaviour, samples made with the same binder type were tested following the exact same protocol (NBN EN 1339 2003) but adding water instead of the 3% NaCl solution (Figure S3). By doing so, it is possible to confirm that no damage (descaled material) is observed, even after 90 cycles of freeze–thaw. Therefore, hydraulic pressure cannot fully explain a freeze–thaw resistance four times worse for CEM II/C-M than for CEM III/A in this study.

Crystallization pressure and the so-called glue-spalling by [42] are other types of mechanisms that could induce the failure which appears to happen more often in site conditions and could occur even when the elements are not fully saturated [39].

Crystallization pressure created by the propagation of ice in the pore structure is directly proportional to the variation in temperature plotted in Equation (3) and it is given by Equation (6).

$$\mathrm{P} \cong -\frac{2\,\gamma_{CL}}{(r_p - \delta)} \tag{6}$$

where P is the local pressure (when the crystal of ice enters the capillary pore with radius r_p, it sucks water from it, inducing a negative pressure). If this equation is plotted for pores with a radius between 5 nm and 50 nm (characteristic of cementitious materials), it can be observed that pores with a diameter greater than 40 nm (Figure 13) do not result in a local pressure that exceeds the overall tensile strength of our materials (approximated to be 10% of the compressive strength value [39,43]). It can be seen here that the smaller the pore radius, the higher the pressure. However, as previously explained, gel pores should not contain ice in the current testing conditions (−15 °C in CEN/TS 12390-9). As for hydraulic pressure, the difference in porosity within the 10–30 nm and 10–100 nm ranges between CEM II/C-M and CEM III/A does not account for the fourfold lower freeze–thaw resistance observed in CEM II/C-M. Please, note that it does not imply that there is no effect of crystallization pressure on the freeze–thaw damage. Instead, it suggests that crystallization pressure alone cannot be the main reason of the differences observed in the freeze–thaw damage (Figure 10).

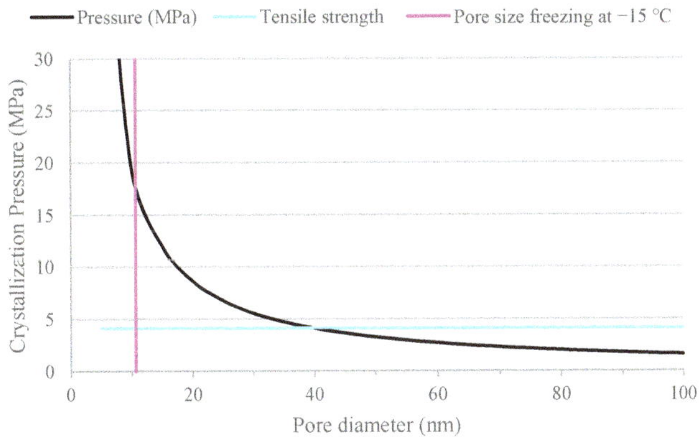

Figure 13. Ice crystallization pressure as a function of the pore diameter.

The fact that the difference in scaling damage between the CEM II/C-M and the CEM III/A is a factor of four, whereas the capillary porosity is 'only' 22% higher and the capillary imbibition rate is 60% higher, initially led us to think that the most viable explanation for the inferior freeze–thaw resistance of the CEM II/C-M was an increase in total porosity and pore connectivity. This type of analysis is not uncommonly found in the literature of cementitious materials. However, if the difference in porosity was indeed what caused CEM II/C-M to perform four times worse than CEM III/A, the theoretical pressure could be roughly estimated by multiplying pressure values in Figure 13 and the number of pores in each pore range (Figure 12) and integrating such a curve. The integration of the curves should, in principle, yield a value at least four times higher for CEM II/C-M compared to CEM III/A. However, within the pore diameter range of 10 to 40 nm—where the individual pore pressure exceeds the local tensile strength (as shown in Figure 13)—the integration of both curves indicates that CEM II/C-M exhibits only about a 40% increase relative to CEM III/A.

The inconsistencies shown in this section indicate that the effect of hydraulic pressure and crystallization pressure would hardly explain the differences shown in Figure 10, implying that salt scaling (very likely glue-spall mechanism [38]) is most likely to explain the differences in freeze–thaw of the samples. This is clearly supported by the fact that samples tested at 0% NaCl (freeze–thaw cycles according to NBN EN 1339 2003 with water instead of NaCl solution) resulted in negligible descaled material for both concrete types.

Further research is needed to better understand the mechanisms driving the significant differences in scaling damage observed for CEM II/C-M and CEM III/A. Previous studies [15,44] have demonstrated that microcrack initiation and interfacial porosity strongly influence freeze–thaw deterioration. Therefore, the role of the interfacial transition zones and possible microcrack development must be investigated in future work. A future study will also assess whether the inclusion of entrained air and/or a reduction in the water-to-binder ratio can more effectively enhance the freeze–thaw resistance of CEM II/C-M cement incorporating low-grade calcined clays. Although the use of CEM II/C-M did not show an important decrease in the overall compressive strength, it could be because the dilution effect provided by the addition of limestone had a significant effect on the tensile strength or fatigue behaviour of the novel mix. A lower local tensile strength and/or fatigue resistance of the mixes, combined with the increased porosity and pore connectivity discussed in previous sections, are expected to directly influence the scaling resistance under the glue-spall mechanism.

3.3.5. Global Warming Potential

The GWP per strength unit (at 7, 28, and 90 days) of 1 m^3 of concrete for mixtures with CEM III/A and CEM II/C-M is shown in Figure 14. Additionally, a concrete made with CEM III/B from [22], using a similar granular skeleton and materials but with a reduced water-to-binder ratio (0.45 instead of 0.50) to achieve the 42.5 N strength class at 28 days, is included to illustrate the impact of higher GGBFS replacement factors. While reducing cement content decreases the GWP, it also reduces strength. Therefore, the coefficient shown in Figure 14 accounts for performance by considering GWP per unit of compressive strength (GWP/f'c).

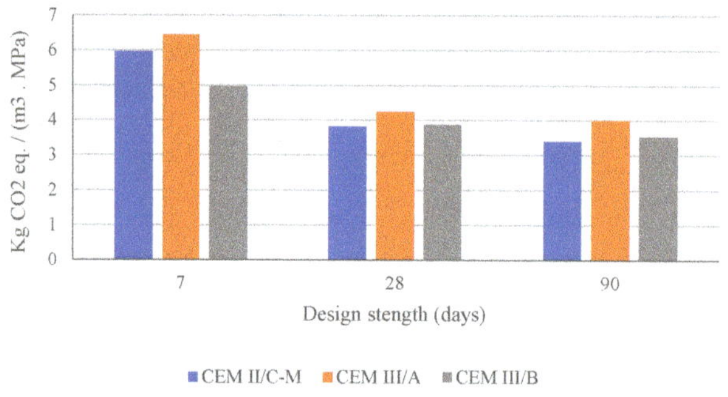

Figure 14. GWP per strength unit (7, 28, and 90 days) of 1 m^3 of concrete for mixtures with CEM III/A, CEM II/C-M, and CEM III/B from Ref. [22].

A lower GWP/f'c coefficient indicates fewer CO_2 emissions per unit of compressive strength. Figure 14 demonstrates that, at 7 days, CEM III/B (with a reduced water-to-binder ratio) is more favourable in terms of GWP/f'c compared to CEM II/C-M. However, when 28 days is used as the design parameter, as common in practice, CEM II/C-M emerges as the more convenient solution, a trend that persists at 90 days.

The analysis highlights that, although reducing cement content is a valid strategy for lowering GWP, the performance of supplementary materials in the mix must also be considered. For instance, while the production-scale GWP of CEM III/B is lower (164 kg CO_2-eq/m^3 compared to 173 kg CO_2-eq/m^3 for CEM II/C-M), the environmental-performance coefficient favours the CEM II/C-M mixture due to its superior strength development. Nevertheless, mixes with lower cement content than those studied here can still provide superior environmental alternatives when a lower strength class (e.g., 32.5 MPa) is sufficient for the application [31].

4. Conclusions

This study evaluated binder mixes based on LC3 blends with minor additions of pozzolanic lava and/or untreated recycled fines as replacements for CEM III/A. The experimental results showed that lava did not meaningfully increase early age strength; its contribution to mortar compressive strength became only slightly noticeable after 28 days, consistent with its late pozzolanic activity and limited reactivity shown by the reactivity test. Low-grade calcined clays contributed to strength development relatively quickly (within 28 days), but further strength gain after 28 days was limited at the mortar scale.

When calcined clays, limestone, and lava were used to replace GGBFS in mortar mixes, both drying and autogenous shrinkage increased relative to CEM III/A. However, mortars made with CEM III/A partially substituted with 10% limestone showed similar shrinkage values to CEM II/C-M, indicating that a small limestone addition may lead to comparable shrinkage behaviour.

At the concrete scale, the CEM II/C-M mix reached the design compressive strength of 42.5 MPa and matched the 90-day compressive strength of CEM III/A, although its early age compressive strength was slightly lower than that of CEM III/A. Durability testing revealed an increase in capillary and open porosity of about 20% for concrete made with CEM II/C-M compared to the CEM III/A reference. This higher porosity, together with a less favourable pore size distribution and possibly lower local tensile strength or fatigue resistance, corresponds with a significant increase in freeze thaw scaling damage for CEM II/C-M.

Finally, the CEM II/C-M formulations delivered a lower GWP per unit f′c at both 28 and 90 days compared to CEM III/A. This shows that not only a successful replacement of GGBFS in terms of strength has been found but also the CEM II/C-M mix containing locally available calcined clays and pozzolanic lava resulted in a more efficient use of materials in the Flanders region. A future perspective is as follows: in specific (non-)reinforced applications within the Flanders region (excluding vibro-compressed elements such as small-size pavers, tiles, concrete brinks, or blocks), freeze–thaw resistance might be a critical performance parameter in horizontal surfaces able to accumulate water/brine after spreading de-icing salts (such as, paver flags, concrete slabs for sidewalks, or pavements). Future research will seek for recommendations to enhance the freeze–thaw resistance of CEM II/C-M mixtures while maintaining a low GWP/f′c coefficient in these elements. The carbonation of the material and the quality of the finished surface may influence the freeze–thaw damage of horizontal elements, and will also be assessed.

Supplementary Materials: The following supporting information can be downloaded at: https://www.mdpi.com/article/10.3390/ma18215048/s1, Figure S1: TGA and Differential mass loss curve (obtained from TGA measurement) performed on the raw clay (before calcination) for determination of kaolinite content according to [45]; Figure S2: Full particle size distribution curve obtained by optical particle analysis; Figure S3. Freeze thaw tested on 45GGBFS-10LL concrete samples following slightly different testing processes. NaCl 3% refers to the regular protocol described in NBN EN 1339 2003, sat. + NaCl 3% includes saturation of the samples in water before the test using NaCl 3% and NaCl 0% includes the use of tap water instead of the NaCl 3% solution.

Author Contributions: Conceptualization, J.M.E.; Methodology, J.M.E., K.D., V.G., and L.D.B.; Formal Analysis, J.M.E., L.D., V.G., and L.D.B.; Investigation, J.M.E. and L.D.; Resources, K.D. and N.D.B.; Data Curation, L.D., V.G., and L.D.B.; Writing—Original Draft, J.M.E.; Writing—Review and Editing, L.D., K.D., V.G., L.D.B., and N.D.B.; Visualization, J.M.E. and L.D.; Supervision, J.M.E., K.D., V.G., L.D.B., and N.D.B.; Project Administration, N.D.B.; Funding Acquisition, N.D.B. All authors have read and agreed to the published version of the manuscript.

Funding: This research was funded by the Research Foundation-Flanders (FWO-Vlaanderen) through research grant G062720N (J. M. Etcheverry).

Data Availability Statement: The original contributions presented in this study are included in the article. Further inquiries can be directed to the corresponding authors.

Conflicts of Interest: Author Laurent Detemmerman and Krist Degezelle were employed by the company DEVAGRO, Beton and Recyclage Harelbeke. The remaining authors declare that the research was conducted in the absence of any commercial or financial relationships that could be construed as a potential conflict of interest.

References

1. Scrivener, K.L.; John, V.M.; Gartner, E.M. Eco-efficient cements: Potential economically viable solutions for a low-CO_2 cement-based materials industry. *Cem. Concr. Res.* **2018**, *114*, 2–26. [CrossRef]
2. Miller, S.A.; Alexander, M.G. Near-term pathways for decarbonizing global concrete production. *Nat. Commun.* **2023**, *14*, 4574. [CrossRef]
3. Londono-Zuluaga, D.; Gholizadeh-Vayghan, A.; Winnefeld, F.; Avet, F.; Ben Haha, M.; Bernal, S.A.; Cizer, Ö.; Cyr, M.; Dolenec, S.; Durdzinski, P.; et al. Report of RILEM TC 267-TRM phase 3: Validation of the R^3 reactivity test across a wide range of materials. *Mater. Struct.* **2022**, *55*, 142. [CrossRef]
4. Lothenbach, B.; Scrivener, K.; Hooton, R.D. Supplementary cementitious materials. *Cem. Concr. Res.* **2011**, *41*, 1244–1256. [CrossRef]
5. Scrivener, K.; Martirena, F.; Bishnoi, S.; Maity, S. Calcined clay limestone cements (LC3). *Cem. Concr. Res.* **2018**, *114*, 49–56. [CrossRef]
6. Briki, Y.; Avet, F.; Zajac, M.; Bowen, P.; Haha, M.B.; Scrivener, K. Understanding of the factors slowing down metakaolin reaction in limestone calcined clay cement (LC3) at late ages. *Cem. Concr. Res.* **2021**, *146*, 106477. [CrossRef]
7. Lin, R.; Han, Y.; Wang, X. Macro—Meso—Micro experimental studies of calcined clay limestone cement (LC3) paste subjected to elevated temperature. *Cem. Concr. Compos.* **2021**, *116*, 103871. [CrossRef]
8. Zunino, F.; Scrivener, K. Microstructural developments of limestone calcined clay cement (LC3) pastes after long-term (3 years) hydration. *Cem. Concr. Res.* **2022**, *153*, 106693. [CrossRef]
9. Fernandez, R.; Martirena, F.; Scrivener, K.L. The origin of the pozzolanic activity of calcined clay minerals: A comparison between kaolinite, illite and montmorillonite. *Cem. Concr. Res.* **2011**, *41*, 113–122. [CrossRef]
10. Lv, T.; Zhang, J.; Hou, D.; Long, W.; Dong, B. Mechanical-thermal activated dredged sludge as a supplementary cementitious material: Microstructure reconstruction and pozzolanic activity enhancement. *Constr. Build. Mater.* **2025**, *476*, 141256. [CrossRef]
11. Villagrán-Zaccardi, Y.A.; Marsh, A.T.M.; Sosa, M.E.; Zega, C.J.; De Belie, N.; Bernal, S.A. Complete re-utilization of waste concretes–Valorisation pathways and research needs. *Resour. Conserv. Recycl.* **2022**, *177*, 105955. [CrossRef]
12. McCarthy, M.J.; Dyer, T.D. *Pozzolanas and Pozzolanic Materials*, 5th ed.; Elsevier Ltd.: Amsterdam, The Netherlands, 2019. [CrossRef]
13. Dhandapani, Y.; Sakthivel, T.; Santhanam, M.; Gettu, R.; Pillai, R.G. Mechanical properties and durability performance of concretes with Limestone Calcined Clay Cement (LC3). *Cem. Concr. Res.* **2018**, *107*, 136–151. [CrossRef]
14. Shui, K.; Yuan, K.; Sun, T.; Li, Q.; Zeng, W. Calcined Clays for Sustainable Concrete. In Proceedings of the 1st International Conference on Calcined Clays for Sustainable Concrete, Lausanne, Switzerland, 23–25 June 2015; Springer: Dordrecht, The Netherlands, 2015.
15. Wang, R.; Hu, Z.; Li, Y.; Wang, K.; Zhang, H. Review on the deterioration and approaches to enhance the durability of concrete in the freeze–thaw environment. *Constr. Build. Mater.* **2022**, *321*, 126371. [CrossRef]
16. Ram, K.; Flegar, M.; Serdar, M.; Scrivener, K. Influence of Low- to Medium-Kaolinite Clay on the Durability of Limestone Calcined Clay Cement (LC3) Concrete. *Materials* **2023**, *16*, 374. [CrossRef] [PubMed]
17. De Brabandere, L.; Grigorjev, V.; Van den Heede, P.; Nachtergaele, H.; Degezelle, K.; De Belie, N. Using Fines from Recycled High-Quality Concrete as a Substitute for Cement. *Sustainability* **2025**, *17*, 1506. [CrossRef]
18. Han, X.; Wang, B.; Feng, J. Relationship between fractal feature and compressive strength of concrete based on MIP. *Constr. Build. Mater.* **2022**, *322*, 126504. [CrossRef]
19. Suraneni, P.; Hajibabaee, A.; Ramanathan, S.; Wang, Y.; Weiss, J. New insights from reactivity testing of supplementary cementitious materials. *Cem. Concr. Compos.* **2019**, *103*, 331–338. [CrossRef]
20. Snellings, R.; Chwast, J.; Cizer, Ö.; De Belie, N.; Dhandapani, Y.; Durdzinski, P.; Elsen, J.; Haufe, J.; Hooton, D.; Patapy, C.; et al. Report of TC 238-SCM: Hydration stoppage methods for phase assemblage studies of blended cements—results of a round robin test. *Mater. Struct.* **2018**, *51*, 111. [CrossRef]
21. Laveglia, A.; Sambataro, L.; Ukrainczyk, N.; De Belie, N.; Koenders, E. Hydrated lime life-cycle assessment: Current and future scenarios in four EU countries BE IR. *J. Clean. Prod.* **2022**, *369*, 133224. [CrossRef]
22. Etcheverry, J.M.; Laveglia, A.; Villagran-zaccardi, Y.A.; De Belie, N. A technical-environmental comparison of hybrid and blended slag cement-based recycled aggregate concrete tailored for optimal field performance. *Dev. Built Environ.* **2024**, *17*, 100370. [CrossRef]
23. Pillai, R.G.; Gettu, R.; Santhanam, M.; Rengaraju, S.; Dhandapani, Y.; Rathnarajan, S.; Basavaraj, A.S. Service life and life cycle assessment of reinforced concrete systems with limestone calcined clay cement (LC3). *Cem. Concr. Res.* **2019**, *118*, 111–119. [CrossRef]
24. Etcheverry, J.M.; Villagran-zaccardi, Y.A.; Van den Heede, P.; Hallet, V.; De Belie, N. Effect of sodium sulfate activation on the early age behaviour and microstructure development of hybrid cementitious systems containing Portland cement and blast furnace slag. *Cem. Concr. Compos.* **2023**, *141*, 105101. [CrossRef]

25. Chen, X.; Wang, H.; Najm, H.; Venkiteela, G.; Hencken, J. Evaluating engineering properties and environmental impact of pervious concrete with fly ash and slag. *J. Clean. Prod.* **2019**, *237*, 117714. [CrossRef]
26. CEMBUREAU. *Key Facts Report 2024*; CEMBUREAU: Brussels, Belgium, 2024.
27. Zunino, F.; Scrivener, K. The influence of the filler effect on the sulfate requirement of blended cements. *Cem. Concr. Res.* **2019**, *126*, 105918. [CrossRef]
28. Zunino, F.; Scrivener, K. Insights on the role of alumina content and the filler effect on the sulfate requirement of PC and blended cements. *Cem. Concr. Res.* **2022**, *160*, 106929. [CrossRef]
29. Etcheverry, J.M.; Yue, Z.; Krishnan, S.; Villagran-Zaccardi, Y.A.; Heede, P.V.D.; Dhandapani, Y.; Bernal, S.A.; De Belie, N. Phase Evolution of Hybrid Alkali Sulfate-Activated Ground- Granulated Blast Furnace Slag Cements. *ACS Sustain. Chem. Eng.* **2023**, *11*, 17519–17531. [CrossRef]
30. Ayati, B.; Newport, D.; Wong, H.; Cheeseman, C. Low-carbon cements: Potential for low-grade calcined clays to form supplementary cementitious materials. *Clean. Mater.* **2022**, *5*, 100099. [CrossRef]
31. Sun, J.; Zunino, F.; Scrivener, K. Hydration and phase assemblage of limestone calcined clay cements (LC3) with clinker content below 50%. *Cem. Concr. Res.* **2024**, *177*, 107417. [CrossRef]
32. Zaccardi, Y.A.V.; Alderete, N.M.; De Belie, N. Improved model for capillary absorption in cementitious materials: Progress over the fourth root of time. *Cem. Concr. Res.* **2017**, *100*, 153–165. [CrossRef]
33. Hall, C. Capillary imbibition in cement-based materials with time-dependent permeability. *Cem. Concr. Res.* **2019**, *124*, 105835. [CrossRef]
34. Kolias, S.; Georgiou, C. The effect of paste volume and of water content on the strength and water absorption of concrete. *Cem. Concr. Compos.* **2005**, *27*, 211–216. [CrossRef]
35. Castro, J.; Bentz, D.; Weiss, J. Effect of sample conditioning on the water absorption of concrete. *Cem. Concr. Compos.* **2011**, *33*, 805–813. [CrossRef]
36. Sayahi, F.; Emborg, M.; Hedlund, H.; Cwirzen, A. Plastic Shrinkage Cracking of Self-compacting Concrete: Influence of Capillary Pressure and Dormant Period. *Nord. Concr. Res.* **2019**, *60*, 67–88. [CrossRef]
37. Acker, P.; Ulm, F.J. Creep and shrinkage of concrete: Physical origins and practical measurements. *Nucl. Eng. Des.* **2001**, *203*, 143–158. [CrossRef]
38. Valenza, J.J.; Scherer, G.W. Mechanism for salt scaling. *J. Am. Ceram. Soc.* **2006**, *89*, 1161–1179. [CrossRef]
39. Sun, Z.; Scherer, G.W. Effect of air voids on salt scaling and internal freezing. *Cem. Concr. Res.* **2010**, *40*, 260–270. [CrossRef]
40. Chatterji, S. Freezing of air-entrained cement-based materials and specific actions of air-entraining agents. *Cem. Concr. Compos.* **2003**, *25*, 759–765. [CrossRef]
41. Helmuth, R.A.; Powers, T.C. *Theory of Volume Changes in Hardened Port-land-Cement Paste During Freezing*; Portland Cement Association: Skokie, IL, USA, 1959.
42. Scherer, G.W. Crystallization in pores. *Cem. Concr. Res.* **1999**, *29*, 1347–1358. [CrossRef]
43. Tremblay, M.-H.; Lory, F.; Marchand, J.; Scherer, G.W.; Valenza, J.J. Ability of the Glue Spall Model to Account for the De-Icer Salt Scaling Deterioration of Concrete. In *Pap. W4-07.3 in Proceedings of the 12th International Congress on the Chemistry of Cement (ICCC), Montreal, QC, Canada, 8–13 July 2007*; Beaudoin, J.J., Makar, J.M., Raki, L., Eds.; National Research Council Canada: Ottawa, ON, Canada, 2007; ISBN 978-0-660-19695-4.
44. Martínez-Martínez, J.; Benavente, D.; Gomez-Heras, M.; Marco-Castaño, L.; García-Del-Cura, M.Á. Non-linear decay of building stones during freeze-thaw weathering processes. *Constr. Build. Mater.* **2013**, *38*, 443–454. [CrossRef]
45. Snellings, R.; Mertens, G.; Elsen, J. Supplementary cementitious materials. *Rev. Mineral. Geochem.* **2012**, *74*, 211–278. [CrossRef]

materials

MDPI

Article

Assessment of the Potential for Producing Geopolymer-Based Granulates as a Substitute for Natural Aggregates

Magdalena Cempa [1,*], Jerzy Korol [2,*] and Agnieszka Klupa [1]

[1] Department of Environmental Analysis and Circular Economy Technologies, Central Mining Institute–National Research Institute, Pl. Gwarków 1, 40-166 Katowice, Poland; aklupa@gig.eu
[2] Department of Mechanical Research and Materials Engineering, Central Mining Institute–National Research Institute, Pl. Gwarków 1, 40-166 Katowice, Poland
* Correspondence: mcempa@gig.eu (M.C.); jkorol@gig.eu (J.K.)

Abstract: This study presents the development and evaluation of a technology for producing geopolymer-based granulates, which act as sustainable substitutes for natural aggregates by utilizing waste materials. The technology is demonstrated to be energy-efficient compared to other manufactured aggregate processes (such as sintering), as it relies on a cold-bonding process and achieves self-hardening at room temperature. The granulation of geopolymer materials using an intensive counter-current mixer represents an innovative solution in the field of producing substitutes for natural aggregates. Coal fly ash (CFA) was used as the primary aluminosilicate precursor, with composite regrind from decommissioned wind turbine blades (CR) and steelmaking dust (SD) tested as additives. Reactive solids and alkaline activator liquids were mixed and granulated in a single operation using an intensive counter-current mixer; moistening and surface powdering were applied to improve granule sphericity. The granules were cold-cured at room temperature and characterized after 28 days by grain size distribution, crushing resistance, water absorption, abrasion (micro-Deval), SEM/EDS and leaching tests. The results indicate that the additives significantly improved the mechanical performance: PM + PK granules reached crushing strengths > 6 MPa, while CFA + SD granules reached > 11 MPa, exceeding many commercial lightweight aggregates (such as LECA or Lytag), as detailed in the paper. The CFA + CR granulates exhibited a compact microstructure and the effective immobilization of several heavy metals, whereas the CFA + DS samples demonstrated the excessive leaching of Cr, Pb and Mo. The process achieved a high solid-to-liquid ratio (>2.0), reducing activator consumption. Composite regrind is recommended as a promising additive.

Keywords: aggregate; coal fly ash; composite regrind; geopolymer; granulate

1. Introduction

Crushed natural rocks and stones serve as the main coarse aggregates in concrete, with variants such as limestone, basalt and gravel chosen according to regional geology. However, increasing construction activity and the rapid exhaustion of these resources threaten the sustainability of the sector [1–3]. Three technical strategies have been suggested to overcome these challenges: recovering aggregates from demolished concrete [4–7], using coarse waste materials (bottom ash or steel slag) directly as aggregates [8,9], and producing manufactured aggregates from industrial waste or by-products [10,11].

Geopolymers, formed through the alkali activation of aluminosilicate precursors, can replace natural aggregates and may offer beneficial physical, chemical and mechanical properties comparable to those of ordinary Portland cement [12]. Currently, researchers increasingly emphasize waste material reuse and recycling, especially within the field of geopolymers [13,14]. Geopolymers typically require feedstocks rich in silica and alumina (together usually exceeding 50% of the total), since these oxides form the structural backbone of the material. Consequently, choosing the appropriate aluminosilicate precursors is essential when producing geopolymers intended as synthetic aggregates. A range of aluminosilicate sources, including kaolin [15], metakaolin [16], fly ash [17], slag [18], red mud [19], rice husk ash [20] and volcanic ash [21], have been widely investigated for this purpose. Other potential materials, such as granite [21], waste glass powder [22,23] and quarry tailings [24], have also been actively researched as geopolymer precursors.

Apart from selecting appropriate precursors for geopolymer-based artificial aggregates, choosing suitable mixture proportions that effectively drive the geopolymerization process is also crucial. The key parameters include the alkaline solution concentration, the alkali activator ratio and the solid–liquid ratio between the aluminosilicate feedstock and the alkaline activator, which are often expressed as molar ratios. Both potassium hydroxide (KOH) and sodium hydroxide (NaOH) are often used as alkali solutions in the synthesis of geopolymers, generally in combination with $K_2Si_2O_3$ or Na_2SiO_3 [25].

In addition to adjusting the liquid composition of the geopolymer-based artificial aggregates, the solid–liquid (S/L) ratio is a key factor that affects the properties of the obtained materials. The S/L ratio also strongly influences the chemical environment during geopolymerization. In general, increasing the S/L ratio (i.e., increasing the solid content) accelerates geopolymerization, as more Si and Al become available to form geopolymer bonds. An appropriate S/L ratio is therefore essential to promote effective geopolymerization and produce high-performance geopolymer-based artificial aggregates. However, if the S/L ratio becomes too high, there will be insufficient liquid for the reaction, which can result in poorly formed bonds [26–28].

Geopolymer production generally involves three stages: (i) mixing reactive solids and activator liquids, (ii) shaping (pelletization, molding, crushing or hand shaping), and (iii) curing (cold bonding, sintering and autoclaving) [29].

Cold bonding is a hardening technique for manufactured aggregates that involves air drying in an enclosed space for periods such as one, three or seven days at relatively low temperatures (under 100 °C). Cold-bonded aggregates are often cured with water or covered with plastic sheets until they gain adequate strength. Given that it does not require additional heating, cold bonding is an energy-efficient method commonly used with cementitious materials such as coal fly ash, iron ore tailings and granulated blast-furnace slag; the aggregate strength is primarily developed through cement hydration and/or pozzolanic reactions [30]. Gesoglu et al. [31] noted that producing lightweight fly ash aggregates via cold bonding consumes less energy and is more cost-effective than sintering.

The reported physical and mechanical data for the obtained geopolymer-based artificial aggregates reveal a wide range of physical and mechanical properties depending on the raw materials and processing (Table 1).

With the progressive depletion of natural resources and growing environmental constraints, the construction sector needs to develop alternative aggregate materials. The geopolymer granules developed in this study respond to these challenges by using industrial waste (fly ash, steel mill dust, and composite regrind) as secondary raw materials. This solution combines the circular economy aspect with the real potential to replace natural aggregates in construction and insulation applications.

Unlike commercial lightweight aggregates, which require thermal sintering processes, the proposed technology allows comparable mechanical parameters to be obtained under room temperature curing conditions, which significantly reduces the energy consumption of the process.

Table 1. Research on the physical and mechanical properties of geopolymer-based artificial aggregates [32].

References	Artificial Aggregates Produced	Physical Properties			Mechanical Properties		
		Specific Gravity	Density, kg/m^3	Water Absorption, %	AIV, %	ACV, %	CS, MPa
[33]	FA-RM	n.d.	1007–1132	9.80–12.10	n.d.	n.d.	1.46–6.18
[34]	FA-SF	1.700	738	18.98	10.24	n.d.	2.03–12.00
[35]	FA-GGBS	n.d.	764–878	18.73–28.30	25.00–39.00	n.d.	n.d.
[36]	FA-SF	1.800	710	17.95	10.03	n.d.	3.34–4.54
[37]	FA-BT, FA-MK, FA-GGBS	1.68–1.89	848–983	13.01–21.26	31.96–50.47	n.d.	14.51–22.81
[38]	FA	n.d.	1450–1500	22.00–23.00	n.d.	n.d.	n.d.
[39]	FA	2.058	n.d.	7.07	28.31	23.96	n.d.
[40]	CFA, FFA	2.40–2.45	n.d.	5.51–6.05	n.d.	n.d.	n.d.

Notes: AIV—Aggregate Impact Value; ACV—Aggregate Crushing Value; CFA—Class C fly ash; FFA—Class F fly ash; FA—fly ash; GGBS—ground granulated blast furnace slag; MK—metakaolin; n.d.—no data; RM—red mud; SC—Crushing Strength; SF—silica fume.

The growing pressure for sustainable development in the construction sector is motivated not only by environmental considerations but also by economic ones. The rising costs of extracting and transporting natural aggregates, combined with increasing fees for industrial waste storage, are a key driver for the search for technologies that enable their valorization. The aim of the developed technology is to transform waste, which has so far been a financial burden, into a fully fledged market product in the form of geopolymer granules that can replace natural aggregates.

Although the processes of alkaline activation, granulation, and low-temperature curing are well known, this study introduces a new dimension by: (i) using wind turbine blade grinding (PK) and steel mill dust (PS) as functional additives in geopolymer granules, (ii) applying a simultaneous activation and granulation step in an intensive counter-rotating mixer at S/L > 2.0, which reduces activator consumption, and (iii) simultaneously evaluating strength and leachability against environmental limits.

Therefore, the aim of this study was not only to develop a method for industrial waste disposal, but above all to demonstrate that it can serve as a basis for a 'sustainable substitute' for natural aggregates. In the context of this study, a 'sustainable substitute' is defined as a material that (1) is based on waste raw materials, (2) is produced in a low-energy-intensive process, and (3) has final physical and mechanical properties at least equal to, and optimally superior to, commercially available natural or sintered aggregates.

2. Materials and Methods

2.1. Materials

Coal fly ash from the combustion of hard coal in a fluidized coal boiler was used as the main precursor material to produce a geopolymer granulate (CFA). Steelmaking dust (SD) and composite regrind (CR) were used as additives for the geopolymer granule production (Figure 1). The steelmaking dust originated from waste gas dedusting during metallurgical processes. The composite regrind was obtained from wind turbine blades that had been taken out of service.

A composite regrind from decommissioned wind turbine blades is typically produced by mechanical grinding or milling, resulting in fine powders or shredded fibers. These materials are then used as fillers or reinforcements in new composites, such as polypropylene or geopolymer matrices, or as additives in cementitious materials. Mechanical recycling is the most cost-effective and widely adopted method, although it often leads to some reduction in mechanical properties compared to the original materials. However, optimized processes such as solid-state shear milling and surface modification can significantly increase the strength and modulus of the resulting composites, making them suitable for high-performance applications in construction and infrastructure [35,41–45].

Figure 1. Photographs of materials: (**a**) coal fly ash (CFA), (**b**) composite regrind (CR), and (**c**) steelmaking dust (SD) used as additives in the production of geopolymer granules. Note: the photos are illustrative of the bulk materials' appearance; detailed particle size characterization is provided in Tables 6 and 7 and Figures 5 and 6.

2.2. Chemical Reagents

A binary combination of a sodium hydroxide solution (NaOH) and a sodium silicate solution (Na_2SiO_3) was used as the alkaline activator. Sodium hydroxide was supplied by the local company Avantor/POCH® (Gliwice, Poland) in the form of white pellets. The sodium hydroxide was characterized by a molar mass equal to 40 g/mol and a purity not less than 99.4%. The sodium silicate solution, also known as water glass or liquid glass, was obtained from Chempur® (Piekary Śląskie, Poland), a manufacturing company, in the form of a thick liquid. According to the producer, the density of the liquid sodium solution at 20 °C was 1.47 g/cm^3, and its silica molar ratio (SiO_2/Na_2O) was 2.51.

2.3. Experimental Procedure

Various mixtures were prepared as part of the study, as presented in Table 2. It should be emphasized that the presented experiment plan (Table 2) was not a systematic parametric analysis, but rather a process of technological optimization. The goal was to develop a complete method for producing granules with the desired properties. The modifications introduced to the process, such as additional moisturizing (M) or powdering (P), were aimed at improving the sphericity and grain size of the product. At the same time, they forced the correction of other parameters, especially the S/L ratio, in order to maintain the appropriate workability of the mixture during granulation. Therefore, the individual recipes, compared later in this paper, differ in several variables at the same time, reflecting the iterative pursuit of an optimized final product.

Table 2. Mixture proportions for forming geopolymer-based granulates.

Exp. Index	Solid: Liquid	Solid Compound	CFA:SD or CFA:CR	$NaOH:Na_2SiO_3$	M	P
ST1	1.50	CFA	-	3.0	no	no
ST2A	2.00	CFA	-	3.0	no	no
ST2	1.90	CFA	-	3.0	yes	no
ST3	1.75	CFA	-	3.0	no	yes
ST5	1.55	CFA + CR	0.67	3.0	yes	no
ST7	1.96	CFA + CR	0.50	3.0	yes	yes
ST10	2.54	CFA + SD	0.40	3.0	no	no

Note: CFA—coal fly ash, CR—composite regrind, SD—steelmaking dust, M—moisturizing, P—powdering, Solid = CFA + CR + SD + P, Liquid = NaOH + Na_2SiO_3 + M.

The reactive solids and activator liquids were mixed and shaped into granules by means of an intensive counter-current mixer with a nominal capacity of 30 L. Granulation, i.e., the controlled agglomeration of highly dispersed raw materials, is a technique aimed at either preparing fine-grained substances for further technological operations or giving them the shape and properties characteristic of the final product [46,47]. During the processing of the tested materials, the mixer was equipped with a star-type paddle agitator (Figure 2). The design of this agitator enabled the disintegration and fragmentation of the agglomerated raw materials. The high linear speed of the agitator, which was about 30 m/s, allowed this stage to be carried out by operating the bowl and agitator in counter-current mode. The next stage of the process was homogenization to ensure a uniform composition of the mixture throughout the entire volume of the batch as well as its granulation in the final stage. Moistening and powdering of the material surface was applied during granulation to obtain granules with shapes as close as possible to spherical.

Figure 2. Production of geopolymer-based granulates in an intensive counter-current mixer.

It is worth noting that the activation and granulation process was carried out in a single stage in an intensive counter-current mixer, which made it possible to obtain a high solid-to-liquid ratio (S/L > 2.0) during curing at room temperature.

The fresh geopolymer mixture was spread onto metal trays and left under room conditions. After 28 days, selected samples were subjected to a series of mechanical, physical and chemical tests.

2.4. Methods of Material Characterization

The main chemical components of the input materials and the geopolymer-based granulates were determined via wavelength dispersive X-ray fluorescence spectroscopy (WDXRF) by means of a ZSX PRIMUS II analyzer (Rigaku, Tokyo, Japan) equipped with a 4 kW X-ray Rh tube.

The trace chemical components of the input materials and the geopolymer-based granulates as well as the water leaching extract were determined via inductively coupled plasma optical emission spectroscopy (ICP–OES) using an Elmer Optima 5300 analyzer (PerkinElmer, Waltham, MA, USA). The analysis of the solid materials was performed after they were mineralized in a mixture of acids (30% HCl, 65% HNO_3, water) prepared according to the relevant standard. The water extracts were prepared at a ratio of 1 kg:10 L with accuracies of ±0.01 g and 0.05 mL, respectively [48]. The water extract pH was determined via the potentiometric method per standard [49].

Diffractometric tests were performed using the powder method (DSH) in Bragg–Brentano geometry with a Bruker D8 DISCOVER diffractometer (Bruker Corp., Billerica, MA, USA), CuKα radiation, Ni filter, and LYNXEYE_XE detector (Bruker Corp., Billerica, MA, USA).

The mineral composition was determined and calculated based on standards licensed in PDF-5+ 2024 RDB ICDD (International Centre for Diffraction Data) and databases: ICSD (Inorganic Crystal Structure Database) and NIST (National Institute of Standard and Technology). DIFFRAC v.5.2 and TOPAS v.4.2 software from Bruker AXS was used for recording and diagnostics. Quantitative calculations of crystalline phases and amorphous substances were performed based on the Rietveld method, using ZnO or Al_2O_3 standards, depending on the sample matrix.

Analyses of grain morphology and chemical composition in the microareas were carried out by scanning electron microscopy (SEM) and X-ray energy dispersion spectroscopy (EDS) using an SU3500 SEM (Hitachi, Ltd., Tokio, Japan) operating in conjunction with an UltraDry EDS Detector (Thermo Fisher Scientific Inc., Waltham, MA, USA) under the following conditions: 15 keV acceleration voltage, BSE detector, magnification from 100× to 1000×. The images were taken after spraying the samples with gold.

Grain size and shape analysis via the optical method with image analysis using a G3S-ID analyzer (Malvern Panalytical Ltd., Malvern, UK) was carried out under the following conditions: air as the dispersion medium; dispersion pressure of 4.0 bar; ×20, ×10 and ×5 lenses; diascopic light.

The crushing resistance of the geopolymer-based granulates was determined according to the standard [50]. Per the standard, crushing resistance tests can be performed for aggregates with a grain size of 4 mm to 22 mm and a bulk density greater than 150 kg/m^3. Therefore, fractions below 4 mm and above 10 mm were removed from the prepared samples for testing individual granulates to homogenize the samples for comparative testing. Individual large granules in the tested material volume could distort the reading of the force required for the piston to penetrate the tested material in accordance with the standard. The tested granules were subjected to mechanical testing consisting of uniaxial compression of the samples in a steel cylinder in a strength testing machine with a maximum force of 30 kN (Figure 3). The crushing resistance (SC) of individual aggregates was calculated via the following equation:

$$SC = \frac{L+F}{A}\ \mathrm{N/mm^2} \tag{1}$$

where SC is the crushing resistance (MPa), L is the force exerted by the piston in newtons(N), F is the force required to depress the piston (N), and A is the piston area (mm^2).

(a) (b) (c)

Figure 3. Crushing resistance testing of geopolymer-based granulates: (**a**) granulated sample in a steel cylinder–before testing, (**b**) granulate compression, (**c**) granulate after compression.

The water absorption testing of the geopolymer-based granulates was performed according to standard [51]. Granules with grain sizes ranging from 4 mm to 8 mm were placed in baskets and flooded with water to a height of approximately 5 cm above the tested sample level. The tested material was then left in the water until the samples were completely saturated (24 h). After this time, the granules were removed, wiped with a damp absorbent cloth, and weighed (m_1). The next step involved drying to a constant mass (at 105 ± 5 °C). After removal from the dryer, the samples were weighed (m).

The aggregate water absorption was determined via the following formula:

$$W_A = \frac{m_1 - m}{m} \cdot 100\% \tag{2}$$

where m is the mass of the sample dried to a constant mass and m_1 is the mass of the sample completely saturated with water.

The aggregate abrasion resistance was determined via the micro-Deval wet method per standard [52]. Granules with grain sizes ranging from 4 mm to 8 mm were used for the tests. The samples were rinsed in water and dried to a constant weight at 105 ± 5 °C. Samples weighing 500 g each were placed in the device container, and 2.8 dm^3 of water were added. The device rotated the containers at a speed of 100 rpm. The set number of revolutions was 12,000 ± 10 revolutions. After the device was stopped, the containers were opened, and their contents were transferred to a protective sieve with a mesh size of 10 mm. The protective sieve was placed on another sieve with a mesh size of 1.6 mm. The material and steel balls on the sieves were rinsed with a stream of clean water. The balls were carefully separated from the material on the protective sieve. The remaining mineral material was transferred to a sieve with a mesh size of 1.6 mm and rinsed with water. The material remaining on the sieve was transferred to a tray, dried at 105 ± 5 °C and weighed. The micro-Deval coefficient was calculated via the following formula:

$$Mde = \frac{500 - m}{5} \tag{3}$$

where Mde is the micro-Deval coefficient, and m is the mass of the fraction retained on the 1.6 mm sieve [g].

Size distribution was determined via wet sieving methods according to standard [53] using a laboratory vibrating screen and standardized laboratory sieves (diameter—200 mm; height—50 mm; mesh sizes 1.6 mm, 2 mm, 4 mm, 8 mm and 16 mm) produced in accordance with the standard.

3. Results and Discussion

3.1. Characterization of the Materials Used as the Main Precursor for Producing the Geopolymer-Based Granulate

The main components of the coal fly ash (CFA) were silicon (SiO_2, 30.0 wt%) and calcium (CaO, 20.9 wt%). The contents of aluminum (Al_2O_3) and iron (Fe_2O_3) were 22.2 wt% and 5.4 wt%, respectively (Table 3). The total content of SiO_2, Al_2O_3 and Fe_2O_3 was 57.6 wt%. The tested ash belonged to the high-calcium ash group [54,55].

Table 3. Main chemical composition (via XRF) of the materials used for the geopolymer-based granulate production.

Sample Index	CaO	SiO_2	Al_2O_3	Fe_2O_3	MgO	Na_2O	K_2O	TiO_2	ZnO	MnO	P_2O_5	SO_3	Cl	Br	C	H
	wt%															
CFA	20.90	30.00	22.20	5.39	1.93	2.71	2.06	1.09	0.04	0.09	0.16	8.25	0.77	0.01	4.08	<0.01
CR	19.60	30.70	9.10	0.59	1.14	0.32	0.30	1.06	<0.01	0.02	0.04	0.02	0.02	<0.01	34.10	2.88
SD	8.85	2.09	0.73	37.20	2.46	<0.01	1.37	0.06	36.20	2.46	0.13	1.42	1.80	0.03	2.62	<0.01

The ash also contained metals that are known for environmental toxicity and adverse effects on human health, such as arsenic, cadmium, lead, chromium, nickel, copper, zinc, tin, cobalt, molybdenum, antimony, barium, manganese and selenium [56,57]. Among the metals listed, the highest contents were identified for Zn (451 mg/kg), Ba (317 mg/kg) and Mn (304 mg/kg). The lead content did not exceed 200 mg/kg, the chromium and copper contents did not exceed 70 mg/kg, the nickel and cobalt contents did not exceed 40 mg/kg, and the contents of the remaining elements were less than 4 mg/kg. The total content of the analyzed elements was 1439 mg/kg (Tables 4 and 5). The analyzed coal fly ash consisted mainly of mineral phases such as quartz (31.5%) and anhydrite (14.0%). Amorphous substances accounted for 35.5% (Figure 4). The ash grains were irregular in shape. The percentage of grains with a size (CE parameter) less than 10 μm was 96.2%. The percentage of grains below 1 μm was 14.0%. The grain size did not exceed 150 μm (Figures 5 and 6, Tables 6 and 7). The maximum grain sizes (CE parameter) for the composite regrind and steelmaking dust were 450 μm and 400 μm, respectively (Table 5).

Table 4. Trace chemical composition (via ICP–OES) of the materials used for the geopolymer-based granulate production.

Sample Index	As	Ba	Cd	Co	Cr	Cu	Mo	Ni	Pb	Sb	Se	Sn	V	Sr	Li	W
CFA	0.33	317	2.17	10.8	67.1	64.6	1.56	37.2	175	3.09	2.04	2.25	94.1	245	82.8	<4
CR	0.29	33.8	<1	81.0	16.1	5.54	10.5	7.90	1.32	<1	3.16	9.31	7.88	35.2	22.2	<4
SD	<0.1	236	341	37.0	3604	1142	272	1031	9518	118	17.4	181	59.6	68.0	20.1	14.2

Table 5. Water leaching results (via ICP–OES) for selected elements from the geopolymer-based granulate (expressed as mg/kg of solid dry material) and pH of the water leachate.

Sample Index	Mn	Cd	Co	Cr	Cu	Ni	Pb	Zn	As	Se	Mo	V	Ti	Sb	Ba	Sr	Li	pH
	mg/kg s.d.m.																	-
CFA	<0.002	<0.005	<0.01	<0.05	<0.01	<0.02	<0.05	<0.05	<0.05	<0.05	0.509	<0.05	<0.005	0.703	3.46	53.7	1.34	12.58
CR	0.461	<0.005	2.35	<0.05	<0.01	<0.02	<0.05	0.224	<0.05	<0.05	0.854	<0.05	<0.005	<0.05	0.08	0.31	0.311	7.68
SD	<0.002	<0.005	<0.01	25.3	0.11	<0.02	133	8.10	<0.05	1.40	90.9	<0.05	0.053	0.25	12.3	11.1	3.02	12.40

The main chemical components of the composite regrind (CR) were carbon (C, 34.1 wt%) and silicon (SiO_2, 30.7 wt%). Like the coal fly ash, this material was characterized by a high calcium content (CaO, 19.6 wt%). The sum of SiO_2 and Al_2O_3 was 39.8 wt%. The contents of the other elements did not exceed 3 wt%. The chemical composition of the steelmaking dust (SD) was different from that of the coal fly ash. The main components were iron (Fe_2O_3, 37.2 wt%) and zinc (ZnO, 36.2 wt%). The sum of SiO_2 and Al_2O_3 was 2.8 wt% (Table 3).

The amorphous phase in the composite regrind accounted for 84.0 wt% of the material. The identified crystalline components included gibbsite, hexahydrite, quartz and feldspar. The main mineral components in the steelmaking dust were franklinite (34.0 wt%) and zincite (24.0 wt%), while others included magnetite (14.0 wt%), quartz (4.5 wt%) and hematite (4.0 wt%). The amorphous phase constituted 17.5 wt% of the steelmaking dust (Figure 4).

The total contents of metals such as arsenic, cadmium, lead, chromium, nickel, copper, zinc, tin, cobalt, molybdenum, antimony, barium, manganese and selenium were 193 mg/kg and 326 g/kg, respectively, for the composite regrind and steelmaking dust. The high metal content in the steelmaking dust was primarily due to the high levels of zinc (291 g/kg) as well as manganese (18.9 g/kg), lead (9.5 g/kg) and chromium (3.6 g/kg). The total content of the remaining elements (arsenic, cadmium, nickel, copper, tin, cobalt, molybdenum, antimony barium, and selenium) was 3375 mg/kg.

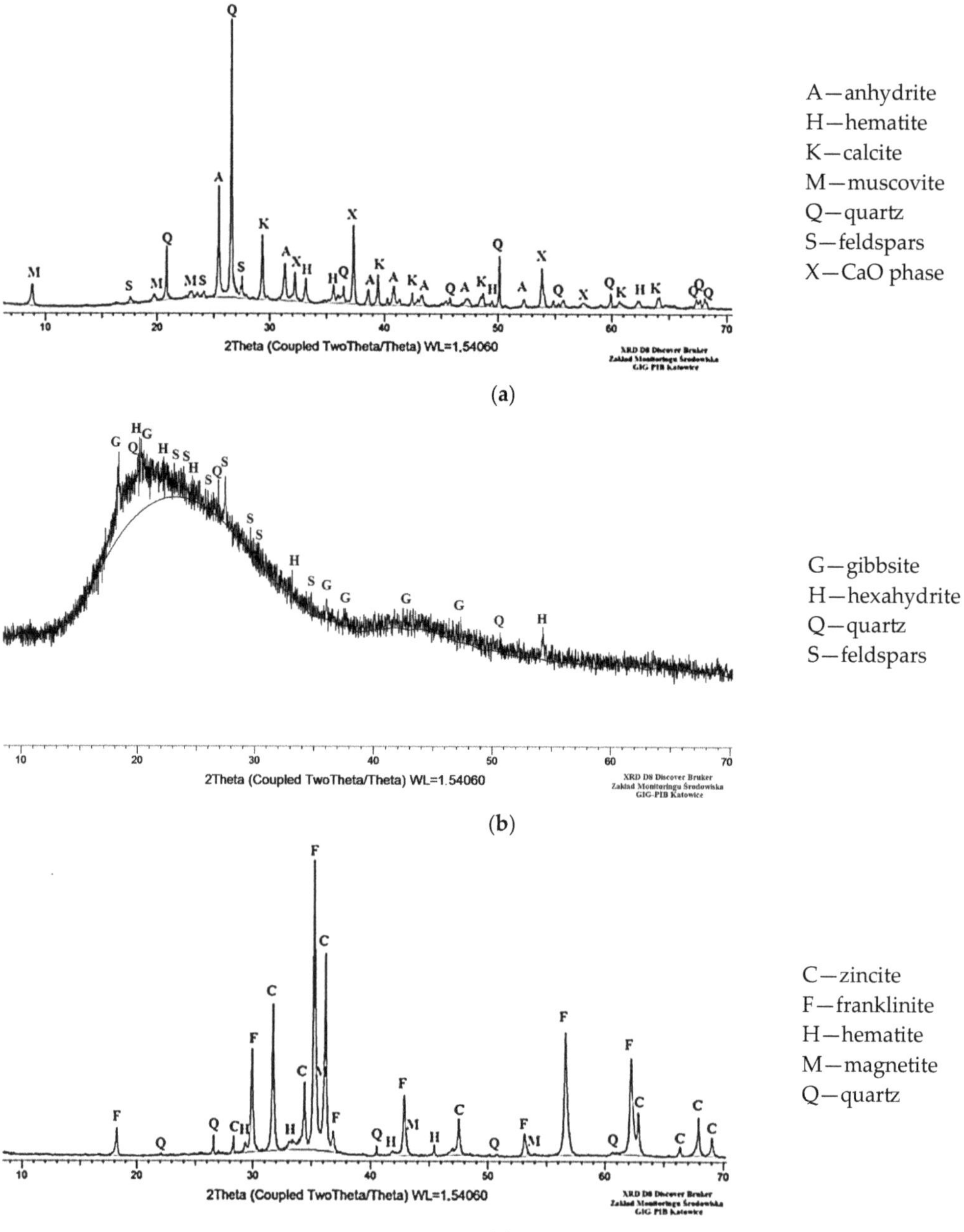

Figure 4. Mineral composition (via XRD) of the main precursor used for geopolymer-based granulate production: (**a**) coal fly ash (CFA), (**b**) composite regrind (CR), and (**c**) steelmaking dust (SD).

Figure 5. SEM images with grains of the coal fly ash (from the combustion of hard coal in a fluidized coal boiler, CFA) used as the main precursor for geopolymer-based granulate production.

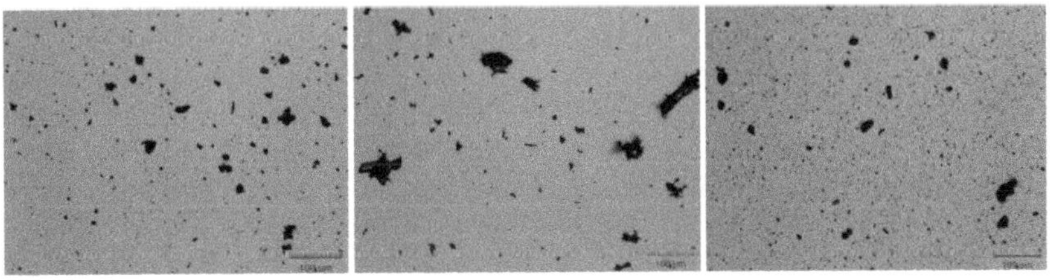

(a) (b) (c)

Figure 6. Photos of selected grains: (a) coal fly ash (CFA), (b) composite regrind (CR), and (c) steelmaking dust (SD) used as the main precursor in the production of geopolymer granules based on ash.

Table 6. Particle size (number distribution) of the materials used for the geopolymer-based granulate production.

CE Diameter	CFA	CR	SD
μm		%	
<5	84.7	68.0	93.7
5–10	11.5	16.4	5.87
10–20	3.18	10.51	0.65
20–50	0.632	4.71	0.11
50–100	0.020	0.312	0.004
100–200	0.001	0.034	<0.001
>200	n.d.	0.008	<0.001

Note: n.d.—not detected.

Table 7. Particle size (volume distribution) of the materials used for the geopolymer-based granulate production.

CE Diameter	CFA	CR	SD
μm		%	
<5	4.48	0.19	12.3
5–10	11.0	1.14	13.6
10–20	23.7	5.21	13.7
20–50	41.3	21.5	18.9
50–100	13.9	13.8	8.53
100–200	5.64	13.5	5.44
200–400	n.d.	26.3	27.6
>400	n.d.	18.4	n.d.

Note: n.d.—not detected.

Analyzing the input materials for the leachability of the abovementioned elements revealed that the coal fly ash (CFA) and composite regrind (CR) were moderately stable, whereas the steelmaking dust (SD) was characterized by a high mobility of selected elements. The coal fly ash mainly leached barium (3.46 mg/kg), antimony (0.70 mg/kg) and molybdenum (0.51 mg/kg). The composite regrind primarily leached cobalt (2.35 mg/kg) and molybdenum (0.85 mg/kg), as well as manganese (0.46 mg/kg), zinc (0.22 mg/kg) and barium (0.08 mg/kg). Despite the high zinc content, molybdenum was the main element leached from the steelmaking dust (91 mg/kg), accounting for 33% of all the leached elements and 33% of the total molybdenum content. In addition, lead (133 mg/kg), chromium (25.3 mg/kg), barium (12.3 mg/kg), zinc (8.1 mg/kg), selenium (1.4 mg/kg), antimony (0.25 mg/kg) and copper (0.11 mg/kg) were leached as well.

3.2. Grain Composition and Granulate Shape

Clear differences between individual samples can be observed in the analysis of the geopolymer granule grain composition (Figure 7). The grain fraction with a size of 4–8 mm was dominant in samples ST1 and ST2, accounting for 74.9% and 64.9% of the total mass, respectively. The opposite trend was observed in samples ST2A and ST3, where the grain fraction below 2 mm was clearly dominant, reaching 67.9% and 65.4% of the total mass, respectively, whereas the share of grains above 4 mm did not exceed 8.4%.

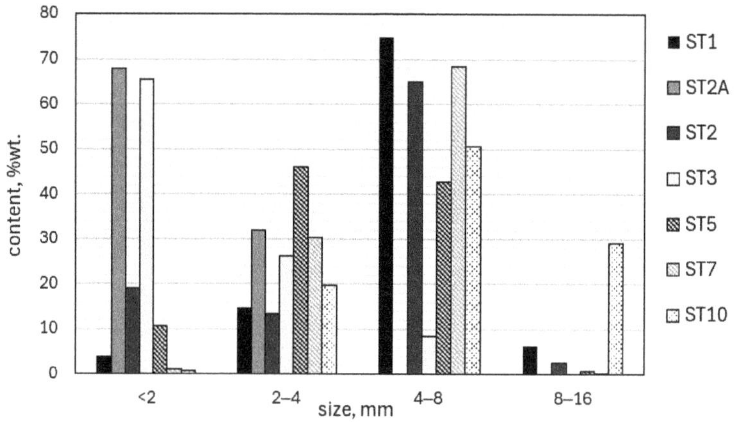

Figure 7. Grain size composition of geopolymer-based granulate.

In the case of samples ST5 and ST7 (based on coal fly ash and composite regrind), 2–8 mm was the main fraction size; specifically, in the case of sample ST5, the 2–4 mm and 4–6 mm fractions were similar and amounted to 46.0% and 42.7% of the total mass, respectively, whereas the 4–8 mm fraction was dominant in the case of sample ST7 (68.3% of the total mass). The most diverse grain size distribution was found in sample ST10, which contained all the analyzed fractions, including the largest grains of 8–16 mm, which accounted for 29% of the total mass. Figure 8 presents photographs of the granules in the 2–4 mm and 4–8 mm grain size fractions.

Fine granules with a moist surface formed during the homogenization stage of the ST1 sample components, which conjoined during the maturation stage, creating agglomerates with uneven surfaces. In experiments ST2 and ST2A, the solid-to-liquid ratio was increased to 2.00 and 1.90, respectively (Table 2). The application of staged moistening enabled the formation of spherical granules larger than 4 mm in experiment ST2. In experiment ST3, the use of surface powdering made it possible to produce fine (2–4 mm) spherical granules.

Figure 8. Photos of the produced geopolymer-based granulate.

3.3. Mechanical and Physical Properties of the Produced Geopolymer-Based Granulates

The results of crushing resistance, water absorption and abrasion resistance tests performed for the geopolymer-based granulates after 28 days of curing are depicted in Figure 9. The use of additives such as composite regrind (CR) and steelmaking dust (SD) had a significant effect on the mechanical properties of the geopolymer granulates. The crushing resistance of the (CFA) geopolymer granulate (based on fly ash) did not exceed 2 MPa. The use of composite regrind as an additive improved the mechanical properties of the material. In the case of sample ST5, there was a 2.9-fold increase in value (CS = 5.6 MPa), and in the case of sample ST7 (where powdering was used in addition to moistening), a 3.4-fold increase in value was recorded (CS = 6.7 MPa). In the case of test ST10, where steelmaking dust was used as an additive, there was a 5.7-fold increase in crushing resistance (CS = 11.1 MPa) and a reduction in water absorption to 11.5%. Water absorption was high in the other tests, exceeding 32%. In summary, the strength test results indicated that the obtained aggregate strength was significantly greater than that of most aggregates available on the market.

The crushing strengths of various commercial aggregates are as follows: Liapor: 0.7–10 MPa; Arlita: 0.98 MPa; Lytag: 0.43 MPa; LECA and Ardelite: 0.09 MPa; and LECA Gniew: 0.7–4.0 MPa [58–60]. Compared to the results published for other wastes used to produce geopolymer granules, the obtained values of CS amounted 6.7 MPa (CFA + CR) and 11.1 MPa (CFA + SD) are among the hardest for materials cured under ambient conditions with a limited amount of activator. It should be emphasized that although the strength of the granules obtained (especially ST10) is close to the parameters of some dense (natural) aggregates, the main reference group for us remains manufactured lightweight aggregates (LWA). This is due to the fact that our cold bonding technology is a direct, low-energy alternative to the energy-intensive sintering processes used in the production of LWA. In addition, the high water absorption of most of our granules (above 32%) confirms their porous structure, typical of lightweight aggregates, and fundamentally different from the almost zero water absorption of dense aggregates. Our material therefore fills a market niche, offering strength significantly superior to typical LWAs, while maintaining a low-energy production regime.

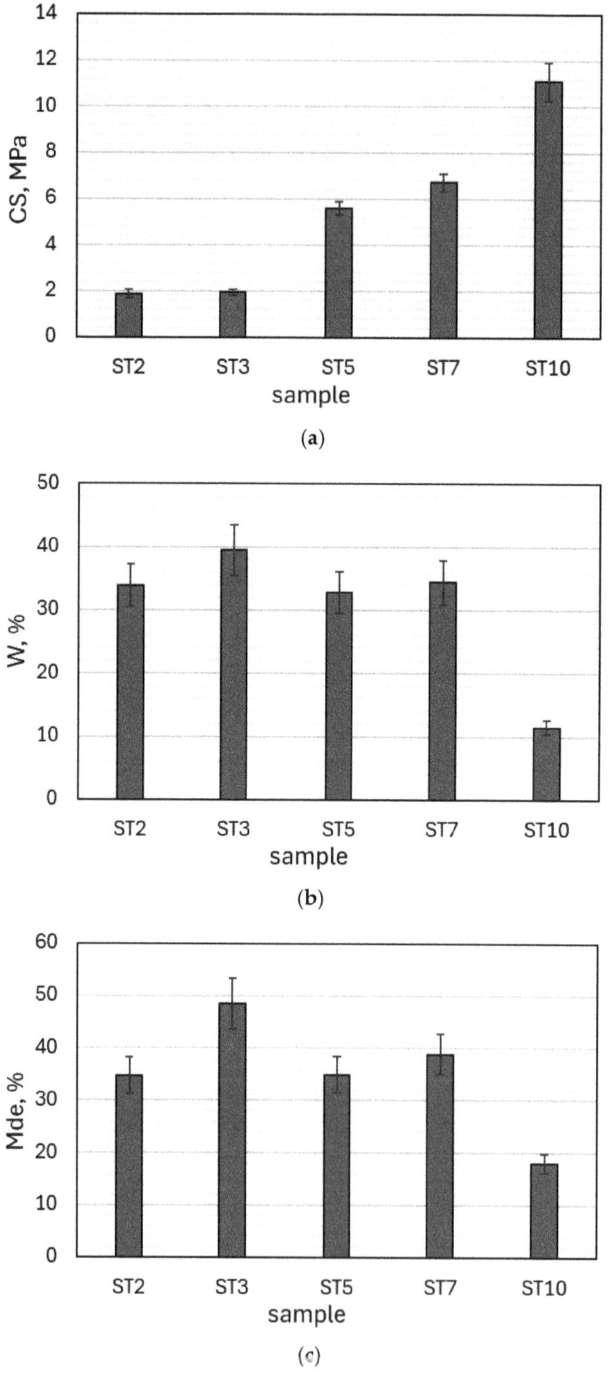

Figure 9. Results of mechanical and physical property testing for geopolymer-based granulates: (**a**) crushing resistance, (**b**) water absorption, and (**c**) abrasion resistance.

3.4. Chemical and Mineral Composition of the Geopolymer-Based Granulates

Tables 8 and 9 present the chemical and mineral compositions of the geopolymer-based granulates after 28 days of curing. Table 10 presents the results of selected metal leachability from the tested granulates. Samples ST2 and ST3 (coal fly ash-based) were dominated by typical silicate minerals and calcium phases such as quartz, feldspar, portlandite, anhydrite and muscovite, and exhibited a balance between the contents of amorphous and crystalline substances. The ratios of the amorphous to crystalline contents were 0.94 and 1.02 for samples ST2 and ST3, respectively. In samples obtained on the basis of the coal fly ash (CFA) and composite regrind (CR), a gradual decrease in quartz content from 31.5 wt% (content in the CFA) to 10.5 wt% (content in sample ST7) and an increase in the amorphous content from 35.5 wt% (content in CFA) to 70.0 wt% (content in ST7) was observed.

Table 8. Mineral composition of the geopolymer-based granulate.

Phase	Sample Index				
	ST2	ST3	ST5	ST7	ST10
	wt%				
Quartz (SiO_2)	26.5	22.5	12.5	10.5	5.5
Franklinite ($ZnFe_2O_4$)	n.d.	n.d.	n.d.	n.d.	32.0
Zincite (ZnO)	n.d.	n.d.	n.d.	n.d.	15.0
Calcite ($CaCO_3$)	5.5	5.5	3.5	3.5	4.5
Kaolinite (Al4[$Si_4O_{10}(OH)_8$])	n.d.	n.d.	n.d.	n.d.	n.d.
Feldspars (potassium + plagioclases)	5.0	4.0	2.0	4.0	1.0
Anhydrite ($CaSO_4$)	4.0	4.0	5.0	4.0	4.0
Portlandite ($Ca(OH)_2$)	3.0	3.0	2.0	2.0	n.d.
Hematite (Fe_2O_3)	2.0	2.0	2.0	2.0	2.0
Apatite (Ca[(F/OH/Cl/($PO_4)_3$)	2.0	3.0	4.0	2.0	n.d.
CaO substance	1.0	1.0	1.0	2.0	n.d.
Muscovite ($KAl_2[AlSi_3O_{10}(OH)_2]$)	1.0	2.0	n.d.	n.d.	n.d.
Magnetite ($FeFe_2O_4$)	1.0	2.0	n.d.	n.d.	9.0
Chlorite ($(Mg,Fe)_6[(AlSi_3)O_{10}](OH)_8$)	n.d.	n.d.	n.d.	n.d.	n.d.
Crystalline substance (C)	51.0	49.0	32.0	30.0	73.0
Amorphous substance (A)	48.0	50.0	67.0	70.0	25.5
A:C phase ratio	0.94	1.02	2.09	2.33	0.35

Note: n.d.—not detected.

Table 9. Main chemical composition (via XRF) of the geopolymer-based granulate.

Sample	CaO	SiO_2	Al_2O_3	Fe_2O_3	MgO	Na_2O	K_2O	TiO_2	ZnO	MnO	P_2O_5	SO_3	Cl	Br	C	H
	wt%															
ST2	30.10	22.60	10.70	9.60	0.83	11.40	2.08	1.48	0.11	0.09	0.11	5.87	0.63	0.02	3.42	0.55
ST3	30.20	22.00	9.36	10.30	0.70	13.20	1.89	1.37	0.12	0.12	0.13	5.40	0.64	0.02	3.66	0.51
ST5	18.60	29.50	9.14	2.89	0.91	20.20	0.91	0.79	0.03	0.03	0.06	2.20	0.33	0.00	14.30	<0.01
ST7	19.70	28.00	8.11	2.81	0.80	17.70	0.94	0.98	0.03	0.04	0.08	1.60	0.31	0.01	17.30	1.38
ST10	14.10	11.80	4.12	27.20	1.89	4.20	1.42	0.28	24.40	1.65	0.31	2.42	1.28	0.03	3.22	<0.01

The ratios of the amorphous to crystalline contents were 2.09 and 2.33 for samples ST5 and ST7, respectively. This significant increase in the amorphous-to-crystalline ratio (A:C) in samples ST5 and ST7 provides direct mineralogical evidence of a higher degree of substrate reaction and a more effective geopolymerization process. This suggests that the CR additive did not act merely as an inert filler, but may have actively participated in the reaction or facilitated the dissolution of aluminosilicates from fly ash, leading to the precipitation of a larger volume of amorphous geopolymer binder. It is this amorphous phase that is the main carrier of mechanical properties, which directly explains the significant increase in the strength of these granules observed in Section 3.3.

Table 10. Water leaching results (via ICP–OES) of selected elements from the geopolymer-based granulate (expressed as mg/kg of solid dry material) and pH of the water leachate.

Sample	Mn	Cd	Co	Cr	Cu	Ni	Pb	Zn	Se	Mo	Sn	V	Ti	Sb	Ba	Sr	Li	W	pH
	mg/kg s.d.m.																		-
ST2	0.003	0.031	<0.01	0.16	0.027	<0.02	<0.05	<0.05	1.2	0.83	0.056	12.3	0.019	0.15	0.005	0.13	11.5	1.87	12.84
ST3	0.013	0.069	<0.01	0.17	0.066	0.035	<0.05	<0.05	0.9	0.63	<0.05	9.4	0.010	0.3	<0.002	0.16	8.9	1.18	12.74
ST5	0.035	0.16	0.29	0.21	0.26	0.13	<0.05	<0.05	0.88	3.2	0.097	10.6	0.023	0.45	0.041	0.26	0.27	1.27	11.73
ST7	0.0029	0.14	0.23	0.2	0.11	0.12	<0.05	<0.05	0.76	3.8	<0.05	9.1	0.019	0.57	0.036	0.2	2.1	1.64	11.31
ST10	<0.002	0.016	<0.01	23.4	0.21	<0.02	6.4	42.2	4.7	105	0.12	8.7	<0.005	0.93	0.56	0.16	3.3	12.3	12.60
Limit [1]	nd	0.04	nd	0.50	2.00	0.40	0.50	4.00	0.10	0.50	nd	nd	nd	0.06	20.00	nd	nd	nd	-

Note: [1] Permissible leaching limit values for the acceptance of waste at inert waste landfills [61].

Sample ST10 (based on coal fly ash and steelmaking dust) exhibited a different mineral composition than the other geopolymer samples. The main components included zinc and iron minerals, i.e., franklinite (32.0 wt%), zincite (15.0 wt%), magnetite (9.0 wt%) and hematite (2.0 wt%). Quartz accounted for 5.5 wt%, while the amorphous phase accounted for 25.5 wt%. The ratio of amorphous to crystalline content was 0.35, therefore crystalline phases dominated in this material, as in the steelmaking dust (SD).

Figure 10 compares the leachability of selected metals from the input materials and the granules obtained on their basis. These values were compared with the guidelines for the acceptance of waste for disposal in inert waste landfills [61]. Exceeded permissible leaching limits were observed in the following ranges:

- geopolymer granules based on coal fly ash (ST3, ST2)—the permissible leaching limits were exceeded for cadmium (only in ST3), selenium, molybdenum and antimony;
- geopolymer granules based on coal fly ash and composite regrind (ST5 and ST7)—the permissible leaching limits were exceeded for cadmium, selenium, molybdenum and antimony;
- geopolymer granules based on coal fly ash and steelmaking dust (ST10)—the permissible leaching limits for most metals were exceeded: chromium, lead, zinc, selenium, molybdenum and antimony.

Alkaline activators, which increase leachate pH, were used in the geopolymer synthesis. A change in pH has a significant effect on metal immobilization [45]. The following trends were observed: (i) barium, antimony, lead—higher leachability of these metals from the input materials than the geopolymer-based granules; (ii) molybdenum, cadmium, chromium, copper, nickel, zinc, selenium—higher leachability of these metals from the geopolymer-based granules than the input materials.

As pH decreases, metals are released from soil particles due to reduced adsorption and increased dissolution, especially for metals like cadmium and zinc, which are highly mobile even at neutral or slightly alkaline pH. Some metals (notably Pb and Zn) show amphoteric behavior: their leaching is high at both low and high pH, but minimized at intermediate pH values. The effect of pH on metal leaching is also influenced by material properties (buffer capacity, organic matter, mineral composition) and the specific metal involved.

The leachability of chromium, zinc, and molybdenum is significantly higher with CFA + SD geopolymer granules (ST10) than with the input materials. The leachability of lead from sample ST10 is lower than from its synthesis precursors. The pH of leachate from steelmaking dust was 12.40, while the pH of the leachate from ST10 granulates was 12.60, so the alkaline activators used may have a key influence on the mobilization/immobilization of metals.

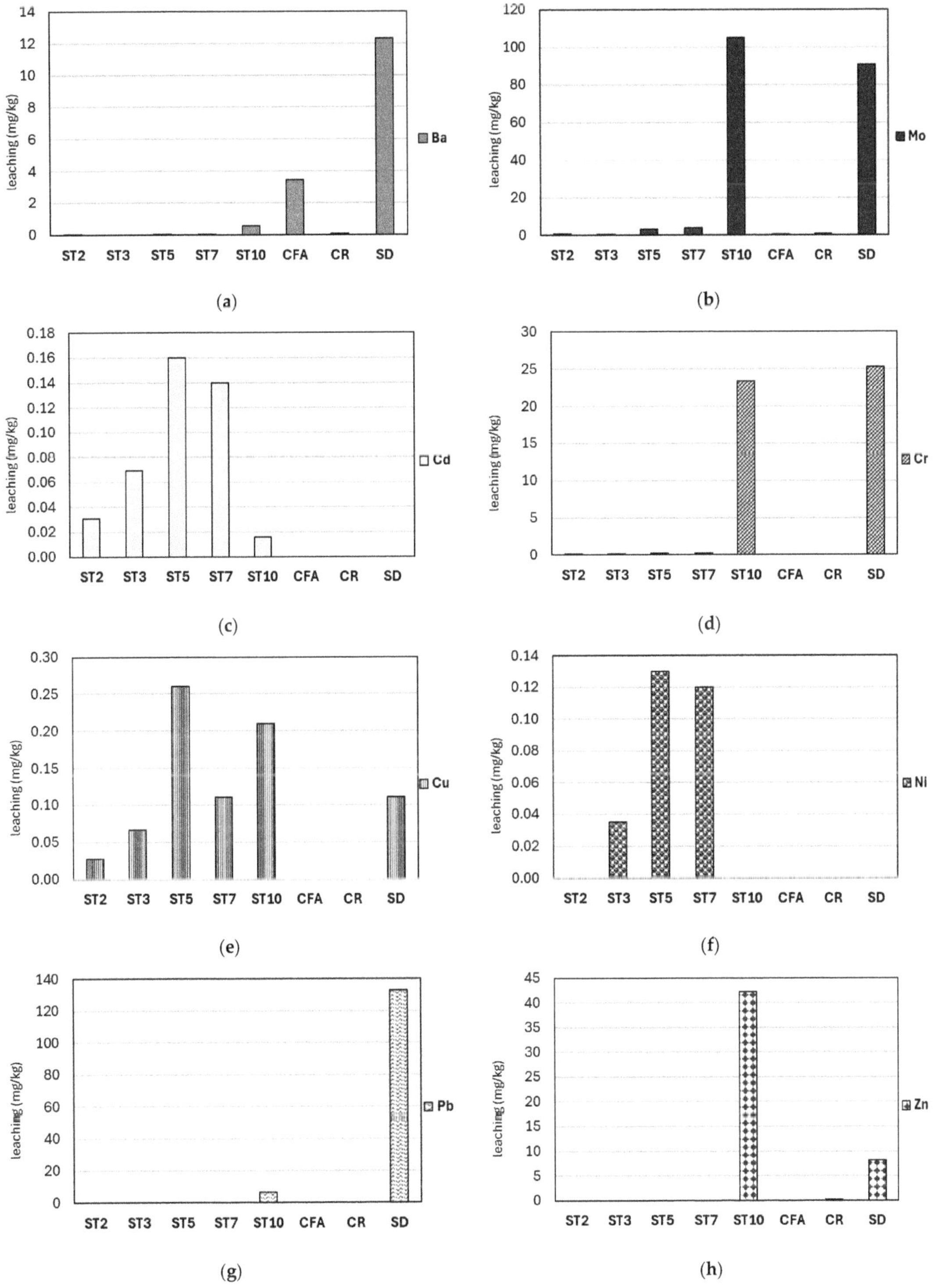

Figure 10. *Cont.*

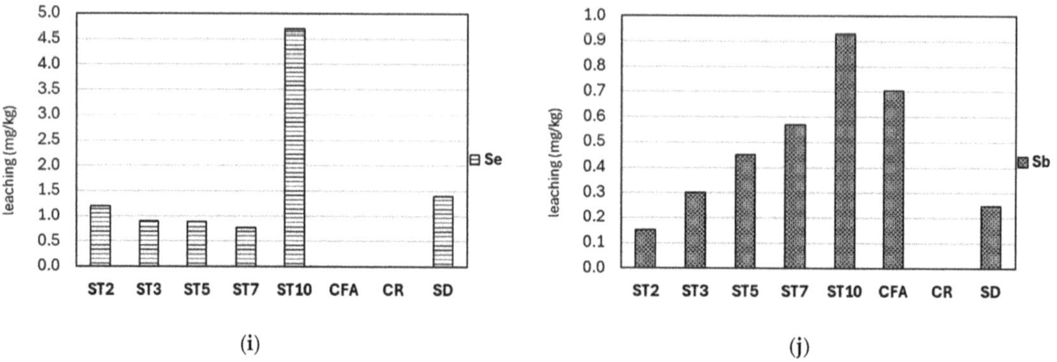

Figure 10. Leaching content of selected metals: (**a**) barium, (**b**) molybdenum, (**c**) cadmium, (**d**) chromium, (**e**) copper, (**f**) nickel, (**g**) lead, (**h**) zinc, (**i**) selenium, (**j**) antimony from the input materials (CFA, CR, SD) and the geopolymer-based granulate (ST2, ST3, ST5, ST7, ST10); Note: Explanations of abbreviations are provided in Table 2.

NaOH can selectively leach zinc from franklinite, especially at high concentrations and elevated temperatures, while iron remains largely insoluble. This allows for efficient zinc recovery with minimal iron dissolution. ZnO is highly soluble in NaOH, forming soluble zincate ions ($[Zn(OH)_4]^{2-}$). The leaching process is chemically controlled and efficient, with high zinc recovery rates [62]

NaOH, especially when combined with oxidants or at high temperatures, efficiently extracts chromium from chromite and spinel phases as water-soluble chromate. The process is enhanced by oxidizing agents (e.g., H_2O_2) and high alkali concentrations [63]. NaOH is very effective for leaching molybdenum from minerals like wulfenite, with extraction rates above 99% at elevated temperatures and high alkali concentrations [64]. NaOH is highly effective for leaching lead from oxide and some secondary minerals, under optimal conditions (high temperature, high NaOH concentration). However, in some matrices (e.g., fly ash), NaOH can also stabilize lead, reducing its leachability by forming less soluble compounds [65].

The observed high leachability of Cr, Zn, and Mo from ST10 granules is not only a technical disqualification of this variant, but also a key scientific discovery. It demonstrates the fundamental duality and challenge of alkaline activation of metallurgical waste. The strongly alkaline environment necessary for the dissolution of aluminosilicates is also an aggressive chemical agent. As already mentioned, these conditions selectively attack the mineral phases present in SD, such as franklinite and zincite, leading to the mobilization of heavy metals. This illustrates a fundamental process relationship: conditions optimal for geopolymerization (high pH) may also be conditions conducive to the leaching of amphoteric contaminants (Pb, Zn) or those that form soluble oxyanions (Cr, Mo).

3.5. Geopolymer-Based Granulate Surface Morphology

Figures 11 and 12 present SEM images of geopolymer granules based on coal fly ash (CFA) as well as geopolymer granules based on coal fly ash and composite regrind (CFA + CR) after 28 days of curing. The chemical composition of the selected micro-areas corresponding to Figure 12 is presented in Table 11.

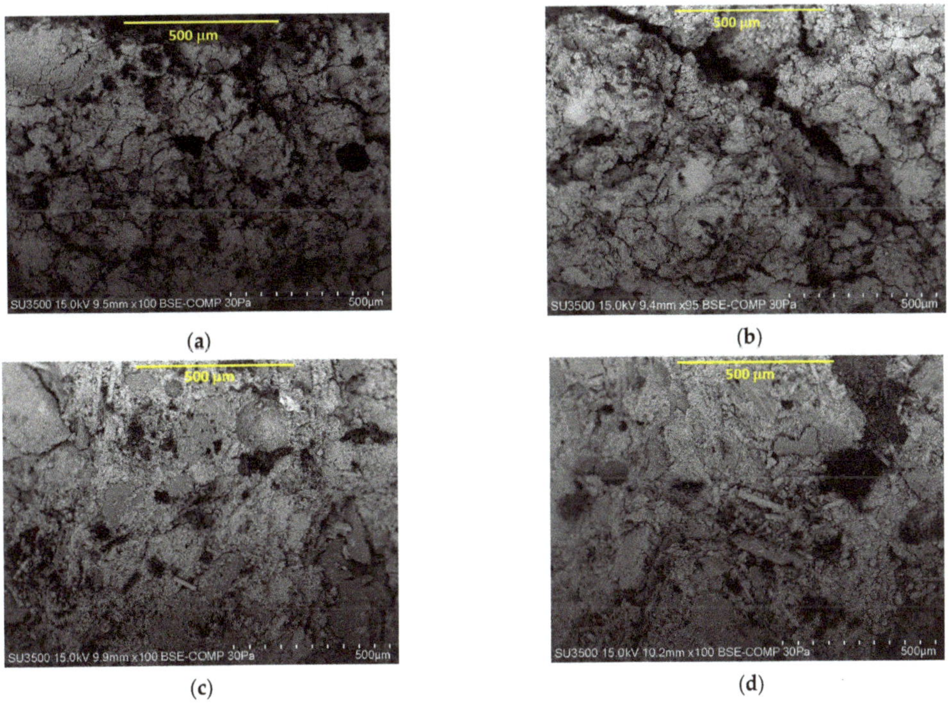

Figure 11. Surface area comparison for the CFA geopolymer granulate (represented by sample ST3, (**a**,**b**)) and CFA + CR geopolymer granulate (represented by sample ST7, (**c**,**d**)).

Pławecka et al. [45] assessed whether mechanically ground waste from decommissioned wind turbine blades can be used as a filler (0%, 5%, 15%, 30% by dry mass) in fly ash-based geopolymer composites to reduce waste and substitute conventional fillers. The compressive strength of the materials with waste fraction additions was lower than that of the reference sample. Low-to-moderate additions (around 5–15%) can yield materials with acceptable mechanical properties. Studies have demonstrated that the main drawback was high porosity and that reducing the porosity (improved mixing, vibration or modified processing) could improve strength.

This study has demonstrated that the structure of the CFA + CR geopolymer granules (based on fly ash and composite regrind) (Figure 11c,d) was more compact and dense; the pores were smaller, and the material was better bound than the CFA geopolymer granules (based on fly ash) (Figure 11a,b). The CFA + CR geopolymer structure was characterized by the presence of fibers whose lengths ranged from a few micrometers up to several hundred micrometers. An EDS analysis indicated that the main components of the fibers included silicon, calcium and aluminum. The presence of titanium, sodium and magnesium was identified as well (Table 11). This analysis showed that the presence of composite regrind (Si–Ca–Al) fibbers promotes the formation of structural bridges and surface sealing, which confirms the observed increase in strength and decrease in water absorption compared to ash granules alone. Microstructure analysis (Figures 11 and 12) provides an explanation for the strength increase mechanism observed in Section 3.3. While the base CFA granulate (Figure 11a,b) exhibits a porous structure with numerous voids and loosely bound grains, the structure of the CFA + CR granulate (Figure 11c,d) is fundamentally different. It is clearly more compact and dense, and, crucially, permeated with fibers from the composite milling. As shown by EDS analysis (Table 11, e.g., areas P1, P2 for ST5 and ST7), these fibers (with a Si-Ca-Al composition) are not a passive filler, but are well integrated into the amorphous geopolymer matrix. They act as micro-reinforcement and "structural bridges," transferring stresses and hindering the propagation of microcracks. This direct reinforcement at the micrometric level is the scientific reason for the observed more than 3-fold increase in compressive strength (up to 6.7 MPa).

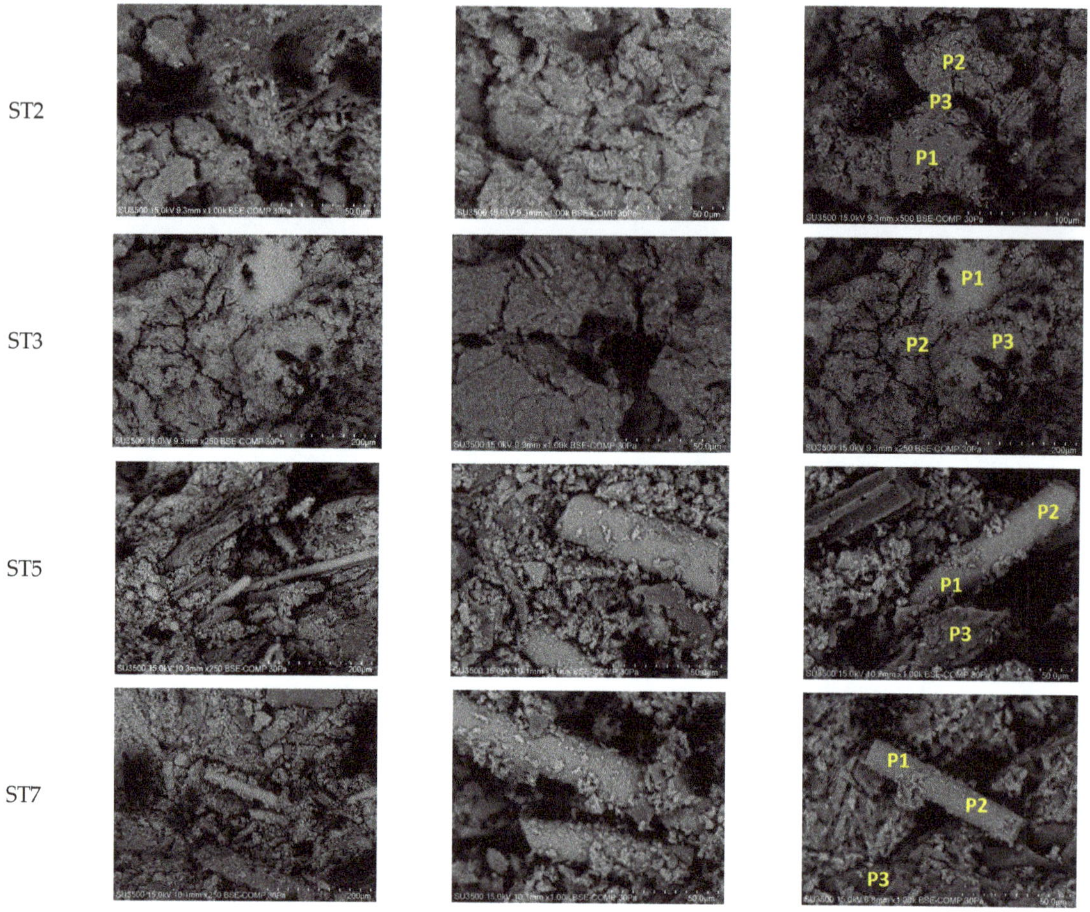

Figure 12. SEM images of selected geopolymer-based granulates ST2, ST3, ST5, ST7; Note: Explanations of abbreviations are provided in Table 2.

Research on geopolymers synthesized from coal fly ash highlights the critical influence of microstructure on the final mechanical properties, durability and environmental performance [66]. This study also showed a clear correlation between the structure of the geopolymer and the mechanical properties of the granules. In this case, the structural cohesion was enhanced by the use of an intensive counter-current mixer for granulate production combined with the presence of fibers.

Table 11. Chemical composition (via EDS) of selected micro-areas of the geopolymer-based granulate.

Sample Index [1]	Micro-Area Index [2]	C	O	Na	Mg	Al	Si	K	Ca	Ti	Fe
		wt%									
ST2	P1	5.1	42.8	1.9	-	9.9	30.3	10.2	-	-	-
	P2	6.5	40.5	8.1	0.8	11.5	20.8	1.5	7.4	-	2.9
	P3	31.4	31.6	7.6	0.5	6.6	13.3	1.0	6.6	-	1.6
ST3	P1	3.8	38.9	1.8	-	1.8	3.8	-	48.7	-	1.2
	P2	4.1	41.2	9.9	0.5	8.7	16.5	0.9	18.2	-	-
	P3	3.8	44.5	3.7	1.3	9.9	15.0	0.4	18.6	0.9	2.0
ST5	P1	9.2	42.7	1.5	0.7	7.0	25.5	-	13.6	-	-
	P2	8.4	42.0	1.1	0.6	7.3	25.6	-	14.5	0.5	-
	P3	47.7	33.1	2.8	-	2.7	8.5	0.6	4.7		-
ST7	P1	6.1	41.5	2.0	0.6	7.2	27.3	-	14.8	0.6	-
	P2	8.8	37.3	1.2	0.5	7.6	27.4	-	15.7	0.7	1.0
	P3	54.6	33.4	1.6	-	1.5	5.2	-	3.7	-	-

Note: [1] Explanations of abbreviations are provided in Table 2; [2] The micro-areas (P1, P2, P3) are marked in Figure 12.

3.6. Justification of Potential as a 'Sustainable Substitute'

The research results confirm that the developed geopolymer granules, especially those with PK added, are a viable and sustainable substitute for natural aggregates.

The substantiation of this thesis is based on proven facts. First, the 'sustainable' aspect results from a triple benefit: (1) the valorization of three different industrial waste streams (PM, PK, PS), which is in line with the principles of the circular economy and responds to the challenges described in the Introduction; (2) the use of energy-efficient 'cold bonding' technology, which allows curing at room temperature; and (3) demonstrated in the process of optimization and reduction in alkaline activator consumption (S/L > 2.0).

Secondly, the 'substitute' aspect has been clearly confirmed by mechanical properties. It has been shown that the addition of PK or PS leads to granules with crush strength (>6 MPa and >11 MPa, respectively) that 'surpasses many commercial lightweight aggregates'. Thus, it has been proven that these materials are not an inferior substitute, but a fully fledged, and in terms of strength often superior, technical product capable of replacing natural aggregates in demanding applications.

3.7. Economic Feasibility and Process Cost Implications

The developed method for producing geopolymer granules was designed with economic factors in mind. Unlike traditional lightweight sintered aggregates, the proposed solution is based on a granulation and curing process at room temperature ('cold bonding'). Eliminating the sintering stage is crucial for profitability. As indicated in the literature, the production of aggregates using the 'cold bonding' method consumes less energy and is more cost-effective than sintering, which translates into lower operating costs.

An additional factor reducing process costs is the achieved high solid-to-liquid ratio (S/L > 2.0). As shown in the Conclusions, this has enabled a significantly reduced amount of costly alkaline activators (NaOH and Na_2SiO_3). Lower consumption of activators, which have been a frequent cost factor in geopolymerization processes, is a direct result of the use of an intensive counter-rotating mixer.

From an economic point of view, the main advantage of the proposed solution is the use of waste raw materials (fly ash, steel mill dust, and composite grinding). As indicated in the Introduction, research on the recycling and reuse of waste materials is crucial for sustainable development. The use of these materials as feedstock minimizes their acquisition costs and is in line with the circular economy strategy.

It has been demonstrated that the production process can be carried out in a single technological operation, using an intensive counter-rotating mixer and an ambient maturation system. Low energy consumption (resulting from cold bonding) and limited demand for activators (resulting from high S/L) make this technology a promising industrial alternative.

In direct comparison with natural aggregates, whose resources are rapidly depleting, the unit cost of producing geopolymer granules may be higher. However, the overall profitability of the process is supported by the elimination of energy costs (compared to sintering) and reduced activator costs (thanks to S/L > 2.0). From an economic point of view, the proposed process therefore represents a viable alternative, particularly in the context of industrial waste valorization.

3.8. Scope of Environmental Analysis and Context of Alkaline Activators

It should be emphasized that the scope of the environmental analysis in this study focused on the valorization of industrial waste (PM, PK, PS), the energy efficiency of the cold bonding process compared to sintering, and the assessment of the washability of contaminants from the final product. The issue of quantitative environmental impact assessment (LCA) related to the production of alkaline activators (NaOH and Na_2SiO_3) itself was beyond the scope of this study. However, this issue was addressed indirectly through process optimization. It was demonstrated that the developed technology, thanks to the use of an intensive counter-rotating mixer, allows operation at a high solid-to-liquid ratio (S/L > 2.0). As indicated in the Conclusions, this results in a "significantly reduced amount" of alkaline activators consumed. The reduction in the consumption of these components is crucial because it directly minimizes the overall environmental footprint associated with their production, while improving the economic profitability of the process.

3.9. Summary

The developed method of granulation and cold bonding allowed for obtaining granules with compressive strength exceeding 6 MPa (CFA + CR) and 11 MPa (CFA +SD), which places them above the range of typical values for commercial lightweight aggregates. These parameters confirm that the proposed material meets the technical criteria for use as an aggregate in structural concrete or prefabricated elements. These studies have shown that the addition of composite grinding (CR) fundamentally modifies the geopolymerization process. This leads to the formation of a compact, fiber-reinforced microstructure, as confirmed by SEM analyses. This explains the mechanism by which above-average compressive strength (>6 MPa) was achieved, higher than that of many commercial lightweight aggregates (such as LECA, Liapor, Lytag). At the same time, the addition of steel dust (SD) resulted in the highest strength (>11 MPa), which is related to the different mineral composition of the final product.

The results indicate also that CFA + SD granules (ST10), despite their highest mechanical strength (11.1 MPa), exceed the permissible values for Cr, Pb, and Mo leaching, which limits their direct use or storage as inert materials. Therefore, this variant should be treated as a material for use in closed structures, e.g., in cement or geopolymer composites, where further binding reactions can ensure more effective immobilization of heavy metals.

In order to improve chemical stability, it is proposed to modify the alkaline activator (increase the silicate modulus, partial replacement of Na with K), introduce pozzolanic additives (metakaolin or GGBS—Ground Granulated Blast Furnace Slag) or apply surface protective coatings of a thin layer of geopolymer or cement grout. Therefore the key scientific conclusion is the identification of the dual role of steel dust (SD) additive. While it significantly improves mechanical parameters, a fundamental material challenge has been identified: the strongly alkaline conditions required for activation also mobilize hazardous heavy metals (Cr, Pb, Mo), leading to exceedances of standards. This points to the need for further research on immobilization in geopolymer matrices based on steel mill waste.

Another important innovative aspect of the developed technology is the high solid-to-liquid ratio (S:L > 2.0), which has made it possible to reduce the consumption of alkaline activators (NaOH and Na_2SiO_3) in the synthesis process. The reduction in the alkaline liquid content by over 30% compared to standard recipes translates directly into lower production costs and carbon footprint, confirming the practical industrial potential of the process.

Due to their favorable mechanical properties and volume stability, these granulates have the potential be used both as lightweight aggregate for concrete and precast elements, as well as ballast or insulation material in road and civil engineering. The use of industrial waste as the primary raw material makes the developed method an economically and environmentally sustainable solution.

4. Conclusions

The conclusions, which are based on the mechanical and physicochemical analysis results for geopolymer-based granulates with a substitution of coal fly ash (CFA) by composite regrind (CR) or steelmaking dust (SD), are as follows:

(1) The use of an intensive counter-current mixer enabled the mixing and granulation of the feedstock in the form of fluidized fly ash, alkaline activators, and additives such as composite regrind or steelmaking dust in a single technological operation. Moistening and/or powdering the ash-based granule surfaces had a significant effect on the quality of the obtained product, i.e., the size and shape of the geopolymer granules.

(2) The developed method enabled the production of geopolymer granules with a significantly reduced amount of water glass and the NaOH solution (solid-to-liquid ratio above 2.0).

(3) The produced geopolymer-based granulates had a compact, stable structure and were capable of self-hardening at room temperature without the need for further processing. The crushing resistance of the CFA + CR (>6 MPa) and CFA + SD (>11 MPa) geopolymer granules was greater than that of most aggregates available on the market.

(4) The method used to produce the CFA + CR geopolymer granules ensured the stabilization of metals such as chromium, copper, nickel, zinc, lead and barium. In the case of the CFA + SD geopolymer granules, the leaching levels of metals such as chromium, zinc and molybdenum were exceeded several times relative to the permissible leaching limits for the acceptance of waste for disposal in inert waste landfills.

(5) Considering both the mechanical and chemical properties of the produced geopolymer granules, the use of composite regrind as an additive is promising. In the next stage of the work, the geopolymer granule production process will be optimized to improve the immobilization of metals such as cadmium, selenium, molybdenum and antimony.

Taking into account both strength and environmental parameters, the CR additive should be considered the most promising component for cold-cured geopolymer granules. Despite its high strength, the DS additive requires additional metal stabilization methods to meet the environmental requirements for inert waste landfill.

The results obtained clearly confirm that geopolymer granules produced under cold curing conditions are a viable alternative to natural aggregates. The developed technology is characterized by low energy consumption, limited use of activators, and the possibility of using a wide range of industrial waste as raw materials. Combined with mechanical properties that exceed those of commercially available lightweight aggregates, these materials have high implementation potential in the construction sector, contributing to the reduction in natural resource exploitation.

From an economic point of view, the developed technology has key advantages for its potential profitability. The process is characterized by low energy consumption, as it is based on an energy-efficient cold bonding method and granule curing at room temperature. At the same time, significantly reduced consumption of expensive alkaline activators (NaOH solution and water glass) has been demonstrated thanks to the high solid-to-liquid ratio (S/L > 2.0) achieved in the process. Raw material costs are minimized by using industrial waste (CFA, CR, DS). The combination of these factors—energy efficiency, reduced consumption of activators, and the use of waste raw materials—makes the proposed technology promising and potentially competitive.

The authors acknowledge that while the term 'energy-efficient' is used in this paper to describe the underlying principles of the geopolymerization process (i.e., low-temperature processing and the valorization of industrial by-products which would otherwise require disposal), this study does not contain a quantitative, comparative evaluation of the energy consumption of the proposed technology versus the conventional extraction and processing of natural aggregates. Natural aggregate production, while mechanically intensive, often involves lower inherent energy input compared to manufacturing processes.

Therefore, to definitively validate the 'energy-efficient' and 'sustainable' nature of these geopolymer-based granulates in the context of substituting natural aggregates, a comprehensive Life Cycle Assessment (LCA) is required. Such an analysis was beyond the scope of this initial technological study but remains a critical objective for future research. This future work must quantify the energy inputs and environmental impacts across the entire value chain—from raw material extraction (for activators) and transport, through processing, to the end-of-life considerations of the final product—and compare these metrics directly against those for natural aggregates.

5. Patents

Patent application (Poland), "Method for Producing Geopolymers Using an Intensive Counter-Current Mixer"/"Sposób Wytwarzania geopolimerów z wykorzystaniem intensywnego mieszalnika przeciwbieżnego" (PW 6/2025).

Author Contributions: Conceptualization, M.C. and J.K.; methodology, M.C. and J.K.; validation, M.C. and J.K.; formal analysis, M.C.; investigation, M.C., J.K. and A.K.; resources, M.C.; data curation, M.C.; writing—original draft preparation, M.C.; writing—review and editing, M.C., J.K. and A.K.; visualization, M.C.; supervision, J.K.; project administration, M.C. and J.K.; funding acquisition, J.K. All authors have read and agreed to the published version of the manuscript.

Funding: This research was funded by the Polish Ministry of Science and Higher Education as part of research work No. 111840225 at the Central Mining Institute–National Research Institute in Poland.

Institutional Review Board Statement: Not applicable.

Informed Consent Statement: Not applicable.

Data Availability Statement: The original contributions presented in this study are included in the article. Further inquiries can be directed to the corresponding authors.

Conflicts of Interest: The authors declare no conflicts of interest.

Abbreviations

The following abbreviations are used in this manuscript:

AIV	Aggregate Impact Value
ACV	Aggregate Crushing Value
CFA	Class C fly ash
CE	Circle Equivalent diameter

FFA	Class F fly ash
FA	Fly Ash
GGBS	ground granulated blast furnace slag
MK	Metakaolin
RM	Red Mud
CFA	Coal fly ash from the combustion of hard coal in a fluidized coal boiler (material used in the experimental section)
CR	Composite regrind (material used in the experimental section)
DS	Steelmaking dust (material used in the experimental section)
CS	Crushing Strength
SF	Silica Fume
Na_2SiO_3	Sodium Silicate
NaOH	Calcium Hydroxide
XRF	X-ray Fluorescence
XRD	dispersive X-ray fluorescence

References

1. Cruz, C.O.; Gaspar, P.; de Brito, J. On the concept of sustainable sustainability: An application to the Portuguese construction sector. *J. Build. Eng.* **2019**, *25*, 100836. [CrossRef]
2. Prasetyo, G.; Sutanto, M.E.; Putra, A.M.; Ichsan, M.; Dwidienawati, D.; Bawono, M. Insights and challenges of sustainable project management implementations in construction sectors: Systematic literature analysis. *J. Infrastruct. Policy Dev.* **2024**, *8*, 8104. [CrossRef]
3. Spassova, A. Construction Industry and Sustainability. In *Interdisciplinary Approaches to Climate Change for Sustainable Growth*; Springer: Cham, Switzerland, 2022; pp. 261–287.
4. Akhtar, A.; Sarmah, A.K. Construction and demolition waste generation and properties of recycled aggregate concrete: A global perspective. *J. Clean. Prod.* **2018**, *186*, 262–281. [CrossRef]
5. Liu, C.; Xu, X.; Lv, Z.; Xing, L. Self-healing of Concrete Cracks by Immobilizing Microorganisms in Recycled Aggregate. *J. Adv. Concr. Technol.* **2020**, *18*, 168–178. [CrossRef]
6. Pawluczuk, E.; Kalinowska-Wichrowska, K.; Bołtryk, M.; Jiménez, J.R.; Fernández, J.M. The Influence of Heat and Mechanical Treatment of Concrete Rubble on the Properties of Recycled Aggregate Concrete. *Materials* **2019**, *12*, 367. [CrossRef] [PubMed]
7. Dimitriou, G.; Savva, P.; Petrou, M.F. Enhancing mechanical and durability properties of recycled aggregate concrete. *Constr. Build. Mater.* **2018**, *158*, 228–235. [CrossRef]
8. Saxena, S.; Tembhurkar, A.R. Impact of use of steel slag as coarse aggregate and wastewater on fresh and hardened properties of concrete. *Constr. Build. Mater.* **2018**, *165*, 126–137. [CrossRef]
9. Keulen, A.; van Zomeren, A.; Harpe, P.; Aarnink, W.; Simons, H.A.E.; Brouwers, H.J.H. High performance of treated and washed MSWI bottom ash granulates as natural aggregate replacement within earth-moist concrete. *Waste Manag.* **2016**, *49*, 83–95. [CrossRef]
10. Tajra, F.; Elrahman, M.A.; Lehmann, C.; Stephan, D. Properties of lightweight concrete made with core-shell structured lightweight aggregate. *Constr. Build. Mater.* **2019**, *205*, 39–51. [CrossRef]
11. Qian, D.; Yu, R.; Shui, Z.; Sun, Y.; Jiang, C.; Zhou, F.; Ding, M.; Tong, X.; He, Y. A novel development of green ultra-high performance concrete (UHPC) based on appropriate application of recycled cementitious material. *J. Clean. Prod.* **2020**, *261*, 121231. [CrossRef]
12. Liew, Y.M.; Heah, C.Y.; Mohd Mustafa, A.B.; Kamarudin, H. Structure and properties of clay-based geopolymer cements: A review. *Prog. Mater. Sci.* **2016**, *83*, 595–629. [CrossRef]
13. Provis, J.L.; Bernal, S.A. Geopolymers and Related Alkali-Activated Materials. *Annu. Rev. Mater. Res.* **2014**, *44*, 299–327. [CrossRef]
14. Sambucci, M.; Sibai, A.; Valente, M. Recent Advances in Geopolymer Technology. A Potential Eco-Friendly Solution in the Construction Materials Industry: A Review. *J. Compos. Sci.* **2021**, *5*, 109. [CrossRef]
15. Okoye, F.; Durgaprasad, J.; Singh, N. Fly ash/Kaolin based geopolymer green concretes and their mechanical properties. *Data Brief* **2015**, *5*, 739–744. [CrossRef] [PubMed]
16. Aboulayt, A.; Gounni, A.; El Alami, M.; Hakkou, R.; Hannache, H.; Gomina, M.; Moussa, R. Thermo-physical characterization of a metakaolin-based geopolymer incorporating calcium carbonate: A case study. *Mater. Chem. Phys.* **2020**, *252*, 123266. [CrossRef]
17. Liu, J.; Doh, J.-H.; Ong, D.E.; Wang, S.; Yang, Y.; Dinh, H.L.; Zi, G. Correlation between dissolubilities of Si, Al, and Fe from aluminosilicate precursor and strength of fly ash-based geopolymer. *Constr. Build. Mater.* **2023**, *393*, 132107. [CrossRef]

18. Huang, G.; Wang, M.; Liu, Q.; Zhao, S.; Liu, H.; Liu, F.; Feng, L.; Song, J. Simultaneous utilization of mine tailings and steel slag for producing geopolymers: Alkali-hydrothermal activation, workability, strength, and hydration mechanism. *Constr. Build. Mater.* **2024**, *414*, 135029. [CrossRef]
19. Ke, X.; Bernal, S.A.; Ye, N.; Provis, J.L.; Yang, J. One-Part Geopolymers Based on Thermally Treated Red Mud/NaOH Blends. *J. Am. Ceram. Soc.* **2015**, *98*, 5–11. [CrossRef]
20. Nawaz, M.; Heitor, A.; Sivakumar, M. Geopolymers in construction—recent developments. *Constr. Build. Mater.* **2020**, *260*, 120472. [CrossRef]
21. Kamseu, E.; Alzari, V.; Nuvoli, D.; Sanna, D.; Lancellotti, I.; Mariani, A.; Leonelli, C. Dependence of the geopolymerization process and end-products to the nature of solid precursors: Challenge of the sustainability. *J. Clean. Prod.* **2021**, *278*, 123587. [CrossRef]
22. Xiao, R.; Huang, B.; Zhou, H.; Ma, Y.; Jiang, X. A state-of-the-art review of crushed urban waste glass used in OPC and AAMs (geopolymer): Progress and challenges. *Clean. Mater.* **2022**, *4*, 100083. [CrossRef]
23. Xiao, R.; Polaczyk, P.; Zhang, M.; Jiang, X.; Zhang, Y.; Huang, B.; Hu, W. Evaluation of Glass Powder-Based Geopolymer Stabilized Road Bases Containing Recycled Waste Glass Aggregate. *J. Transp. Res. Board* **2020**, *2674*, 22–32. [CrossRef]
24. Wang, S.; Yu, L.; Yang, F.; Zhang, W.; Xu, L.; Wu, K.; Tang, L.; Yang, Z. Resourceful utilization of quarry tailings in the preparation of non-sintered high-strength lightweight aggregates. *Constr. Build. Mater.* **2022**, *334*, 127444. [CrossRef]
25. Pavithra, P.; Srinivasula Reddy, M.; Dinakar, P.; Hanumantha Rao, B.; Satpathy, B.K.; Mohanty, A.N. Effect of the Na_2SiO_3/NaOH Ratio and NaOH Molarity on the Synthesis of Fly Ash-Based Geopolymer Mortar. In *Geo-Chicago*; American Society of Civil Engineers: Reston, VA, USA, 2016; pp. 336–344.
26. Rahim, R.H.A.; Azizli, K.A.; Man, Z.; Rahmiati, T.; Ismail, L. Effect of Solid to Liquid Ratio on the Mechanical and Physical Properties of Fly Ash Geopolymer without Sodium Silicate. *Appl. Mech. Mater.* **2014**, *625*, 46–49. [CrossRef]
27. Aziz, I.H.; Abdullah, M.M.A.B.; Salleh, M.M.; Azimi, E.A.; Chaiprapa, J.; Sandu, A.V. Strength development of solely ground granulated blast furnace slag geopolymers. *Constr. Build. Mater.* **2020**, *250*, 118720. [CrossRef]
28. Ma, H.; Zhu, H.; Yi, C.; Fan, J.; Chen, H.; Xu, X.; Wang, T. Preparation and Reaction Mechanism Characterization of Alkali-activated Coal Gangue–Slag Materials. *Materials* **2019**, *12*, 2250. [CrossRef]
29. Qian, L.-P.; Xu, L.-Y.; Alrefaei, Y.; Wang, T.; Ishida, T.; Dai, J.-G. Artificial alkali-activated aggregates developed from wastes and by-products: A state-of-the-art review. *Resour. Conserv. Recycl.* **2022**, *177*, 105971. [CrossRef]
30. Tang, P.; Florea, M.; Brouwers, H.J.H. Employing cold bonded pelletization to produce lightweight aggregates from incineration fine bottom ash. *J. Clean. Prod.* **2017**, *165*, 1371–1384. [CrossRef]
31. Gesoğlu, M.; Güneyisi, E.; Öz, H.Ö. Properties of lightweight aggregates produced with cold-bonding pelletization of fly ash and ground granulated blast furnace slag. *Mater. Struct.* **2012**, *45*, 1535–1546. [CrossRef]
32. Almadani, M.; Razak, R.A.; Abdullah, M.M.A.B.; Mohamed, R. Geopolymer-Based Artificial Aggregates: A Review on Methods of Producing, Properties, and Improving Techniques. *Materials* **2022**, *15*, 5516. [CrossRef]
33. Tian, K.; Wang, Y.; Hong, S.; Zhang, J.; Hou, D.; Dong, B.; Xing, F. Alkali-activated artificial aggregates fabricated by red mud and fly ash: Performance and microstructure. *Constr. Build. Mater.* **2021**, *281*, 122552. [CrossRef]
34. Saleem, N.; Rashid, K.; Fatima, N.; Hanif, S.; Naeem, G.; Aslam, A.; Fatima, M.; Aslam, K.; Miral, M. Appraisal of geopolymer lightweight aggregates sintered by microwave radiations. *J. Asian Concr. Fed.* **2020**, *6*, 37–49. [CrossRef]
35. Rehman, M.U.; Rashid, K.; Haq, E.U.; Hussain, M.; Shehzad, N. Physico-mechanical performance and durability of artificial lightweight aggregates synthesized by cementing and geopolymerization. *Constr. Build. Mater.* **2020**, *232*, 117290. [CrossRef]
36. Aslam, A.; Rashid, K.; Naeem, G.; Saleem, N.; Hanif, S. Microwave Assisted Synthesis and Experimental Exploration of Geopolymer Lightweight Aggregate. *Tech. J.* **2020**, *25*, 8–15.
37. Gomathi, P.; Sivakumar, A. Synthesis of Geopolymer Based Class-F Fly Ash Aggregates and its Composite Properties in Concrete. *Arch. Civ. Eng.* **2014**, *60*, 55–75. [CrossRef]
38. Risdanareni, P.; Schollbach, K.; Wang, J.; De Belie, N. The effect of NaOH concentration on the mechanical and physical properties of alkali activated fly ash-based artificial lightweight aggregate. *Constr. Build. Mater.* **2020**, *259*, 119832. [CrossRef]
39. Arjun, K.; Darshan, B.U. Synthesis of Geopolymer Coarse Aggregates Using Class-F Fly Ash and Studies on Its Physical Properties. In *Sustainable Construction and Building Materials: Select Proceedings of ICSCBM 2018*; Springer: Singapore, 2019; pp. 155–165.
40. Parvathy, S.S.; Sharma, A.K.; Anand, K.B. Comparative study on synthesis and properties of geopolymer fine aggregate from fly ashes. *Constr. Build. Mater.* **2019**, *198*, 359–367. [CrossRef]
41. Fonte, R.; Xydis, G. Wind turbine blade recycling: An evaluation of the European market potential for recycled composite materials. *J. Environ. Manag.* **2021**, *287*, 112269. [CrossRef] [PubMed]
42. Mamanpush, S.H.; Li, H.; Englund, K.; Tabatabaei, A.T. Extruded Fiber-Reinforced Composites Manufactured from Recycled Wind Turbine Blade Material. *Waste Biomass Valorization* **2020**, *11*, 3853–3862. [CrossRef]
43. Pu, Z.; Yang, S.; Wang, Q. Recycling of waste wind turbine blades for high-performance polypropylene composites. *J. Appl. Polym. Sci.* **2024**, *141*, e55474. [CrossRef]

44. Tan, B.; Wang, X.; Pu, Z.; Yang, S.; Nie, M. Fabrication of High-Strength Waste-Wind-Turbine-Blade-Powder-Reinforced Polypropylene Composite via Solid-State Stretching. *Sustainability* **2025**, *17*, 840. [CrossRef]
45. Pławecka, K.; Przybyła, J.; Korniejenko, K.; Lin, W.-T.; Cheng, A.; Łach, M. Recycling of Mechanically Ground Wind Turbine Blades as Filler in Geopolymer Composite. *Materials* **2021**, *14*, 6539. [CrossRef]
46. Izak, P.; Serkowski, S. Energooszczędna technologia przygotowania granulatów ceramicznych. *Cz. II—Właściwości granulatu. Ceram. Mater.* **2011**, *63*, 552–557.
47. Serkowski, S.; Guenter, H. Energooszczędna technologia przygotowania granulatów ceramicznych. Cz. I—Przemiał i granulacja. *Ceram. Mater.* **2011**, *63*, 266–272.
48. *PN-EN 12457-4:2006*; Characterization of Waste—Leaching—Compliance Testing for Leaching of Granular Waste Materials and Sludge. Polish Committee for Standardization: Warsaw, Poland, 2006.
49. *PN-EN ISO 10523:2012*; Water Quality—Determination of pH. Polish Committee for Standardization: Warsaw, Poland, 2012.
50. *PN-EN 13055-1*; Lightweight Aggregates—Part 1: Lightweight Aggregates for Concrete, Mortar and Grout. Polish Committee for Standardization: Warsaw, Poland, 2002.
51. *PN-EN 1097-6*; Tests for Mechanical and Physical Properties of Aggregates—Part 6: Determination of Particle Density and Water Absorption. Polish Committee for Standardization: Warsaw, Poland, 2013.
52. *PN-EN 1097-1*; Tests for Mechanical and Physical Properties of Aggregates—Part 1: Determination of the Resistance to Wear (Micro-Deval). Polish Committee for Standardization: Warsaw, Poland, 2023.
53. *ISO 3310-1:2016*; Test Sieves—Technical Requirements and Testing—Part 1: Test Sieves of Metal Wire Cloth. International Organization for Standardization: Geneva, Switzerland, 2016.
54. Jiang, T.; Jin, Y. Phase Analysis of Alkali-Activated Slag Hybridized with Low-Calcium and High-Calcium Fly Ash. *Sustainability* **2022**, *14*, 3767. [CrossRef]
55. Yin, Y.; Yin, J.; Zhang, W.; Tian, H.; Hu, Z.; Ruan, M.; Xu, H.; Liu, L.; Yan, X.; Chen, D. FT-IR and micro-Raman spectroscopic characterization of minerals in high-calcium coal ashes. *J. Energy Inst.* **2018**, *91*, 389–396. [CrossRef]
56. Briffa, J.; Sinagra, E.; Blundell, R. Heavy metal pollution in the environment and their toxicological effects on humans. *Heliyon* **2020**, *6*, e04691. [CrossRef]
57. Rouhani, A.; Gutiérrez, M.; Newton, R.A.; Al Souki, K.S. An overview of potentially toxic element pollution in soil around lead–zinc mining areas. *Environ. Rev.* **2025**, *33*, 1–16. [CrossRef]
58. Skotniczny, G.; Kozioł, M.; Korol, J.; Poneta, P. Production and Evaluation of Synthetic Lightweight Aggregates Based on Mixture of Fluidized Bed Fly Ash and Post-Mining Residues. *Materials* **2022**, *15*, 660. [CrossRef] [PubMed]
59. Declaration of Performance. Available online: https://www.liapor.com/en/products/declaration-of-performance.html (accessed on 2 June 2025).
60. Leca. Technical Data Sheet. Available online: https://www.leca.pl/sites/leca.pl/files/pdf/Leca-KERAMZYT-geotechniczny-8-10-20-RX-270121.pdf (accessed on 2 June 2025).
61. Dz.U.2015.1277 Rozporządzenie Ministra Gospodarki z Dnia 16 Lipca 2015 r. w Sprawie Dopuszczania Odpadów do Składowania na Składowiskach. Available online: https://sip.lex.pl/akty-prawne/dzu-dziennik-ustaw/dopuszczanie-odpadow-do-skladowania-na-skladowiskach-18219829 (accessed on 18 July 2025).
62. Sánchez, F.; de la Torre, E.; Guevara, A.; Moreno, K.; Aragón-Tobar, C.F. Comparative Study of Steel Mill Dust Leaching with Phosphoric Acid and Sodium Hydroxide. *Metals* **2024**, *14*, 1087. [CrossRef]
63. Peng, H.; Liu, Z.; Tao, C. A green method to leach vanadium and chromium from residue using NaOH-H2O2. *Sci. Rep.* **2018**, *8*, 426. [CrossRef]
64. Keng, X.I.E.; Wang, H.; Wang, S. Direct leaching of molybdenum and lead from lean wulfenite raw ore. *Trans. Nonferrous Met. Soc. China* **2019**, *29*, 2638–2645. [CrossRef]
65. Ghasemi, S.M.S.; Azizi, A. Alkaline leaching of lead and zinc by sodium hydroxide: Kinetics modeling. *J. Mater. Res. Technol.* **2018**, *7*, 118–125. [CrossRef]
66. Izquierdo, M.; Querol, X.; Davidovits, J.; Antenucci, D.; Nugteren, H.; Fernández-Pereira, C. Coal fly ash-slag-based geopolymers: Microstructure and metal leaching. *J. Hazard. Mater.* **2009**, *166*, 561–566. [CrossRef] [PubMed]

materials

MDPI

Article

Preliminary Assessment of Commercial Antibacterial Finishes on Knitted Fabrics from Recycled Cotton and Polyester

Muhammad Sajid Faheem [1], Hafsa Jamshaid [1], Rajesh Kumar Mishra [2,*], Adeel Abbas [3,*], Miroslav Muller [2], Jiri Urban [2] and Michal Penc [2]

1 Department of Textiles, School of Engineering & Technology, National Textile University, Faisalabad 37610, Pakistan; sajid.faheem@ntu.edu.pk (M.S.F.); hafsa@ntu.edu.pk (H.J.)
2 Department of Material Science and Manufacturing Technology, Faculty of Engineering, Czech University of Life Sciences Prague, Kamycka 129, Suchdol, 165 00 Prague, Czech Republic; muller@tf.czu.cz (M.M.); urbanjiri@tf.czu.cz (J.U.); penc@tf.czu.cz (M.P.)
3 School of Fashion and Textiles, RMIT University, Melbourne, VIC 3000, Australia
* Correspondence: mishrar@tf.czu.cz (R.K.M.); s4068392@student.rmit.edu.au (A.A.)

Abstract: Recycled materials are employed in several areas to reduce the environmental burden. Therefore, recycling textiles is a vast domain being investigated by several researchers. Functionalization of textiles using active nanomaterials is a growing area of interest nowadays. However, functionalized textiles with antibacterial properties still do not employ the incorporation of recycled materials. The huge consumer need for different types of functional textiles necessitates a focus on recycled textiles in this area. Hence, this research focuses on the development of metal oxide nanoparticle-functionalized antibacterial textiles using recycled cotton and polyester yarns. Two different finishes have been used with a 50:50 ratio. Using a two-fold approach, antibacterial finishes were applied during both the yarn and fabric stages to analyze the differences in performance. Agar plate qualitative antibacterial analysis revealed efficient antibacterial function both before and after commercial laundering. However, thermo-physiological comfort properties were found to be variable for recycled yarn types and finishing stages. The air permeability of fabric-finished specimens was found to be about 47% lower than yarn-finished specimens owing to reduced porosities. However, the overall moisture management capability (OMMC) index was found to be 21% higher than that of yarn-finished specimens. Among serviceability parameters, bursting strength was found to decrease with increasing concentration, as the finishing treatments made the yarns crispier which eventually compromised yarn strength.

Keywords: recycled textiles; metal oxide nanoparticles; antibacterial activity; functional textiles; thermal comfort; serviceability

1. Introduction

Recycling textiles has gained much popularity in recent times owing to its crucial role in decreasing the environmental burden [1]. Manufacturing products using virgin fiber materials increases the environmental burden; hence, different techniques of mechanical and chemical recycling have been introduced into the market [2]. Such techniques enhance the possibilities of the service life of textile materials through reuse in some other products [3]. However, textile fibers usually lose their mechanical strength during recycling which demands the adoption of appropriate fabrication methods that can handle these delicate recycled materials effectively [4]. Knitting technology can convert single yarns into fabric, with lower effort and more convenience [5]. The weaving process involves higher mechanical stresses during the preparatory stage (warping, sizing, and drawing), as well as during manufacturing (shedding, picking, and beating up). Knitting can handle soft, delicate, and low-strength yarns due to its simpler manufacturing processes. Hence, the knitting process is more suitable for recycled yarns. Knitted fabrics are also preferable for their form fitting, stretching and recovery, and thermo-physiological comfort attributes [6]. This is why knitted fabrics are a primary choice for comfortable clothing [7]. However, such recycled yarn knitted fabrics can also potentially be employed in some functional applications where slightly lower mechanical attributes are also acceptable, and functional performance is more dominant over casual clothing characteristics [8]. Antibacterial knitwear is used for resisting bacterial infections caused due to body sweat and can be engineered using recycled materials [9]. The lower number of twists in recycled yarns make them ideal for knitting, and such attributes also work for enhanced absorption of functional finishes and efficient handling of body perspiration to maintain the body heat balance [10]. Moreover, recycled yarns are often made by blending recycled fibers with some virgin materials which also serve to tailor functional and thermo-physiological comfort characteristics.

In recent years, increasing cases of deaths caused by viral infections such as SARS (severe acute respiratory syndrome), norovirus, and avian influenza have been reported [11–17]. Findings have been reported about the integration of recycled fibers with some functional materials for antibacterial characteristics [18]. Researchers have engineered antibacterial nonwoven felts by incorporating antibacterial properties into recycled polyester using nano zinc oxide during a spun bonding process [19]. Other researchers have worked on different chemical fixation methods for imparting antibacterial properties to recycled polyester nonwoven felts [20]. On the other hand, some researchers have investigated the properties of yarns made from waste polyester bottles and silica aerogels to produce antibacterial knitwear. However, such fabrics have decreased thermal conductivity and enhanced hydrophobicity [21]. Some others have developed recycled polyester and silver nanocomposite (SNC) polyester blended yarns to engineer antibacterial fabrics. Increasing SNC polyester content improved the antibacterial performance, but surface textures and touch characteristics were compromised [22].

The above-mentioned techniques are viable to design functional textiles with recycled materials. These methods involve chemical recycling and the aforementioned processing methods. However, there is still a huge amount of mechanically recycled yarn being produced on an industrial scale, and such yarns also require some fruitful directions of employment in their second service lives. Scientific studies on these materials are limited to certain conventional attributes. Researchers performed a comparative analysis of the thermal comfort attributes of knitted fabrics engineered using different blends of recycled cotton and polyester [23]. It has been found that the thermo-physiological comfort properties of recycled cotton/polyester knitted fabrics are comparable to those of virgin materials, proposing the use of recycled fibers in knitwear [24]. Another analysis proved that the bursting strength of recycled polyester-based knitwear is equivalent to that of virgin polyester knitted fabric [25]. However, 2.3% higher microfiber leaching from recycled polyester knitted fabrics was observed [26]. Similarly, several researchers have focused on comparisons of the thermal comfort and mechanical performance of recycled fiber knitted fabrics with virgin ones [27–29]. While the point of discussion remains the same, the potential of such knitted fabrics towards functional applications, i.e., antibacterial textiles has not been explored. The previous literature also lacks studies identifying the influence of varying blend ratios of recycled and virgin cotton/polyester fibers on antibacterial properties. Hence, this research focuses on the development of single jersey knitted fabrics using recycled and virgin yarns with blends of cotton and polyester.

Metal oxide nanoparticles are materials of growing interest for drug delivery. Many types of nanoparticles, e.g., zinc oxide (ZnO), magnesium oxides, and titanium dioxides (TiO_2), etc., have shown antibacterial properties against a broad spectrum of microorganisms. However, there have been no studies on the efficacy of the combined use of recycled textiles and metal oxide nanoparticles to provide antibacterial properties. Hence, in this study, copper, magnesium, and silver finishes were applied through dip coating to incorporate antibacterial attributes into recycled yarns and knitted fabrics. Thermo-physiological comfort and mechanical characteristics were also assessed to check the suitability of the designed antibacterial knitwear in maintaining clothing–body heat equilibrium and service life factors. This study consists of a two-fold approach. On one hand, the effect of the blending percentage of virgin and recycled fibers was studied; on the other hand, the effect of an antibacterial finish on the functional properties was investigated. In addition to this, the fabric was treated with antibacterial finish to check the effects of finishes on the mechanical and thermo-physiological comfort properties of fabric samples.

2. Materials and Methods

2.1. Materials

Cotton and polyester are among the most recycled fibers through mechanical recycling methods. Mechanically recycled cotton and polyester yarns were manufactured with different blend ratios as shown in Table 1. Recycled cotton and cotton/polyester yarns were sourced from Combined Fabrics Limited (Lahore, Pakistan). The recycled polyester was purchased from a local yarn market in Faisalabad, Pakistan. The study also aimed to analyze the influence of antibacterial finishes on the physical attributes of yarns. Hence, the physical characteristics of employed yarns have been described in Table 2. The results presented are averaged from 25 measurements with standard deviation. Two different antibacterial finishes, i.e., copper and silver (CS) nano-finish (50:50) and magnesium and silver (MS) nano-finish (50:50), with 1% and 3% concentrations were employed to analyze the antibacterial function. These finishes were supplied by ZAMVIR, Tech. Fab. International (Faisalabad, Pakistan). Figure 1 shows the antibacterial performances of the employed finishes.

Table 1. Composition of recycled yarns.

Sr. No.	Yarn Type	Blend Ratio
1	Virgin Cotton/Recycled Cotton	60:40
2	Recycled Cotton/Recycled Polyester	63:37
3	Recycled Polyester	100

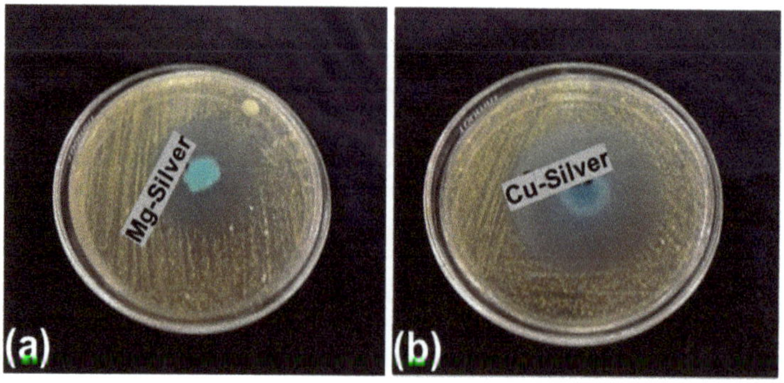

Figure 1. Antibacterial performance of employed finishes: (**a**) magnesium–silver; (**b**) copper–silver.

Table 2. Physical properties of recycled yarns (average ± standard deviation).

Sr. No.	Physical Property	Yarn Type		
		Recycled Cotton (C)	Recycled Cotton/Polyester (PC)	Recycled Polyester (P)
1	Count (Ne)	24 ± 2	26 ± 2.1	25 ± 2.1
2	Twist Per Inch (2.5 cm) (TPI)	17.7 ± 1.2	18.3 ± 1.2	17.2 ± 1.1
3	Count Lea Strength Product (CLSP)	2592 ± 70	3595 ± 85	2210 ± 62
4	Single Yarn Strength (N)	291.5 ± 11.2	438.8 ± 18.5	341.7 ± 12.1
5	Elongation (%)	3.44 ± 0.04	17.36 ± 0.52	7.08 ± 0.64
6	Tenacity (cN/Tex)	12.83 ± 0.23	20.06 ± 0.31	13.88 ± 0.3
7	Coefficient of Variation (CV%)	8.94 ± 0.35	8.54 ± 0.22	9.66 ± 0.37

2.2. Fabrication Method and Experimental Plan

The study consisted of a two-step experimental plan. In the first step, yarns were treated with finishes to analyze the antibacterial performance and the impact of the finishes on physical attributes of yarns. In the second step, these yarns were converted into knitted fabrics, and their antibacterial, thermo-physiological comfort, and mechanical properties were analyzed. In addition to these, knitted fabrics were engineered using recycled cotton yarn, and the finish was applied onto the fabric instead of the yarn to analyze the performance difference between coated yarn and coated fabric specimens. Figure 2 highlights the overall research methodology. A detailed experimental plan is also shown in Table 3. Yarns were converted into knitted fabrics using single jersey structures consisting of all knit stitches with plain architecture on a sample knitting machine (Karlmeyer, Obertshausen, Germany). The machine supports a yarn range of 18 Ne to 36 Ne. Hence, all three yarns smoothly functioned during the knitting process.

Table 3. Experimental plan and specimen codes.

Sr. No.	Specimen Description	Finish Concentration	Specimen Code
1. 2.	Recycled Cotton Yarn–Copper and Silver Nano-finish *	1% 3%	CY-CS1% CY-CS3%
3. 4.	Recycled Cotton Yarn–Magnesium and Silver Nano-finish *	1% 3%	CY-MS1% CY-MS3%
5. 6.	Recycled Polyester/Cotton Yarn–Copper and Silver Nano-finish *	1% 3%	PCY-CS1% PCY-CS3%
7. 8.	Recycled Polyester/Cotton Yarn–Magnesium and Silver Nano-finish *	1% 3%	PCY-MS1% PCY-MS3%
9. 10.	Recycled Polyester Yarn–Copper and Silver Nano-finish *	1% 3%	PY-CS1% PY-CS3%
11. 12.	Recycled Polyester Yarn–Magnesium and Silver Nano-finish *	1% 3%	PY-MS1% PY-MS3%
13. 14.	Recycled Cotton Fabric–Magnesium and Silver Nano-finish **	1% 3%	CF-MS1% CF-MS3%

* Antibacterial finish was applied on yarns and then knitting was performed. ** Antibacterial finish was applied on knitted fabrics.

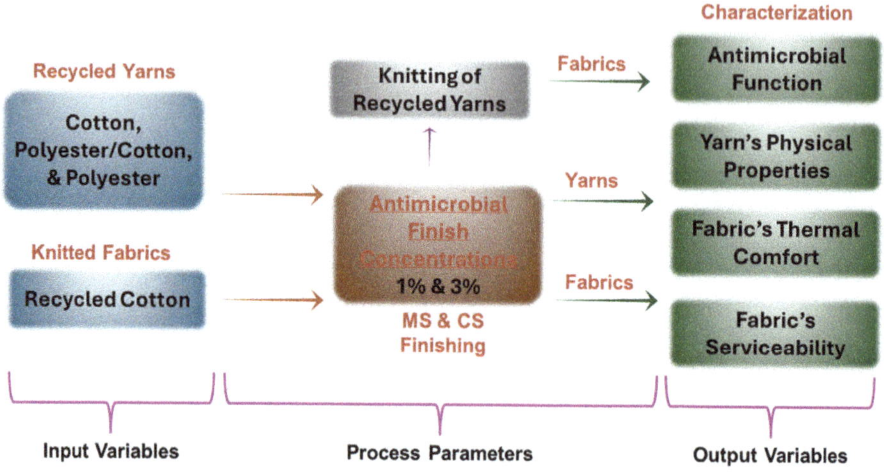

Figure 2. Overall research methodology.

Pretreatments and Antibacterial Finishing Process

All recycled yarns were processed in the form of yarn lea through a wet pretreatment (solomatic bleaching) process. A textile bleaching process that combines scouring and bleaching into a single step was used to remove any inherent impurities and colorants from the fiber surfaces. Scouring and bleaching were carried out through a batch-wise/exhaust method using specific recipes of hydrogen peroxide (H_2O_2) and sodium hydroxide (NaOH)/sodium carbonate (Na_2CO_3) with a wetting agent and a stabilizer with a material-to-liquor ratio 1:15, at a temperature of 95–100 °C for 30 min, followed by hot and cold rinsing, as well as neutralization using 1 g/L acetic acid. All of the reagents were procured from Havel Composites (Prague, Czech Republic). Different concentrations of bleaching agent and alkali/base were kept for each yarn type owing to unique blend ratios. For recycled cotton yarn, 7 g/L H_2O_2, 3 g/L NaOH, 1 g/L wetting agent, and 2 g/L stabilizer were used. For recycled polyester/cotton yarn, 5 g/L H_2O_2, 2 g/L NaOH, 1 g/L wetting agent, and 1.5 g/L stabilizer were used. For recycled polyester yarn, 2 g/L Na_2CO_3, 1 g/L wetting agent, and 2 g/L detergent were used. These concentrations were based on supplier specifications. The wet pretreatments cleaned the fiber surfaces, facilitating more penetration and adhesion for the antibacterial finish. The antibacterial nano-finishes were also applied on pretreated leas of yarn and knitted fabric samples through a batch-wise/exhaust method. Aqueous solutions of the antibacterial nano-finishes were prepared in distilled water for two different concentrations: 1% and 3% *w*/*v*, respectively. Pres and knitted fabric samples were immersed in the prepared solutions and heated to raise the temperature. The process was continued for 30 min at 80 °C for uniform application and fixation of the finish followed by hydro-extracting without rinsing and drying at 100 °C.

2.3. *Characterization*

2.3.1. Physical and Mechanical Properties of Yarn

The physical properties of yarns provide insights into various attributes determining behaviors during fabric manufacturing and in-service life. Twist per Inch/2.5 cm (TPI) describes the amount of twist per unit length and was measured using a digital TPI tester and the ASTM D1422 standard untwist/retwist test method [30]. TPI plays an important role in the softness and crispiness of yarn; hence, it was crucial to analyze. The count lea (120 yards/109.73 m of yarn) strength product (CLSP) is an important mechanical attribute of yarn that predicts the tensile properties of yarn when it is converted into fabric form in the future. CLSP was evaluated on a lea strength tester using the ASTM D1578 standard test method [31]. However, for CLSP evaluation, the yarn needed to be converted into lea form; hence, a warping reel was employed for lea formation by following the procedure described under the ASTM D1578 standard method. Single yarn strength, on the other hand, is also an important mechanical attribute of yarn, providing detailed comprehension of breaking force and elongation. Hence, single yarn strength was determined using an Uster Tensorapid (Uster, Switzerland) following the ASTM D2256 standard test method [32]. All measurements were performed 25 times per each sample. The results were found to be statistically significant with a coefficient of variation lower than 5%.

2.3.2. Fourier Transform Infrared Spectroscopy

The surface modification of knitted fabric samples was assessed by their Fourier transform infrared (FTIR) spectra. The FTIR spectra of fabric samples were recorded on an attenuated total reflectance–Fourier transform infrared (ATR-FTIR) spectrometer (spectrum TWO, PerkinElmer, Waltham, MA, USA), at room temperature in the spectra range between 4000 and 600 cm^{-1}, using the ATR reflection technique on an adapter with a crystal of ZnSe. The spectra were collected as a result of 32 running scans at a resolution of 1 cm^{-1}.

2.3.3. Scanning Electron Microscopy (SEM)

The surface morphology of knitted fabric samples was examined by scanning electron microscopy (Thermo Scientific Phenom Pure G6 Desktop SEM (GH Eindhoven, The Netherlands). Before SEM analysis, the surface of the fabric sample was sputter-coated with gold (a conductive layer). The images were taken at different suitable accelerating voltages and magnifications with a slow scanning speed to obtain higher-quality images.

2.3.4. Antibacterial Efficacy

The antibacterial efficacy of the applied nano-finishes on the knitted fabric samples against a fresh bacterial strain of *Staphylococcus aureus* (*S. aureus*) was evaluated qualitatively using the agar diffusion method according to the test method AATCC 147 [33]. The swatches of treated fabrics with known dimensions were sterilized (in an autoclave at 121 °C for 15 min) and placed in the center of agar plates and inoculated with the test bacteria. Then, the plates were incubated in a thermostat environment at 37 ± 2 °C for 24 h. After the incubation period, the diameter of the growth-free zone (zone of inhibition) around the fabric and the bacterial growth under/around the fabric were identified.

2.3.5. Washing Durability

The finished knitted fabrics were laundered/washed in an Electrolux-wascator (a front-loading, horizontal rotating drum-type standard reference washing machine, (Stockholm, Sweden), according to the standard test method ISO 6330 [34]. The standard reference detergent (20 g) and a washing temperature of 40 °C were used as per ISO 6330. The washing was conducted for up to 15 wash/laundry cycles, each of which comprised standard washing and rinsing steps. In order to inspect the durability of the antibacterial nano-finish films/coatings to washing, all of the washed fabric samples were subsequently evaluated for antibacterial performance after 1, 5, 10, and 15 cycles following the standard test method AATCC 147.

2.3.6. Thermo-Physiological Comfort and Serviceability of Knitted Fabrics

Thermo-physiological comfort attributes are vital to ensure the prolonged wearing of textile materials. Knitted fabrics unable to maintain thermal equilibrium between wearer and clothing are also not suitable for functional applications. Hence, thermo-physiological comfort properties including air permeability and moisture management were characterized.

Air permeability is the property of a fabric to allow air to pass through it. Fabrics used in bodywear require reasonably good air permeability. It is directly linked to breathability and comfort. On the other hand, low air permeability is also required in particular fields, including protective and safety clothing. Air permeability was measured under the ISO 9237 international standard test method [35]. The commercial air permeability tester FX 3300 Lab Air IV testing instrument (Textest AG, Schwerzenbach, Switzerland) was used, working according to ISO 9237. A laminar flow of air was calculated in mm/s through a specific area of the fabric at a pressure of 100 Pa. The area of measurement was 100 cm^2 and the measurements were repeated 10 times for each sample to calculate the average. The results were found to be statistically significant with a coefficient of variation lower than 5%.

Overall moisture management capability was characterized in terms of the OMMC index by following the AATCC 195 standard test method [36]. The OMMC index provides insights into overall moisture handling performance in a single value instead of individual parameters, i.e., water vapor permeability, wetting time, wetting speed, etc. OMMC analyzes the management and transport of liquid moisture through the surface and the bulk of the fabric. It was tested on a moisture management tester, MMT (SDL ATLAS, Rock Hill, SC, USA). The samples tested using the MMT device have a standardized size of 80 mm $\times$ 80 mm. Each measurement lasted 120 s. During the first 20 s, a water drop was applied to the upper surface of the sample. The water drop was applied onto the center of the upper side of the sample. The applied drop was transported through the sample in 3 directions: on the upper surface of the fabric; through the fabric from the upper surface to the lower surface; on the lower surface of the fabric. Considering these factors, the various moisture management parameters listed below were evaluated.

1. Wetting time of top surface (s) denoted as WTT;
2. Wetting time of bottom surface (s) denoted as WTB;
3. Absorption rate of top surface (%/s) denoted as TAR;
4. Absorption rate of bottom surface (%/s) denoted as BAR;
5. Maximum wetted radius for top surface (mm) denoted as MWRT;
6. Maximum wetted radius for bottom surface (mm) denoted as MWRB;
7. Spreading speed on top surface (mm/s) denoted as SST;
8. Spreading speed on bottom surface (mm/s) denoted as SSB;
9. Accumulative one-way transport index (−) denoted as R;
10. Overall moisture management capability (−) denoted as OMMC.

All of the tests were performed under standard testing conditions, e.g., a temperature of 21 $\pm$ 2 °C and a relative humidity of 65 $\pm$ 2%. Ten measurements were taken for each sample, and the average was calculated. The results were found to be statistically significant with a coefficient of variation lower than 5%.

Bursting strength ensures the serviceability of knitted fabrics, as the fabrics encounter several multiaxial stresses caused by the wearer's body and the environment during their service life. It refers to the force required to break or rupture the fabric under hydraulic or pneumatic pressure. To evaluate the bursting strength of the fabrics, the sample size used was 112 mm^2 using the standard method of ASTM D3786 [37]. The samples were clamped over a rubber diaphragm, and a hydraulic bursting strength tester was used to calculate the bursting strength. During the test, the samples were clamped over the rubber diaphragm and fluid pressure was applied to the samples until bursting. The value of bursting pressure for each sample was measured 10 times, and the average was calculated.

3. Results and Discussion

3.1. Physical and Mechanical Characteristics of Yarns

The specialty finishing process influences the physical and mechanical attributes of yarns. Fibers can swell, the surface can become smooth and slippery, and inter-fiber frictions can vary owing to finish absorption [38]. Hence, it was crucial to analyze the above-mentioned properties. Figure 3a shows the results of TPI for raw and finished yarn specimens.

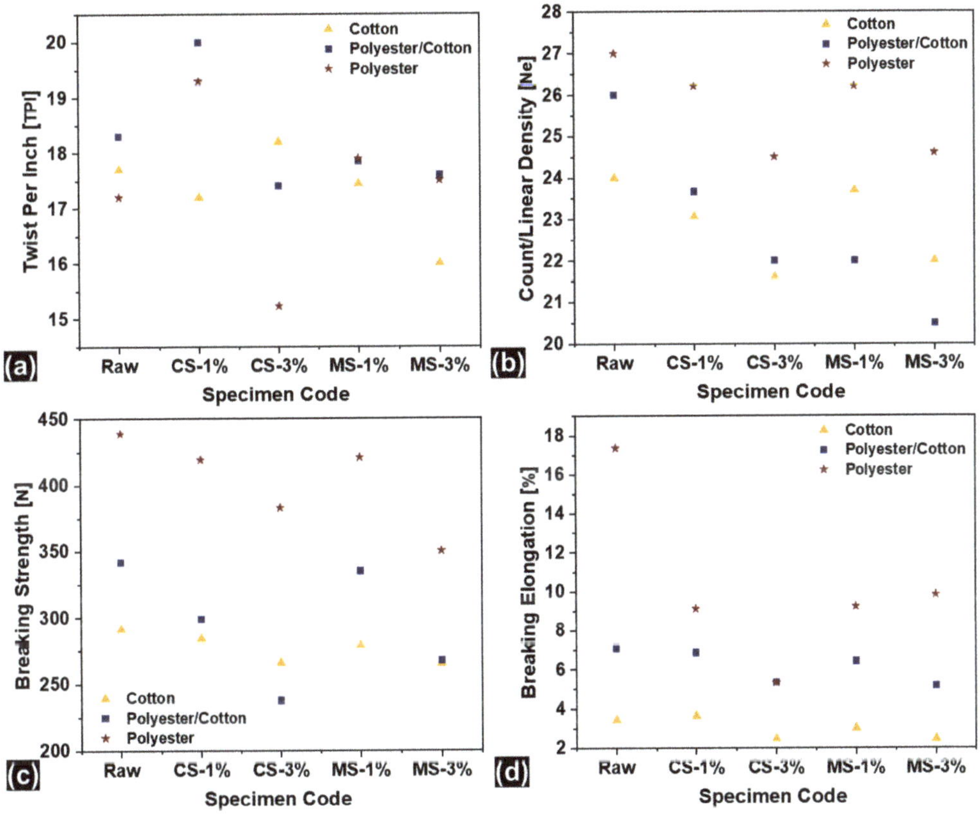

Figure 3. Physical and mechanical properties of recycled yarns, (**a**) twist per inch (2.5 cm); (**b**) count/linear density; (**c**) breaking strength; (**d**) breaking elongation.

Samples of 100% cotton yarn exhibited a different trend from polyester/cotton and 100% polyester yarns. Polyester/cotton and 100% polyester showed increases of about 9.28 ± 0.21% and 12.20 ± 0.33% in TPI, respectively, at a 1% CS finish concentration, while a decrease of 2.82 ± 0.04% was noted for cotton yarn. Cotton, being a natural fiber, has more affinity to fluids as compared to polyester. Hence, the swelling of cotton fibers caused axial expansion of fibers in the yarn cross-section and the amount of TPI was decreased [39]. Polyester, on the other hand, has a different behavior. The finish made the surface of polyester more slippery, and fibers came closer owing to inter-fiber slippage which made the number of TPIs higher [40]. Moreover, 100% recycled polyester had a higher increase in TPI than polyester/cotton blended specimens. However, a further increase in finish concentration resulted in the opposite behavior. The 100% cotton specimen showed about a 2.83 ± 0.05% increase in the number of TPIs. This is because of the inability of the fibers to absorb more fluid and the slippage of fibers due to wet surfaces which led to an increase in TPI. At 3% CS concentration, polyester started absorbing the finish more effectively, and the fibers' swelling caused a 4.91 ± 0.12% and 11.23 ± 0.47% decrease in TPI for polyester/cotton and pure polyester, respectively. The polyester/cotton and 100% polyester yarns exhibited a similar trend for MS finish, but the difference in TPI was not much higher than it was for CS-finished specimens. On the other hand, cotton yarn exhibited a different trend for CS finish. A linear decrease in the number of TPIs was noted: a 1.46 ± 0.02% decrease in TPI was noted at 1% MS concentration, and 3% MS concentration resulted in a further 9.60 ± 0.35% decrease. In fact, all of the yarns were prepared with similar twist levels in order to make them comparable in terms of mechanical and transmission-related performance. Therefore, the results of TPI seem similar. However, the increase in TPI after finishing is slightly variable depending on the yarn type and percentage of finishing. The results of TPI were found to be statistically significant with a coefficient of variation lower than 5%.

Linear density characterization also revealed almost similar trends for cotton and polyester yarns as shown in Figure 3b. The increase and decrease in linear densities is attributed to swollen fibers after deposition of the nanoparticles and a change in the linear density (an indirect yarn count system has been used in this research). Therefore, the yarns became coarser [41]. The same phenomenon was observed in characterized specimens where fibers swelled owing to finish absorption making the yarn coarser. An interesting fact was noted that increasing concentrations of both CS and MS finishes caused an increase in yarn coarseness. Hence, it was assumed that more absorption of the finish occurred in yarns at elevated concentrations. For example, cotton yarn showed a 6.32 ± 0.24% lower linear density for CS-3% as compared to CS-1%. The trends of cotton and polyester specimens were similar. However, polyester/cotton yarn had a different behavior. A linear decrease in yarn fineness/count can be observed from raw specimens towards MS-3% in Figure 3b. This fact confirms the higher affinity of polyester/cotton blended yarn to MS finish which caused about a decrease of 21.15 ± 1.02% in linear density with MS-3% as compared to the raw specimen. The results were found to be statistically significant with a coefficient of variation lower than 5%.

The mechanical properties of yarns also seemed to be affected by the antibacterial finishing process. Figure 3c,d highlight the breaking strength and elongation of mechanically characterized specimens after the single yarn strength test. Most chemical finishing treatments (except softener finishing) cause yarns to be stiffer, and breaking strength and elongation are usually compromised. The applied antibacterial finish worked in a similar manner, causing the breaking strength and elongation of finished specimens to be lower than those of untreated/bleached-only samples. Cotton, being a natural fiber, had different interactions with the antibacterial finishes as compared to polyester. Hence, a smaller change in strength and elongation were observed for cotton yarns before and after finishing. The CS-1% specimen exhibited 2.40 ± 0.02% lower strength than untreated/bleached-only cotton samples, and the increased concentration of CS-3% caused a further 8.64 ± 0.22% decrease in yarn strength. Similar behavior was noted for elongation as well: a maximum difference of 28.48 ± 0.72% was obtained between raw cotton and the CS-3% specimen (Figure 3d). Settlement of unabsorbed antibacterial finish between fibers restricted their movement, making them stiffer, less elastic, and weaker [42]. Polyester, being less hydrophilic than cotton, has more finish between fibers, which is why higher differences in strength and elongation were obtained for polyester and polyester/cotton yarns. Polyester/cotton yarn showed a 12.43 ± 0.23% decrease in strength for CS-1%, which was only 2.40 ± 0.03% for pure cotton. The increasing concentration caused a 30.49 ± 1.08% decrease in strength. Similar behaviors of cotton and polyester were noted for MS-finished specimens, i.e., MS-3% finished polyester offered 20.20 ± 1.04% lower strength than untreated/bleached-only samples of recycled 100% polyester. The facts proved that the mechanical properties of polyester fiber were more adversely affected by antibacterial finishing possibly due to more finish accumulation and curing between fibers. Moreover, the elongation of polyester showed an exponential decrease after finishing, which can be observed in Figure 3d. The results were found to be statistically significant with a coefficient of variation lower than 5%.

3.2. Weight Gain of Knitted Fabrics

In textile chemical finishing processes, there are different desired end uses; the aesthetic and functional properties presented by the textiles depend on the amount of chemical/finish deposited on the textile substrates. Therefore, the weight gain of the knitted fabrics was assessed after treatment with aqueous antibacterial nano-finish solutions of different concentrations. The effect of different concentrations of finishing solutions on the weight gain of knitted fabrics is given in Table 4. The weights (gram per square meter) of untreated/bleached-only samples of cotton, polyester–cotton, and polyester were almost similar ~140 g/m^2. The weight was found to increase, and this effect was observed more strongly when the finish was applied on knitted fabric samples. However, the weight gain was also found to increase with an increase in finish concentration due to the maximum wetting and binding of individual fibers and yarns and ultimately retaining of the finish (nanoparticles) in the fabric structure after hydro-extracting and drying. All measurements were repeated 10 times and the CV% was lower than 5%.

Table 4. Weight gain of knitted fabrics after antibacterial finish application.

Sr. No.	Specimen Code	Finish Concentration (%)	Weight (g/m^2)
1	CY-CS1%	1	143.5 ± 4.08
2	CY-CS3%	3	148.1 ± 5.45
3	CY-MS1%	1	143.1 ± 3.74
4	CY-MS3%	3	147.9 ± 4.44
5	PCY-CS1%	1	143.0 ± 2.71
6	PCY-CS3%	3	147.9 ± 4.01
7	PCY-MS1%	1	142.9 ± 6.04
8	PCY-MS3%	3	147.7 ± 5.41
9	PY-CS1%	1	142.5 ± 2.33
10	PY-CS3%	3	147.4 ± 4.42
11	PY-MS1%	1	142.6 ± 3.12
12	PY-MS3%	3	147.5 ± 6.41
13	CF-MS1%	1	144.0 ± 4.22
14	CF-MS3%	3	149.0 ± 5.48

3.3. Fourier Transform Infrared Spectroscopy

The modification of the knitted fabric surface was investigated by ATR-FTIR spectroscopy to find the molecules present in the samples and their functional groups. The ATR-FTIR spectra of untreated/bleached-only samples and finished knitted fabrics are shown in Figure 4, which affirmed the presence of numerous absorption peaks as an indication of different functional groups.

FTIR spectra of finished yarn and fabric samples were critically analyzed and verified with the literature [43–47] to extract valuable information about the nature of the nanoparticles present in the finish formulations. The antibacterial finishes were produced using nanocluster technology as claimed by the manufacturer/supplier. The present innovative product relates to the emulsion of nanoparticles in stable aqueous dispersions of metal/metal oxides nanoclusters containing many other chemical auxiliaries using green nanochemistry.

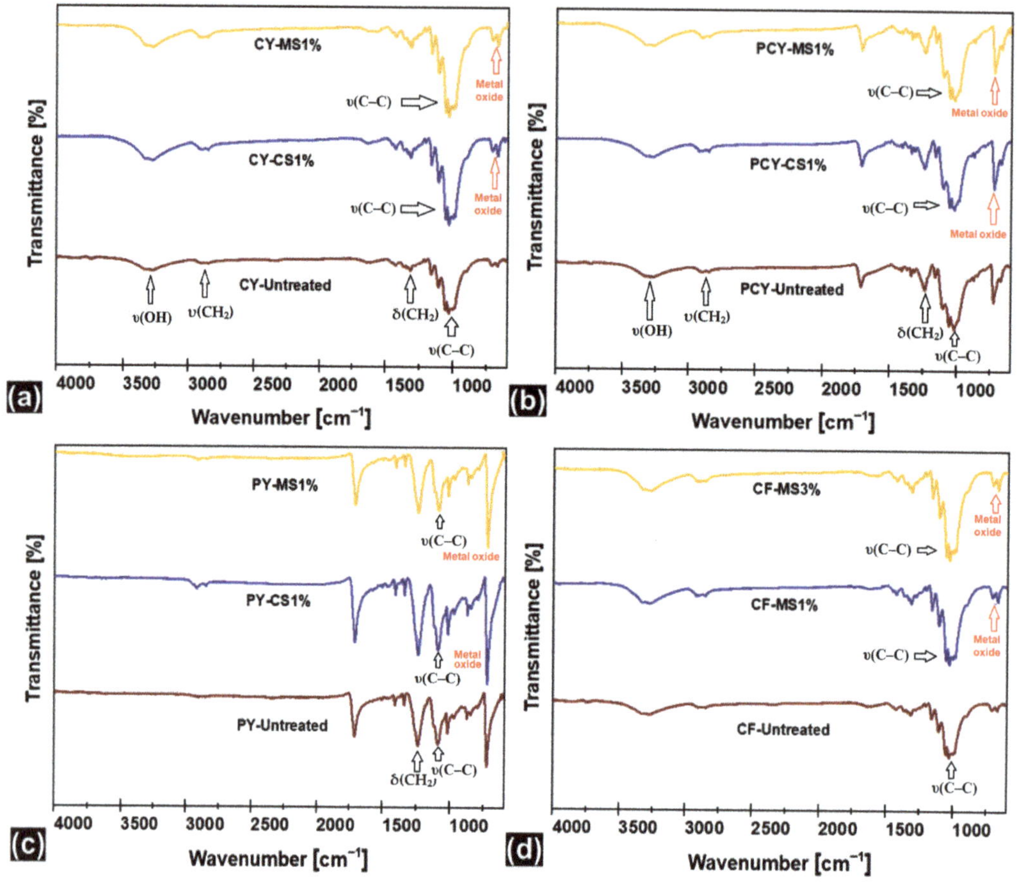

Figure 4. FTIR spectrums of antibacterial finished specimens: (**a**) CY specimens; (**b**) PCY specimens; (**c**) PY specimens; (**d**) CF specimens.

In the case of the untreated/bleached-only sample fabrics of cotton and polyester–cotton, the characteristic vibration modes of cellulose were detected, i.e., υ(OH) at ca. ~3300, υ(CH_2) at ~2900, δ(OH) at ~1640, δ(CH_2) at ~1425, δ(CH) at ~1370, δ(OH) at ~1310, υ(C–C) at ~1020, and υ(OH) at ~894 cm^{-1}, and the characteristic vibration modes of polyester were detected, i.e., υ(C=O) at 1720 cm^{-1}, υ(C-H) at 2920 cm^{-1} and 2880 cm^{-1}, and υ(C-O) at 1250 cm^{-1} [48]. Metal nanoparticles have also their own characteristic vibration peaks. Apart from these, the prominent peaks of lattice vibrations of M—O, M—O—M, and O—M—O (M stands for metal) were observed in the low-energy region of the spectra (~500–1000 cm^{-1}), manifesting the presence of copper, magnesium, and silver metal nanoparticles [49–51]. The FTIR spectra profiles of finished fabrics are very similar to those of untreated/bleached-only samples. However, an increase in the size/intensity of peaks is a direct indication of the amount of antibacterial nano-finish material present.

3.4. Scanning Electron Microscopy

The overall morphology of the knitted fabric surface appearance was inspected by scanning electron microscopy (SEM). Typical SEM images of the untreated/bleached-only samples and finished knitted fabrics are shown in Figure 5, displaying the presence of antibacterial nano-finishes on fabric samples for effective antibacterial activity. The homogeneous distribution of antibacterial nano-finishes on the knitted fabric samples uniformly deposited and modified the surface of the textile substrates (i.e., fibers, yarns, and fabrics) without any accumulation of finish as evidenced from the SEM images. Figure 5a shows the untreated/bleached-only sample without any nano-finish. Figure 5b,c depict SEM images of knitted fabric samples, i.e., finished after knitting and prepared from finished yarn, respectively, which were compared to examine how the finish system was laid on the textile substrate. It can be clearly observed from Figure 5b that the finish is introduced into the yarn as a thin film/coating covering the fibers/yarns but not the surface of the fabric. Contrarily, the fabric that was finished after knitting exhibited a thin film/coating covering the fibers/yarns as well as the surface of the fabric, as evidenced from Figure 5c. Uniform deposition of the finish was found due to the maximum wetting and binding of individual fibers/yarns that also resulted in partial covering or filling of the inter-fiber and inter-yarn spaces/pores of loops/stitches by the finishing of the fabric after knitting. This leads to deterioration in the wearing comfort of fabric samples finished after knitting.

Figure 5. SEM images of specimens: (**a**) CF-untreated/bleached-only sample; (**b**) CY-MS-3%; (c) CF-MS3%.

3.5. Antibacterial Efficacy

The antibacterial nano-finishes were purchased from a company. Therefore, they were first evaluated to ensure their antibacterial activity, and the inhibition zone was clearly observed which confirmed their efficacy (also already presented in Figure 1). Then, on the basis of these findings further work was conducted. The formation of the inhibition zone is the most widely accepted visual fact relating to the antibacterial activity of any finished textile substrates. But sometimes a clear inhibition zone does not appear around the finished fabric samples due to non-/poor leaching properties of the antibacterial finishes. In that situation, the other most important truth/criterion for evaluation of antibacterial activity is the presence/absence of bacterial colonies growing under or around the finished fabric samples. In this study, antibacterial activity was present and observed on the basis of the killing of bacterial strains by the finished fabric samples before and after washing, i.e., the presence/absence of bacterial colonies. The overall antibacterial efficacy results are given in Tables 5 and 6 and Figures 6 and 7.

Table 5. Antibacterial activity of finished knitted fabrics.

Sr. No.	Specimen Code	Bacterial Growth in Sample Contact Area (Yes / No)	Clear Zone of Inhibition (mm)
1.	CY-CS1%	No	0.00
2.	CY-CS3%	No	0.00
3.	CY-MS1%	Yes	0.00
4.	CY-MS3%	No	0.00
5.	PCY-CS1%	No	0.00
6.	PCY-CS3%	No	0.00
7.	PCY-MS1%	Yes	0.00
8.	PCY-MS3%	No	0.00
9.	PY-CS1%	No	0.00
10.	PY-CS3%	No	0.00
11.	PY-MS1%	Yes	0.00
12.	PY-MS3%	No	0.00
13.	CF-MS1%	Yes	0.00
14.	CF-MS3%	No	0.00

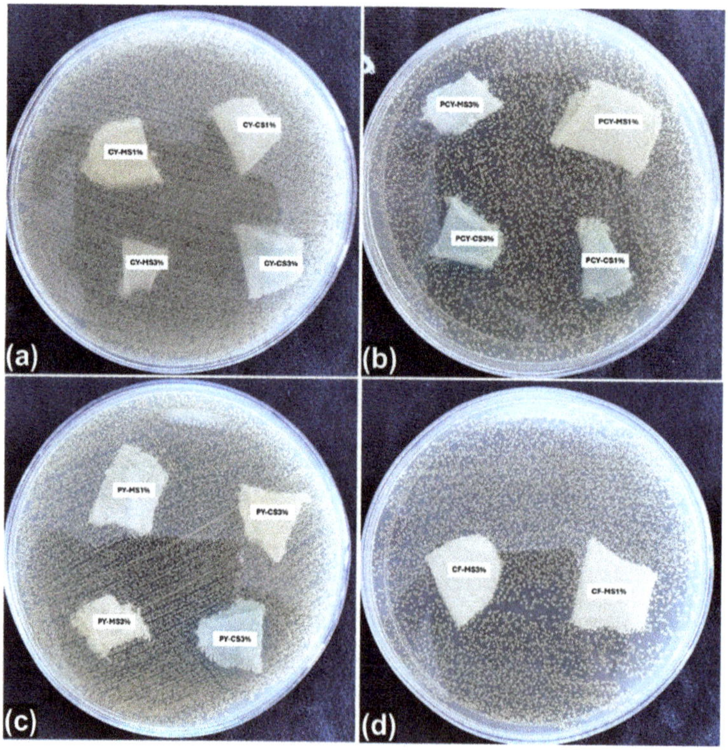

Figure 6. Agar plate images during antibacterial activity: (**a**) CY specimens; (**b**) PCY specimens; (**c**) PY specimens; (**d**) CF specimens.

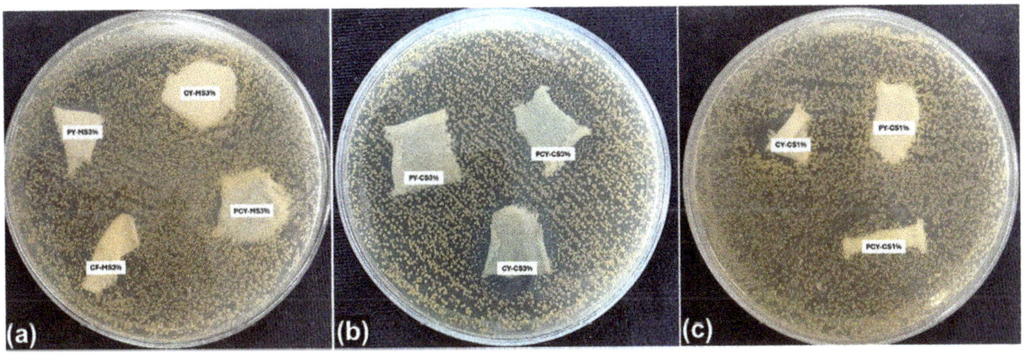

Figure 7. Agar plate images during antibacterial activity: (**a**) MS-3% specimens; (**b**) CS-3% specimens; (**c**) CS-1% specimens.

Table 6. Antibacterial activity of finished knitted fabrics after 1, 5, 10, and 15 washing cycles.

Sr. No.	Specimen Code	Bacterial Growth in Sample Contact Area (Yes / No) After Washing Cycles				Clear Zone of Inhibition (mm) After Washing Cycles			
Washing Cycles		1	5	10	15	1	5	10	15
1	CY-CS1%	No	No	No	Yes	0.00	0.00	0.00	0.00
2	CY-CS3%	No	No	No	No	0.00	0.00	0.00	0.00
3	CY-MS1%	Yes	Yes	Yes	Yes	0.00	0.00	0.00	0.00
4	CY-MS3%	No	No	No	No	0.00	0.00	0.00	0.00
5	PCY-CS1%	No	No	No	Yes	0.00	0.00	0.00	0.00
6	PCY-CS3%	No	No	No	No	0.00	0.00	0.00	0.00
7	PCY-MS1%	Yes	Yes	Yes	Yes	0.00	0.00	0.00	0.00
8	PCY-MS3%	No	No	No	No	0.00	0.00	0.00	0.00
9	PY-CS1%	No	No	No	Yes	0.00	0.00	0.00	0.00
10	PY-CS3%	No	No	No	No	0.00	0.00	0.00	0.00
11	PY-MS1%	Yes	Yes	Yes	Yes	0.00	0.00	0.00	0.00
12	PY-MS3%	No	No	No	No	0.00	0.00	0.00	0.00
13	CF-MS1%	Yes	Yes	Yes	Yes	0.00	0.00	0.00	0.00
14	CF-MS3%	No	No	No	No	0.00	0.00	0.00	0.00

The in vitro antibacterial activity of knitted fabrics was determined by testing against the *Staphylococcus aureus* (*S. aureus*) bacterial strain. After the incubation period, the inhibition zone and bacterial growth were assessed for each fabric sample to compare the antibacterial efficacy. The untreated/bleached-only fabric samples revealed no antibacterial activity against *S. aureus*; the presence of colonies can be seen on the surface, thus indicating bacterial growth. The visual aspects of the antibacterial tests are shown in Figure 6.

The results revealed that most of the finished fabric samples showed an antibacterial effect. The finished fabric samples placed on the bacteria-inoculated surfaces killed all of the bacteria under and around the region of the fabric. It also demonstrated the distinct inhibition of bacterial growth around finished fabric samples for *S. aureus*. The copper–silver-based antibacterial nano-finish exhibited a higher inhibitory activity against *S. aureus* as compared to that of the magnesium–silver-based antibacterial nano-finish. However, a higher inhibition of bacterial growth was obtained for a higher concentration of each antibacterial nano-finish, emphasizing the influence of concentration.

Antibacterial finishes and different metal nanoparticles (such as copper, magnesium, silver, titanium, zinc, etc.) have been validated as potent antibacterial agents, due to their effectiveness in microbial resistance through different proposed mechanisms to oppose microbial growth. They inhibit growth or preferably kill microorganisms by several different mechanisms which act around the cell wall of microorganisms. Thus, cell wall damage, alteration of cytoplasm membrane permeability, alteration of physical or chemical state of nucleic acids and proteins, inhibition of enzyme action, or inhibition of nucleic acid or protein synthesis are all physical and chemical approaches that can be employed by antibacterial finishes to inhibit growth or kill microorganisms [52–54].

The most important properties of textile treatments are durability and washing fastness. Washing procedures usually cause a reduction in antibacterial activity. In general, the effectiveness of the finished material decreases and, in parallel, a reduction in bacterial viability can be observed as shown in Figure 7.

In this study, antibacterial activity was present and observed on the basis of the killing of bacterial strains by the finished fabric samples before and after washing, i.e., the presence/absence of bacterial colonies' growth. Since the antibacterial activity was qualitatively evaluated, only the presence or absence of the bacterial colony was visualized. The results of wash durability after 1, 5, 10, and 15 cycles are given in Table 6.

After 15 washing cycles, there was a change in the bactericidal activity of the knitted fabric samples finished with 1% antibacterial nano-finish concentrations, as summarized in Table 6.

3.6. Air Permeability of Knitted Fabrics

Maintaining adequate air permeability is necessary for antibacterial knitted fabrics along with their functional attributes. Usually, antibacterial knitwear is worn for prolonged times in order to perform its core functions. Hence, if air permeability is not maintained, the thermal equilibrium of the body is disturbed, and a sense of discomfort develops. This is why it was crucial to characterize the developed specimens in terms of air permeability, the results of which are detailed in Figure 8. It is well established that air permeability decreases after the wet finishing of fabrics [55–58]. This is attributed to swelling of the fibers and yarns which causes a shrinkage in the knitted fabric. Therefore, the overall porosity decreases, reducing the ability to allow air passage [51,52].

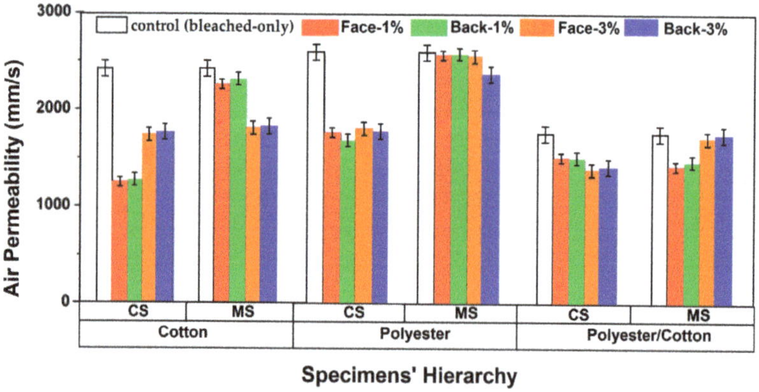

Figure 8. Air permeability of knitted fabrics.

Air permeabilities were separately characterized on the face and back surfaces for the finished samples. However, the obtained values were almost equivalent for all specimens. Changing the finish concentration developed more prominent air permeability variation for cotton knitted fabrics. Increasing the concentration of CS finish increased air permeability by about 41.02 ± 2.01% instead of the increasing areal density/grams per square meter of fabric (Figure 9).

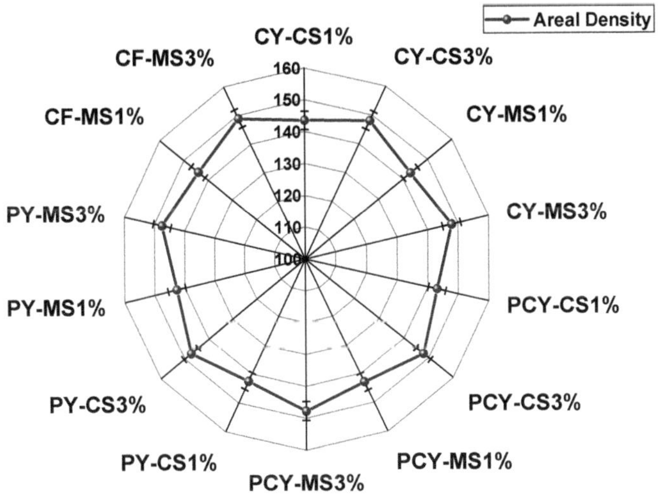

Figure 9. Areal density of knitted fabrics.

However, the trend was found to be different for MS-finished fabrics, where increasing MS concentration to 3% resulted in a 19.77 ± 0.78% decrease in the air permeability of cotton knitted fabrics. The phenomenon was governed by an increasing number of TPIs at CS-3% (Figure 3a) which made the yarns tighter and crispier. Thus, the inter-yarn pores through the loops in the knitted structure became more open and allowed more air to pass through the fabric. The number of TPIs decreased for MS-3% specimens (Figure 3a); hence, air permeability decreased with an increasing concentration of MS finish. However, polyester knitted specimens also exhibited a similar trend to the cotton fiber-based sample. The 3% CS-finished specimen showed a 5.80 ± 0.12% higher air permeability than the 1% CS-finished specimen. The 3% MS-finished fabric also had a 7.92 ± 0.15% lower air permeability as compared to the 1% concentration. Hence, it was evident that the inherent behaviors of antibacterial finishes also affect air permeabilities. However, polyester/cotton blended fabrics exhibited totally opposite behavior. Increasing CS concentration decreased air permeability, and vice versa for MS-finished knitted fabrics. The blending of fibers not only showed a significant influence on changing trends, but also decreased the overall air permeability values. The highest air permeability of the polyester/cotton 3% MS-finished specimen was 26.72 ± 0.91% lower than the 3% MS-finished specimen of polyester fiber. Another factor under consideration was to compare the air permeability of yarn-finished antibacterial fabrics with specimens finished at the fabric stage. Curing of the antibacterial finish occurred between inter- and intra-loop spaces when cotton fabric was being treated. The phenomenon decreased the porosity of the fabric. Hence, the air permeabilities of specimens treated at the fabric stage were found to be much lower than samples from treated yarn. The fabric sample treated with CS-1% exhibited a 47.76 ± 1.58% lower air permeability than CS-1% yarn-finished fabric sample. Such a phenomenon proved that antibacterial treatment at the yarn stage was more effective in maintaining higher air permeability.

Figure 10 shows visual images of untreated/bleached-only fabric samples (Figure 10a), samples treated at the yarn stage (Figure 10b), and samples treated at the fabric stage (Figure 10c).

It is visible that the porosity of the untreated/bleached-only sample is the maximum, followed by the sample made from finished yarns and the sample finished at the fabric stage. The swelling of the fibers and yarns can also be observed in Figure 10b,c.

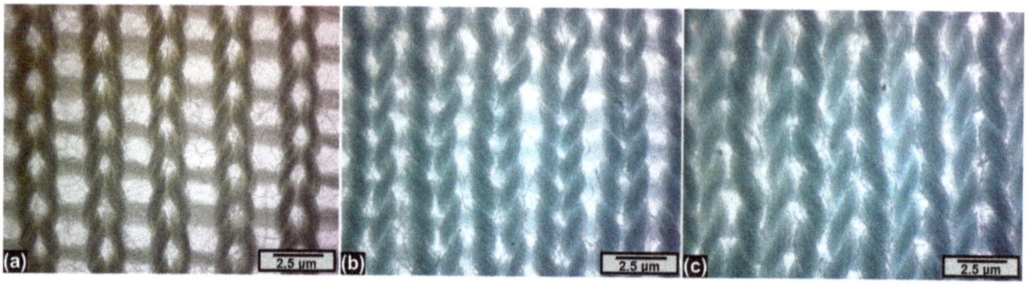

Figure 10. Photographic analysis of specimens: (**a**) unfinished raw fabric; (**b**) yarn-finished fabric; (**c**) finish applied after knitting.

3.7. OMMC of Knitted Fabrics

OMMC is an indexed result of several moisture management parameters, e.g., water vapor permeability, wetting rate, wetting time, etc. Hence, a single value of the OMMC index encompasses all auxiliary moisture management attributes. Antibacterial textile materials are mostly prone to bodily fluids like sweat and wound exudates. Hence, the fabrics should be capable of managing these fluids and transporting them into external environments. The results of all of the moisture management parameters for the untreated/bleached-only and finished samples are given in Table 7.

In general, the OMMC index depends on two major factors: firstly, the hydrophilicity of the material itself to absorb/adsorb the moisture; secondly, the fabric porosities supporting the moisture management. The knitted structures and other knitting parameters were kept constant in this research. Hence, the trends observed in the OMMC results were governed by the fiber type, yarn type, and finish applied. It was observed that the OMMC values for the treated samples were higher than those of the untreated/bleached-only samples. This can be attributed to the micro channels created by the finishing treatment which allow capillary action and transportation of water molecules through the inter-fiber channels. Such observations have also been reported by several researchers [48,55]. In several cases, the finishing treatment with higher concentration also facilitated better moisture management due to higher adsorption of water molecules on the nanoparticles deposited on the fiber/fabric surface [56]. The 100% cotton and 100% polyester knitted fabrics exhibited a decrease in OMMC index with an increasing concentration of CS finish. The cotton specimen showed a 30.88 ± 1.12% decrease, while polyester fabric showed a 10.29 ± 0.25% lower OMMC index for CS-3% as compared to fabrics treated with a 1% concentration of the finish. Increasing the concentration of antibacterial finish decreased the inherent fabric porosities, and CS finish could not handle moisture appropriately which resulted in a compromised OMMC index. The MS-finish-treated specimens showed a different trend of an increasing OMMC index with increasing finish concentration. The MS-3% treated cotton yarn specimen has a 8.19 ± 0.17% higher OMMC index than the 1% concentration of MS, and the same difference was noticed for 100% polyester knitted fabrics. Cotton knitted fabric was also treated at the fabric stage with MS finish for comparison with yarn-finished knitted fabric. The results also proved the affinity of MS finish towards moisture. The fabric-stage-treated cotton fabric exhibited the highest OMMC index of 0.8 among all other specimens; the value was 21.21 ± 0.88% higher than yarn-finished specimens of cotton. Polyester/cotton blended yarn knitted fabrics on the other hand showed a linear increase in OMMC indices with increasing concentrations for both finish types. The inherent moisture handling behavior of the fiber and finishing material remains the dominating parameter up to a 3% concentration of CS finish, leading to an elevated moisture management index. A similar phenomenon was seen for MS-finished fabrics and increased the OMMC index by

1.35 ± 0.04%. The polyester/cotton blended yarn knitted specimens offered overall higher moisture management as compared to other yarn-finished specimens, proving blending to be an efficient solution to impart better moisture management attributes irrespective of the finish type or concentration employed.

Table 7. Moisture management parameters of the control/bleached-only and finished samples. The values are the average of 10 measurements with a CV% less than 5%.

Sr. No.	Specimen Code	WTT (s)	WTB (s)	TAR (%/s)	BAR (%/s)	MWRT (mm)	MWRB (mm)	SST (mm/s)	SSB (mm/s)	R (-)	OMMC (-)
1	CY-control	33.25	34.14	312.11	11.20	3.12	1.24	0.31	0.15	−320.12	0.44
2	CY-CS1%	30.42	32.14	346.42	15.41	3.78	1.52	0.38	0.19	−298.24	0.68
3	CY-CS3%	31.21	33.54	327.56	13.41	3.25	1.33	0.34	0.17	−304.71	0.45
4	CY-MS1%	30.84	32.78	334.74	14.25	3.54	1.42	0.36	0.18	−300.45	0.61
5	CY-MS3%	30.48	32.41	342.81	15.28	3.68	1.51	0.37	0.19	−299.42	0.67
6	PCY-control	33.15	33.89	315.52	12.46	3.26	1.26	0.35	0.16	−322.54	0.56
7	PCY-CS1%	30.53	31.78	347.85	15.52	3.95	1.74	0.39	0.20	−297.94	0.68
8	PCY-CS3%	30.76	32.78	333.76	13.95	3.81	1.44	0.37	0.19	−301.23	0.61
9	PCY-MS1%	30.89	32.63	335.36	14.12	3.79	1.43	0.39	0.17	−312.36	0.59
10	PCY-MS3%	30.57	31.88	345.02	15.46	3.87	1.46	0.38	0.19	−300.56	0.66
11	PY-control	33.44	34.42	312.43	12.48	3.25	1.25	0.36	0.17	−319.57	0.51
12	PY-CS1%	31.16	33.08	337.47	15.45	3.84	1.38	0.40	0.19	−307.47	0.59
13	PY-CS3%	30.14	31.56	350.70	14.91	3.72	1.63	0.42	0.17	−289.73	0.66
14	PY-MS1%	30.35	31.79	335.78	18.75	3.78	1.96	0.45	1.75	−285.41	0.68
15	PY-MS3%	29.24	28.34	355.45	20.45	4.23	2.12	0.58	0.22	−254.28	0.71

The OMMC index values of the developed specimens were compared with the untreated/bleached-only sample. The results are shown in Figure 11.

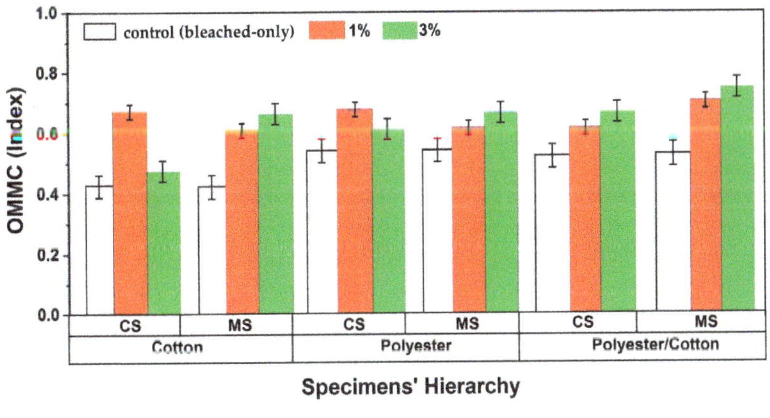

Figure 11. OMMC indices of knitted fabrics.

3.8. Bursting Strength

Bursting strength is a crucial indicator determining the serviceability of knitted fabrics. Knitwear fabrics undergo several multidirectional environmental stresses which can cause ruptures or bursts in the influenced area. Hence, bursting performance was characterized in this research, providing insights into fabric serviceability. Figure 12 highlights the results of the characterized specimens. Bursting strength was measured at each concentration of applied antibacterial finish to analyze the influence of changing the finish type and concentration on serviceability.

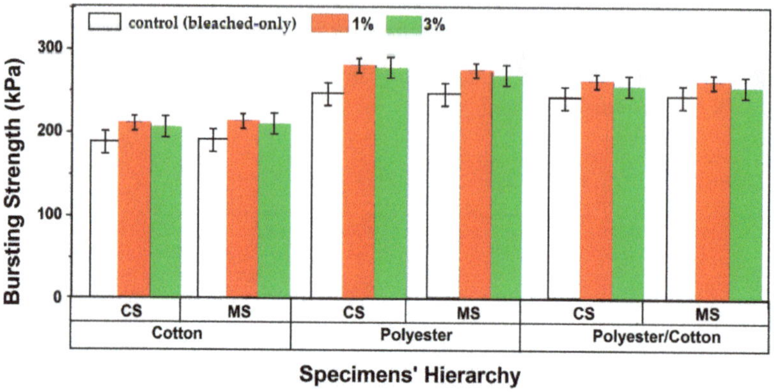

Figure 12. Bursting strength of knitted fabrics.

The bursting strength values of the finished samples were slightly higher than those of the untreated/bleached-only samples [27,48]. This can be attributed to the cross-linking of the nanoparticles on the fiber surfaces, which improves the inter-fiber friction and thus the resistance to bursting of the yarns and fabric [45,57]. On the other hand, a linear trend of a slight decrease in bursting strength was noticed for all specimens with an increase in antibacterial finish concentration from 1% to 3%. This could be due to the overload of nanoparticles which deteriorate the bursting resistance of the yarns and fabric [44,56], though the overall bursting resistance/strength in all finished samples was still higher than that of the untreated/bleached-only sample. The influence of fiber-inherent mechanical properties remained dominant, i.e., polyester was stronger as a synthetic fiber which resulted in the highest bursting strength followed by a decreasing trend towards the polyester/cotton blend and cotton. The highest bursting strength of 280 kPa was observed for recycled polyester fabrics having a 1% concentration of CS finish. However, MS-finished fabrics having a similar concentration resulted in a 1.78 $\pm$ 0.04% lower bursting strength. Bursting strength, being dependent on the mechanical attributes of yarns, corresponded to the single yarn strength results shown in Figure 3c. The strength of MS-finished polyester yarns was relatively lower than that of CS-finished ones. Hence, the bursting strength of the corresponding fabrics samples was observed to be lower. The cotton sample showed about a 0.9 $\pm$ 0.02% higher bursting strength for MS-finished fabrics, and a similar performance was noticed for polyester/cotton blended yarn knitted fabrics. The contribution of both polyester and cotton in a single yarn balanced the bursting performance. The trend of decreasing bursting strength with increasing concentrations remained similar for all yarn types owing to a lower single yarn strength for 3% finished specimens (Figure 3c). The fabric-finished cotton specimens exhibited a higher bursting strength as compared to the yarn-finished cotton fabrics. This phenomenon can be attributed to the enhanced compactness and links between interloping zones supported by the antibacterial finish. Also, the swelling of yarns visible in Figure 10b,c is responsible for the compactness and increase in bursting resistance. Finish application at the fabric stage improved the bursting performance by 2.8 $\pm$ 0.11% and 2.4 $\pm$ 0.05% for 1% and 3% finish concentrations, respectively.

4. Conclusions

In this research, two unique approaches were successfully combined to develop recycled textiles for value-added medical textile products. The antibacterial applications do not require extreme mechanical attributes. Hence, the recycled yarns were found to be suitable for such applications.

There was an increase in TPI after finishing depending on the yarn type and the percentage of finishing. The swelling of fibers after the deposition of nanoparticles also increased the linear density. The chemical finishing treatments caused the yarns to be stiffer, and breaking strength and elongation were compromised. The applied antibacterial finish worked in a similar manner, causing the breaking strength and elongation of finished specimens to be lower than those of untreated/bleached-only samples. The results were found to be statistically significant with a coefficient of variation lower than 5%.

Using a two-fold approach, the antibacterial finishes were applied at both the yarn and fabric stages, i.e., before and after knitting. Microscopic and SEM analysis revealed that finish application at the fabric stages contributed to a reduction in porosities. This phenomenon did not affect the antibacterial activity. The antibacterial effect was improved by increasing the concentration from 1% to 3%. The antibacterial finish was still effective after 15 laundering/washing cycles.

The thermo-physiological comfort and serviceability factors were influenced by the finish concentration. The air permeability of finished fabrics was lower than that of untreated/bleached-only samples. For 100% cotton fabrics, increasing the concentration of CS finish increased the air permeability by about 41.02 ± 2.01% instead of increasing the areal density/grams per square meter of fabric. For recycled polyester specimens, increasing the concentration of CS finish from 1% to 3% also increased the air permeabilities, and the opposite was seen for MS finish. However, the polyester/cotton blend showed an inverse trend. The specimens finished at the fabric stage showed lower air permeabilities than yarn-finished specimens. The decreased air permeability was attributed to reduced porosities but worked efficiently for OMMC enhancement, as moisture channels/capillaries were enhanced for the transportation of water through the fibrous material. Therefore, treated samples showed a higher OMMC as compared to untreated/bleached-only samples. In some cases, finishing treatments with a higher concentration also facilitated better moisture management due to better adsorption of water molecules on the nanoparticle surface. Bursting strength was improved by 2.8 ± 0.11% and 2.4 ± 0.05% for 1% and 3% finish concentrations, as the finish created linkages between fibers and more force was required to burst the knitted fabrics multiaxially.

The limitations of the study include the lack of a detailed characterization of the nanoparticles deposited onto the textile material. Such information will be necessary to further understand the influence of specific functional nanoparticles on antibacterial performance.

Author Contributions: Conceptualization, M.S.F., H.J., R.K.M., A.A., M.M., J.U. and M.P.; methodology, M.S.F., H.J., R.K.M., A.A., M.M., J.U. and M.P.; software, M.S.F., H.J., R.K.M., A.A., M.M., J.U. and M.P.; testing of mechanical properties, M.S.F., H.J., R.K.M., A.A., M.M., J.U. and M.P.; data analysis, M.S.F., H.J., R.K.M., A.A., M.M., J.U. and M.P.; writing—original draft preparation, M.S.F., H.J., R.K.M., A.A., M.M., J.U. and M.P.; writing—review and editing, M.S.F., H.J., R.K.M., A.A., M.M., J.U. and M.P.; resources, H.J., R.K.M., M.M., J.U. and M.P.; supervision, H.J., R.K.M. and M.M. All authors have read and agreed to the published version of the manuscript.

Funding: The authors are grateful to the internal grant agency of the Faculty of Engineering, Czech University of Life Sciences Prague, Grant Nos. 2025:31140/1312/3104: "Research into the production of composite polymer materials with a focus on improving performance", and 2025:31140/1312/3108: "Research on the recyclability of PUR foam in the application of polymer composite systems".

Institutional Review Board Statement: Not applicable.

Data Availability Statement: The original contributions presented in the study are included in the article. Further inquiries can be directed to the corresponding authors.

Conflicts of Interest: The authors declare no conflicts of interest.

References

1. Huang, X.; Tan, Y.; Huang, J.; Zhu, G.; Yin, R.; Tao, X.; Tian, X. Industrialization of open- and closed-loop waste textile recycling towards sustainability: A review. *J. Clean. Prod.* **2024**, *436*, 140676. [CrossRef]
2. Baloyi, R.B.; Gbadeyan, O.J.; Sithole, B.; Chunilall, V. Recent advances in recycling technologies for waste textile fabrics: A review. *Text. Res. J.* **2024**, *94*, 508–529. [CrossRef]
3. Leal Filho, W.; Ellams, D.; Han, S.; Tyler, D.; Boiten, V.J.; Paco, A.; Moora, H.; Balogun, A.-L. A review of the socio-economic advantages of textile recycling. *J. Clean. Prod.* **2019**, *218*, 10–20. [CrossRef]
4. Jamshaid, H.; Hussain, U.; Tichy, M.; Muller, M. Turning textile waste into valuable yarn. *Clean. Eng. Technol.* **2021**, *5*, 100341. [CrossRef]
5. Abbas, A.; Jamshaid, H.; Mishra, R. Weft-Knitted Structure and Their Effect on Fabric Properties. In *Knitting Science, Technology, Process and Materials*; Springer: Berlin/Heidelberg, Germany, 2024; pp. 81–107. Available online: https://link.springer.com/chapter/10.1007/978-3-031-44927-7_4 (accessed on 8 October 2025).
6. Abbas, A.; Anas, M.S.; Awais, H.; Hussain, M.; Underwood, J.; Ashraf, W. Infrared (IR) imaging based comfort appraisal of wool, acrylic, and Thermolite® plaited gloves; Exploring socio-economic materials. *J. Ind. Text.* **2024**, *54*, 1–22. [CrossRef]
7. Naseer, R.; Abbas, A.; Sarwar, Z.; Saeed, M.H.; Gul, Z.; Anas, M.S. Cross-tuck honeycomb based 3D weft knitted auxetic structures with enhanced adaptability/size changing capability & thermo-physiological comfort properties. *J. Text. Inst.* **2024**, *116*, 1297–1312. [CrossRef]
8. Yang, X.; Fan, W.; Wang, H.; Shi, Y.; Wang, S.; Liew, R.K.; Ge, S. Recycling of bast textile wastes into high value-added products: A review. *Environ. Chem. Lett.* **2022**, *20*, 3747–3763. [CrossRef]
9. Arumugam, V.; Militky, J.; Tunak, M. In-plane shear behavior of 3D spacer knitted fabrics. *J. Ind. Text.* **2016**, *46*, 868–886. [CrossRef]
10. Yang, Y.; Yu, X.; Chen, L.; Zhang, P. Effect of knitting structure and yarn composition on thermal comfort properties of bi-layer knitted fabrics. *Text. Res. J.* **2021**, *91*, 3–17. [CrossRef]
11. Mazzanti, S.; Barchiesi, F.; Pallotta, F.; Luchetti, I.; Giacometti, A.; Brescini, L. Severe Acute SARS-CoV-2 Infection and Long COVID: What Do We Know So Far? New Challenges in Diagnosis and Management. *Diseases* **2025**, *13*, 337. [CrossRef]
12. McBride, R.; Fielding, B.C. The Role of Severe Acute Respiratory Syndrome (SARS)-Coronavirus Accessory Proteins in Virus Pathogenesis. *Viruses* **2012**, *4*, 2902–2923. [CrossRef]
13. Mokrani, D.; Timsit, J.-F. Role of Respiratory Viruses in Severe Acute Respiratory Failure. *J. Clin. Med.* **2025**, *14*, 3175. [CrossRef]
14. Winder, N.; Gohar, S.; Muthana, M. Norovirus: An Overview of Virology and Preventative Measures. *Viruses* **2022**, *14*, 2811. [CrossRef] [PubMed]
15. Omatola, C.A.; Mshelbwala, P.P.; Okolo, M.-L.O.; Onoja, A.B.; Abraham, J.O.; Adaji, D.M.; Samson, S.O.; Okeme, T.O.; Aminu, R.F.; Akor, M.E.; et al. Noroviruses: Evolutionary Dynamics, Epidemiology, Pathogenesis, and Vaccine Advances—A Comprehensive Review. *Vaccines* **2024**, *12*, 590. [CrossRef] [PubMed]
16. Kida, Y.; Okuya, K.; Saito, T.; Yamagishi, J.; Ohnuma, A.; Hattori, T.; Miyamoto, H.; Manzoor, R.; Yoshida, R.; Nao, N.; et al. Structural Requirements in the Hemagglutinin Cleavage Site-Coding RNA Region for the Generation of Highly Pathogenic Avian Influenza Virus. *Pathogens* **2021**, *10*, 1597. [CrossRef] [PubMed]
17. Li, H.; Ren, R.; Bai, W.; Li, Z.; Zhang, J.; Liu, Y.; Sun, R.; Wang, F.; Li, D.; Li, C.; et al. A Review of Avian Influenza Virus Exposure Patterns and Risks Among Occupational Populations. *Vet. Sci.* **2025**, *12*, 704. [CrossRef]
18. Pivsa-Art, S.; Sunyikhan, K.; Pivsa-Art, W. Bicomponent multifilament yarns of recycled poly(ethylene terephthalate) and nano-titanium dioxide for antibacterial carpet. *J. Ind. Text.* **2022**, *51*, 1034S–1047S. [CrossRef]
19. Hunskunatai, D.; Pivsa-Art, W.; Roungpaisan, N.; Boonyod, S.; Yalcinkaya, F.; Pivsa-Art, S. Multi-functional Core/Sheath Bicomponent Nonwoven of Recycled Poly(ethylene terephthalate) Fibres as Thermal Insulators. *J. Polym. Environ.* **2024**, *32*, 2380–2393. [CrossRef]
20. Behera, B.K. Comfort properties of non-conventional light weight worsted suiting fabrics. *Ind. J. Fibre Text. Res.* **2007**, *32*, 72–79.
21. Venkataraman, M.; Wiener, J.; Militky, J.; Kotresh, T.M.; Vaclavik, M. Novel techniques to analyse thermal performance of aerogel-treated blankets under extreme temperatures. *J. Text. Inst.* **2014**, *106*, 736–747. [CrossRef]
22. Addis, S.; Ejegu, H.; Dubale, M.; Mamuye, W. Development of Sustainable and Functional Fabrics from Recycled and Nanocomposite Polyester Fibers. *Adv. Mater. Sci. Eng.* **2021**, *2021*, 7087152. [CrossRef]
23. Marmarali, A.; Sevgi, M. Study on the Comfort Properties of Knitted Fabrics Produced from Conventional and Sustainable Cotton and Polyester Fibres. *Tekstilec* **2023**, *66*, 339–350. [CrossRef]
24. Yılmaz, E.; Özgen Keleş, B. A Comparative Study of Cotton/PES Knitted Fabrics Produced from Recycled Fiber-Based and Virgin Yarns. *Fiber Polym.* **2024**, *25*, 4951–4963. [CrossRef]
25. Seval, U. The bursting strength properties of knitted fabrics containing recycled polyester fiber. *J. Text. Inst.* **2021**, *112*, 1998–2003. [CrossRef]

26. Özkan, İ.; Gündoğdu, S. Investigation on the microfiber release under controlled washings from the knitted fabrics produced by recycled and virgin polyester yarns. *J. Text. Inst.* **2020**, *112*, 264–272. [CrossRef]
27. Telli, A.; Özdil, N. Effect of Recycled PET Fibers on the Performance Properties of Knitted Fabrics. *J. Eng. Fibers Fabr.* **2015**, *10*, 47–60. [CrossRef]
28. Kamble, Z.; Behera, B.K.; Behera, P.K. Influence of cellulosic and non-cellulosic particle fillers on mechanical, dynamic mechanical, and thermogravimetric properties of waste cotton fibre reinforced green composites. *Compos. Part. B Eng.* **2021**, *207*, 108595. [CrossRef]
29. Uyanık, S. A study on the suitability of which yarn number to use for recycle polyester fiber. *J. Text. Inst.* **2019**, *110*, 1012–1031. [CrossRef]
30. *ASTM D1422/D1422M-13*; Standard Test Method for Twist in Single Spun Yarns by the Untwist-Retwist Method. American Standards for Testing and Materials: West Conshohocken, PA, USA, 2020.
31. *ASTM D1578-93*; Standard Test Method for Breaking Strength of Yarn in Skein Form. American Standards for Testing and Materials: West Conshohocken, PA, USA, 2022.
32. *ASTM D2256/D2256M-21*; Standard Test Method for Tensile Properties of Yarns by the Single-Strand Method. American Standards for Testing and Materials: West Conshohocken, PA, USA, 2021.
33. *AATCC Test Method 147-2020*; Measurement of Antibacterial Activity of Textile Materials Using the Parallel Streak Method. Standard of the American Association of Textile Color Chemicals Research: Triangle Park, NC, USA, 2020.
34. *ISO 6330:2012*; Textiles—Domestic Washing and Drying Procedures for Textile Testing. International Standards Organization: Geneva, Switzerland, 2012.
35. *ISO 9237:1995*; Textiles—Determination of the Permeability of Fabrics to Air. International Standards Organization: Geneva, Switzerland, 1995.
36. *AATCC Test Method 195-2011*; Liquid Moisture Management Properties of Textile Fabrics. Standard of the American Association of Textile Color Chemicals Research: Triangle Park, NC, USA, 2011.
37. *ASTM D3786/D3786M—13*; Standard Test Method for Bursting Strength of Textile Fabrics-Diaphragm Bursting Strength Tester Method. American Standards for Testing and Materials: West Conshohocken, PA, USA, 2013.
38. Naebe, M.; Haque, A.; Haji, A. Plasma-Assisted Antimicrobial Finishing of Textiles: A Review. *Engineering* **2022**, *12*, 145–163. [CrossRef]
39. Behera, B.K.; Mishra, R. Artificial neural network-based prediction of aesthetic and functional properties of worsted suiting fabrics. *Int. J. Cloth. Sci. Technol.* **2007**, *19*, 259–276. [CrossRef]
40. Hossain, M.d.S.; Islam, M.; Dey, S.C.; Hasan, N. An approach to improve the pilling resistance properties of three thread polyester cotton blended fleece fabric. *Heliyon* **2021**, *7*, e06921. [CrossRef] [PubMed]
41. Chen, Y.; Su, N.; Zhang, K.; Zhu, S.; Zhu, Z.; Qin, W.; Yang, Y.; Shi, Y.; Fan, S.; Wang, Z.; et al. Effect of fiber surface treatment on structure, moisture absorption and mechanical properties of luffa sponge fiber bundles. *Ind. Crops Prod.* **2018**, *123*, 341–352. [CrossRef]
42. Arumugam, V.; Militky, J.; Davies, L.; Slater, S. Thermal and water vapor transmission through porous warp knitted 3D spacer fabrics for car upholstery applications. *J. Text. Inst.* **2017**, *109*, 345–357. [CrossRef]
43. Dhiman, G.; Chakraborty, J.N. Antimicrobial performance of cotton finished with triclosan, silver and chitosan. *Fash. Text.* **2015**, *2*, 13. [CrossRef]
44. Wu, D.Y.; Wang, S.S.; Wu, C.S. Textile Fabrics Containing Recycled Poly(ethylene terephthalate), Oyster Shells, and Silica Aerogels with Superior Heat Insulation, Water Resistance, and Antibacterial Properties. *ACS Appl. Polym. Mater.* **2021**, *3*, 3175–3184. [CrossRef]
45. Roeges, N.P.G. *A Guide to the Complete Interpretation of Infrared Spectra of Organic Structures*; (Katholieke Industriele Hogeschool O-VI); Wiley: New York, NY, USA, 1994; p. x + 340. ISBN 0-471-93998-6.
46. Othman, A.M.; Elsayed, M.A.; Al-Balakocy, N.G.; Hassan, M.M.; Elshafei, A.M. Biosynthesized silver nanoparticles by Aspergillus terreus NRRL265 for imparting durable antimicrobial finishing to polyester cotton blended fabrics: Statistical optimization, characterization, and antitumor activity evaluation. *Biocatal. Agric. Biotechnol.* **2021**, *31*, 101908. [CrossRef]
47. Venkataraman, M.; Jasikova, D.; Kotresh, T.; Militky, J. Thermodynamics of aerogel-treated nonwoven fabrics at subzero temperatures. *J. Ind. Text.* **2014**, *45*, 387–404. [CrossRef]
48. Pugazhendhi, A.; Prabhu, R.; Muruganantham, K.; Shanmuganathan, R.; Natarajan, S. Anticancer, antimicrobial and photocatalytic activities of green synthesized magnesium oxide nanoparticles (MgONPs) using aqueous extract of Sargassum wightii. *J. Photochem. Photobiol. B* **2019**, *190*, 86–97. [CrossRef]
49. Vasantharaj, S.; Sathiyavimal, S.; Saravanan, M.; Senthilkumar, P.; Gnanasekaran, K.; Shanmugavel, M.; Manikandan, E.; Pugazhendhi, A. Synthesis of ecofriendly copper oxide nanoparticles for fabrication over textile fabrics: Characterization of antibacterial activity and dye degradation potential. *J. Photochem. Photobiol. B* **2019**, *191*, 143–149. [CrossRef]

50. Hashem, M.; Ibrahim, N.A.; El-Sayed, W.A.; El-Husseiny, S.; El-Enany, E. Enhancing antimicrobial properties of dyed and finished cotton fabrics. *Carbohydr. Polym.* **2009**, *78*, 502–510. [CrossRef]
51. Ruparelia, J.P.; Chatterjee, A.K.; Duttagupta, S.P.; Mukherji, S. Strain specificity in antimicrobial activity of silver and copper nanoparticles. *Acta Biomater.* **2008**, *4*, 707–716. [CrossRef]
52. Abbas, A.; Anas, M.S.; Azam, Z. In vitro experimental study on effect of fiber denier, yarn doubling, and specialty treatments on physical and thermal behaviors of knitted sports socks. *J. Eng. Fibers Fabr.* **2023**, *18*, 155892502211489. [CrossRef]
53. Yuksekkaya, M.E.; Celep, G.; Dogan, G.; Tercan, M.; Urhan, B. A Comparative Study of Physical Properties of Yarns and Fabrics Produced from Virgin and Recycled Fibers. *J. Eng. Fibers Fabr.* **2016**, *11*, 68–76. [CrossRef]
54. Khan, S.U.; Hassan, T.; Wasim, M.; Khan, M.Q.; Salam, A.; Hassan, S.; Abbasi, A.M.R.; Mustafa, T. Valorization of recycled PET for yarn manufacturing and knitwear fabrics used for apparel applications. *Polym. Bull.* **2023**, *80*, 2779–2799. [CrossRef]
55. Crina, B.; Blaga, M.; Luminita, V.; Mishra, R. Comfort properties of functional weft knitted spacer fabrics. *Tekst. Konfeksiyon* **2013**, *23*, 220–227.
56. Shalaby, S.E.; Al-Balakocy, N.G.; Elshakankery, M.H.; Beliakova, M.K. Production of antimicrobial nonwoven fabrics based on recycled polyethylene terephthalate (R-PET) fibers on industrial scale. *J. Eng. Fiber Fabr.* **2024**, *19*, 15589250241276861. [CrossRef]
57. Lam, Y.L.; Kan, C.W.; Yuen, C. Developments in functional finishing of cotton fibres—Wrinkle-resistant, flame-retardant and antimicrobial treatments. *Text. Prog.* **2012**, *44*, 175–249. [CrossRef]
58. Gulati, R.; Sharma, S.; Sharma, R.K. Antimicrobial textile: Recent developments and functional perspective. *Polym. Bull.* **2022**, *79*, 5747–5771. [CrossRef]

materials

MDPI

Article

Transforming Waste into Value: The Role of Recovered Carbon Fibre and Oil Shale Ash in Enhancing Cement-Based Structural Composites

Regina Kalpokaitė-Dičkuvienė [1] Inga Stasiulaitiene [2], Arūnas Baltušnikas [1] and Samy Yousef [3,*]

[1] Laboratory of Materials Research and Testing, Lithuanian Energy Institute, Breslaujos 3, LT-44403 Kaunas, Lithuania; regina.kalpokaite-dickuviene@lei.lt (R.K.-D.)

[2] Department of Environmental Technology, Faculty of Chemical Technology, Kaunas University of Technology, LT-50254 Kaunas, Lithuania

[3] Department of Production Engineering, Faculty of Mechanical Engineering and Design, Kaunas University of Technology, LT-51424 Kaunas, Lithuania

* Correspondence: ahmed.saed@ktu.lt

Abstract: Economic and technological factors necessitate the use of alternative fuels during oil shale combustion, a process that generates substantial amounts of solid waste with varying ash compositions. This study evaluates the potential of two such waste materials: (i) fly ash derived from the combustion of oil shale (a fine particulate residue from burning crushed shale rock, sometimes combined with biomass), and (ii) short carbon fibres recovered from the pyrolysis (a process of decomposing materials at high temperatures in the absence of oxygen) of waste wind turbine blades. Oil shale ash from two different sources was investigated as a partial cement replacement, while recycled short carbon fibres (rCFs) were incorporated to enhance the functional properties of mortar composites. Results showed that carbonate-rich ash promoted the formation of higher amounts of monocarboaluminate (a crystalline hydration product in cement chemistry), leading to a refined pore structure and increased volumes of reaction products—primarily calcium silicate hydrates (C–S–H, critical compounds for cement strength). The findings indicate that the mineralogical composition of the modified binder (the mixture that holds solid particles together in mortar), rather than the fibre content, is the dominant factor in achieving a dense microstructure. This, in turn, enhances resistance to water ingress and improves mechanical performance under long-term hydration and freeze–thaw exposure. Life cycle assessment (LCA, a method to evaluate environmental impacts across a product's lifespan) further demonstrated that combining complex binders with rCFs can significantly reduce the environmental impacts of cement production, particularly in terms of global warming potential (−4225 kg CO_2 eq), terrestrial ecotoxicity (−1651 kg 1,4-DCB), human non-carcinogenic toxicity (−2280 kg 1,4-DCB), and fossil resource scarcity (−422 kg oil eq). Overall, the integrative use of OSA and rCF presents a sustainable alternative to conventional cement, aligning with principles of waste recovery and reuse, while providing a foundation for the development of next-generation binder systems.

Keywords: cement-based structural composites; metakaolin; oil shale ash; recycled carbon fiber; waste wind turbine blades

1. Introduction

The cement industry is a major partner in European construction activities. This is evident in the large European cement market, valued at over USD 19.02 billion in 2023.

It had a production volume of 181.17 million metric tons in 2022. With future development ambitions, the European market is expected to reach USD 31 billion by 2033 and 202.83 million metric tons by 2028 [1,2]. However, high cement production leads to large carbon dioxide emissions, estimated at 7% of global emissions and 4% within the European Union [3]. The climate impacts per ton range from 832 to 1075 kg CO_2-equivalents [4]. In addition to CO_2, cement production releases hydrochloric acid, fluorhydric acid, and air pollutants. These emissions cause acid rain [5–7]. To address these challenges, major European cement producers plan to use more renewable electricity and partially replace cement in building materials [8]. In this context, oil shale ash (OSA) has recently been utilised as a substitute for cement. OSA is produced by burning oil shale, which generates approximately 40–70% ash, a significant amount in the context of pollution [9]. This is due to its silicon dioxide (SiO_2) and aluminium oxide (Al_2O_3) content, which reacts with calcium hydroxide ($Ca(OH)_2$) from cement hydration. This reaction forms more calcium silicate hydrates (CSH), improving the physical and mechanical properties of cement matrices [10–12]. OSA is widely available as a solid residue from oil shale combustion, with limited engineering uses or usually landfilled [13]. Using OSA as a building material helps reduce environmental problems and increases its economic value [14]. Accordingly, OSA has been used as a partial cement substitute in concrete, cement paste, mortar, and pavement [15–18]. Results showed that adding 10–30% OSA to composites slightly lowers their performance compared to cement-based materials. Reductions are in the ranges of 18–42% for compressive strength, 27–37% for splitting strength, 3–61% for flexural strength, and 3–61% for modulus of elasticity [10]. Additional supplementation is needed to offset these reductions.

Metakaolin (MK) is a supplementary material rich in silicate and aluminate, promoting pozzolanic reactions [19]. It improves pore structure and reduces permeability and porosity, resulting in enhanced mechanical strength, durability, and corrosion resistance. MK is synthesised from natural sources [20–22]. When incorporated into the binder, MK reacts with portlandite (CH) to generate additional CSH through a synergistic effect [23,24]. Despite OSA's and MK's strong environmental and sustainability potential, their elasticity needs improvement [25,26]. Reinforcing cementitious composites with short fibres is common for this purpose. Many fibre types are used, classified as synthetic or natural fibres (like wood) [27–32]. Among them, synthetic carbon fibres have high strength, elasticity, and thermal and chemical stability. These properties slow down failure and improve mechanical performance [33]. However, their high cost limits widespread adoption [34]. Alternatives must be found to produce or recycle them more cheaply. Thus, recycled fibres from fibre-reinforced polymer (FRP) waste are gaining attention. They are less expensive and offer acceptable performance [35–37]. However, cementitious matrices made with recycled wet fibres (resin-blended) degrade over time. This is due to the alkaline activity of cement, which affects long-term performance [38,39]. Therefore, the resin part of recycled wet fibres should be removed before using them as fillers. The pyrolysis process has recently been utilised to decompose the resin in FRP waste [40,41]. It has proven highly efficient.

Wind turbine blade waste (WTB) consists of FRP with high fibre content, about 70 wt.%. Its abundance is estimated at 43 million tons by 2050 [42,43]. Pyrolysis and its catalytic process can decompose WTB resin into high-value chemicals, such as phenol and aromatics [44–46]. The fibres remain as solid residue due to high thermal resistance. The purity of recycled fibres improves after water washing and oxidation. Studies show this method recovers pure glass and carbon fibres. These fibres enhance compressive strength and reduce mortar absorption and water content [47–49]. Thus, WTB offers a cheap, renewable source for short carbon fibres (rCFs) via pyrolysis with post-treatment oxidation. This study aims to enhance the environmental and mechanical properties of cement mortar by incorporating eco-friendly additives. It uses OSA, MK, and rCFs from WTB pyrolysis. The effects on hydration phase composition, pore size, specific surface area, compressive and flexural strength, freeze–thaw cycling, and water absorption were observed over a period of up to 120 days. The study also evaluated the environmental performance of advanced composites compared to conventional materials.

2. Experimental

2.1. Materials

In this research, Portland cement (PC: CEM Grade I 42.5R) conforming to EN 197-1 [50] was used. MK powder was obtained by calcining kaolin at 800 °C for 2 h. Two batches of oil shale ash (OSA) were supplied by the Eesti and Auvere power plants in Estonia [9,51]. Ashes were taken from electrostatic precipitators [9]. Sand in a (0.0–2.0) mm fraction and a superplasticiser (conforming to EN934-3 [52]) were also used. Detailed information on the chemical and phase composition of the raw oil shale can be found in [9,53]. At Eesti Plant, OSA1 was produced by burning only oil shale. This OSA had a high sulfate ($CaSO_4$) content and a specific gravity of 2946 kg/m^3. At Auvere Plant, OSA2 with a high carbonate ($CaCO_3$) content and a specific gravity of 2756 kg/m^3 was produced by burning oil shale mixed with biomass [9]. The oxide composition of PC, MK, and OSA binders is summarised in Table 1 [9,51,53]. According to our previous study [9], both ashes contain calcium-bearing minerals, such as anhydrite, larnite, calcite, lime, and portlandite, together with a high content of orthoclase and quartz. More detailed analysis is presented in [9]. The rCF used in this study was extracted from WTB, which is composed of carbon fibre-reinforced unsaturated polyester resin, through pyrolysis and post-treatments. The pyrolysis process was performed on chopped WTB at 500 °C in a laboratory reactor to decompose the resin fraction. Afterwards, rCF was separated and purified from the solid residue by sieving and oxidation in air at 450 °C as detailed in [47].

Table 1. Chemical composition of binder components.

Sample	CaO	SiO_2	Al_2O_3	Fe_2O_3	TiO_2	P_2O_5	K_2O	MgO	Na_2O	SO_3	MnO	LOI
PC	63.72	13.82	3.85	1.75	0.18	0.08	0.33	1.01	0.23	2.02	0.12	2.66
MK	0.16	47.80	40.1	0.39	1.06	0.51	0.49	-	-	-	-	0.51
OSA1	31.95	23.51	7.54	4.46	0.52	0.18	4.23	4.60	0.42	9.50	0.06	6.42
OSA2	38.77	22.07	7.90	3.93	0.53	0.15	3.84	3.53	0.16	4.52	0.05	9.75

2.2. Design of the Experiments

Figure 1 presents the current workflow diagram for the composite preparation and characterisation in this research. Usta et al. 2020 highlighted the potential of OSA as a binder [11], prompting the use of 6% MK and 20% OSA as cement replacements in this study. The control sample (M) used PC (94%) and MK (6%) to assess the effect of the two ashes. The process began with preparing cement pastes (water-to-cement ratio of 0.5) using various binders (A–C), omitting inert rCF and sand to isolate the impact of OSA on hydration products via microscopic analysis (D). Next, sand and rCF were added in different proportions to create rCF-reinforced OSA/MK/PC mortars (E–G) to study their mechanical properties (H). Finally, the environmental performance of the composites was assessed (I). Each stage will be detailed in the following sections, in the order listed below.

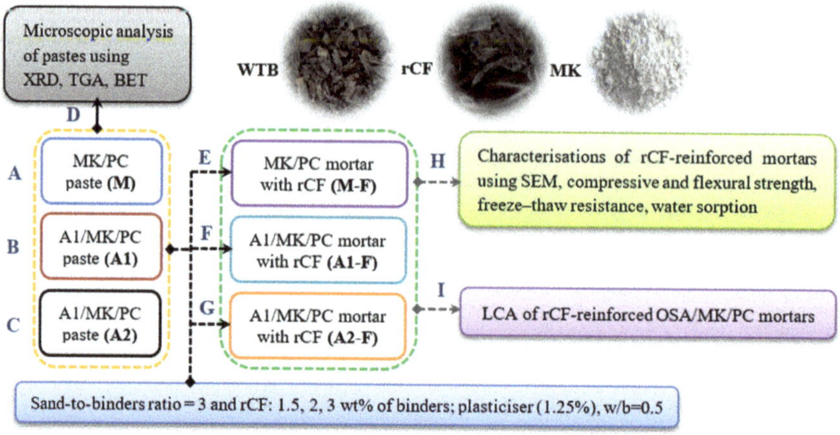

Figure 1. Flowchart of the present work, including the sample preparation and characterisation.

2.3. Preparation of Composites

To prepare the cement mortars, rCF (F: 1.5, 2, and 3 wt.% of binders) and a sand-to-binder ratio of 3 were added to the homogeneous mixtures of M, A1, and A2. This produced M-F, A1-F, and A2-F mortars. The rCF ratio was chosen based on the most common ratio used in the literature [54,55]. Since OSA and MK consume more water than cement [10,53,56], a superplasticiser was used to keep the slump within the 155–170 mm range. The M-F mortars were prepared by mixing PC, MK, and sand particles using a planetary mixer for 5 min (dry mixing). Then, water and plasticiser were added, and mixing continued for another 5 min (wet mixing). After that, rCF was added gradually with the specified content. Mixing continued for an additional 5 min at a slow speed. This step helped distribute the fibres evenly without breaking or damaging their surfaces [48]. The homogeneous mixture was poured into cubic (20 × 20 mm) and prismatic (20 × 20 × 100 mm) metal moulds and then shaken. These samples were prepared for compressive and flexural tests [57]. The moulds were cured for 24 h, followed by hydration for 28 and 120 days in lime water. This prevented additional carbonation and ensured consistent results across samples [58]. A1-F and A2-F mortars were made using the same procedures as M-F mortar. However, for these, the binder OSA was added to PC and MK and mixed under dry conditions first. Fibres were added in wet conditions, followed by the preparation of specimens for compressive and flexural tests as described above. After 28 days of hydration, one set of samples underwent freeze–thawing at temperatures of (−18 + 23 °C) [9]. Other samples were tested for mechanical strength and water absorption. Finally, each sample was given a code based on its composition, as shown in Table 2.

Table 2. Compositions and codes of rCF-reinforced cement mortars.

Components	Control (M) Groups			A1 Groups			A2 Groups		
	M-1.5F	M-2F	M-3F	A1-1.5F	A1-2F	A1-3F	A2-1.5F	A2-2F	A2-3F
PC (wt.%)	94	94	94	74	74	74	74	74	74
MK (wt.%)	6	6	6	6	6	6	6	6	6
A1 (wt.%)	-	-	-	20	20	20	-	-	-
A2 (wt.%)	-	-	-	-	-	-	20	20	20
rCFs (wt.%)	1.5	2	3	1.5	2	3	1.5	2	3
Plasticiser (wt.%) of binder content					1.25				
Binder-to-sand ratio					1:3				

2.4. Characterizations of the Fabricated Cement Composites

The effect of different OSAs on the crystalline phases and hydration products of cement paste was investigated using X-ray diffraction (XRD) with a BRUKER D8 ADVANCE diffractometer (Bruker, Germany). The XRD analysis was performed on A1 and A2 pastes for up to 120 days. The content of hydration products and chemically bound water in the pastes was determined using thermogravimetric analysis (TGA: LINSEIS STA PT1600, (Linseis, Germany) in nitrogen ambient at 20 °C/min. The specific surface area and pore size of the pastes were measured using a Brunauer–Emmett–Teller (BET) apparatus AUTOSORB-iQ-K/MP (Quantachrome, USA) based on the nitrogen physisorption approach. The deposition of hydration products on rCF surfaces was investigated using scanning electron microscopy (SEM). The effect of OSA composition type and rCF concentration on the mechanical strength (compressive and flexural) of cement mortars was investigated using a Zwick Roell universal testing instrument (ZwickRoell group, Germany) at a rate of 0.5 mm/min, in accordance with the European guidelines EN 196-1 [59]. The effect of these additives on the water absorption of mortars was tested according to EN 1015-18 standard [48,60]. The sorptivity (A) of the mortars was determined by plotting the linear relationship between capillary water absorption (i) and the fourth root of time (t0.25) for each sample. The slope was calculated to estimate the value of A using the formula (i = A t0.25) [9]. The frost resistance test of the mortars was performed in accordance with LST L 1413.11 standard [61]. The hydrated specimens were taken from saturated lime water, air-dried for 5 min, placed in a freezer at −17 °C for 4 h, and then immersed in a bath of tap water at 20 °C for 20 h. This freeze–thaw (F-T) cycle was repeated 25 times.

2.5. Life Cycle Assessment of rCF-Reinforced OSA/MK/PC Mortar

In this section, the impact of using MK and OSA as partial cement substitutes and rCF as a reinforcing material for cement mortar on its environmental performance was studied. The evaluation was conducted using a life cycle analysis (LCA) with SimaPro software. The analysis followed ISO 14040 and 14044 standards, using a cradle-to-gate approach [62]. Based on these guidelines, the goal was to study the potential environmental impacts of rCF-reinforced OSA/MK/PC mortar production. The study considered all stages of the production process, required materials, and electricity. The functional unit (FU) was defined as the production of 1 m^3 of rCF-reinforced OSA/MK/PC mortar within the European region, which served as the geographical context [63]. This study focused solely on mortar composite production, excluding the construction and end-of-life phases. Inventory data for the evaluation process were gathered from experimental results obtained during preparation and characterisation. These experiments produced mortar composites with maximum strength. Data for main materials—PC, MK, sand, OSA, rCF, water, and plasticiser—were extracted directly from SimaPro's default Ecoinvent database. Both types of OSA were defined in the LCA based on their composition, listed in Table 1, and MK was defined as calcined kaolin. Because rCF is not available in the Ecoinvent database and only CF-reinforced polymer is, rCF was defined as graphite (its main component). According to the results, A1-3F (with high sulfate content) and A2-3F (with high carbonate content) provided the highest strength. The compositions of both were considered the main scenarios and then compared with typical mortar and OSA/MK/PC mortar. Based on these scenarios, an LCA layout for each was constructed, along with the corresponding constraints, as shown in Figure 2. The evaluation continued by identifying the most frequent impact categories, which helped determine the technical and ecological potentials of the new composites. The study assumed that all materials used in production were available on-site, so transportation was not considered. Finally, input data, including materials and electricity consumption, were collected and are presented in Table 3 [64–68].

According to Diaz-Basteres [64], producing one ton of mortar consumes 4 kWh/ton for dry mixing and the same amount for wet mixing. The electricity required to produce 1 m^3 of mortar composite was calculated in this research after adjusting for the specified functional unit.

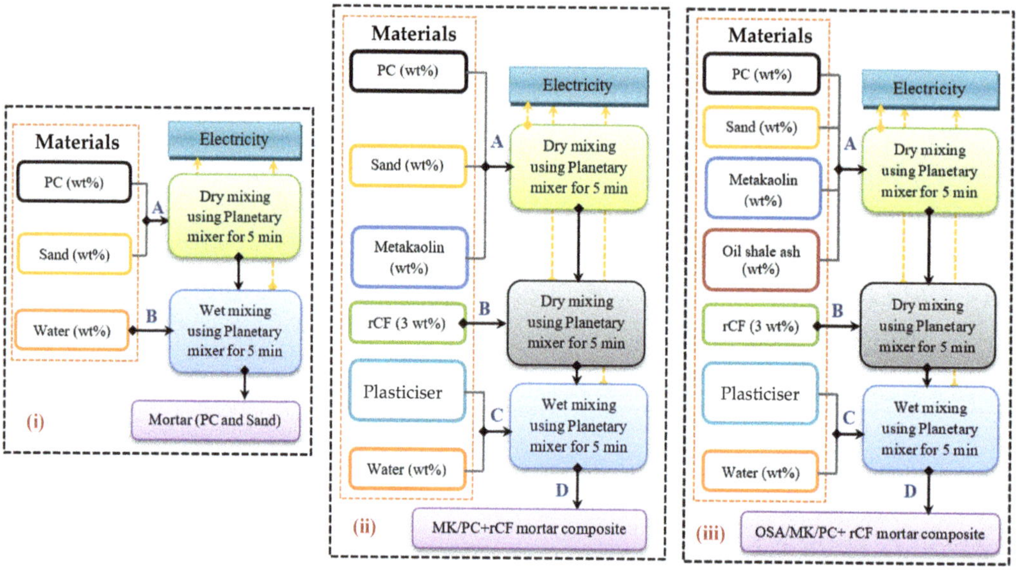

Figure 2. LCA layout of (i) mortar, (ii) OSA/MK/PC mortar composite, and (iii) OSA/MK/PC + rCF mortar composite production scenarios.

Table 3. Inventory data required to produce 1 m^3 (FU) of mortar composites.

Input Materials	Estimated Value per FU		
	Mortar	**MK/PC + rCF Mortar**	**OSA/MK/PC + rCF Mortar (A1-3F and A2-3F)**
Portland cement (PC)	496.2	473.11	297.32
Sand	1488.6	1503.36	1200.67
Metakaolin (MK)	----	28.57	22.82
Oil shale ash (OSA)	----	----	80.54
Recycled short carbon fibre (rCFs)	----	9.05	9.05
Plasticiser	----	8.98	10.08
Water	248.1	255.46	209.40
Total mass	2233	2279	1830
Electricity for dry mixing (kwh/FU) [64]	8.93	8.93	8.93
Electricity for wet mixing (kwh/FU) [64]	8.93	8.93	8.93

3. Results and Discussions

3.1. XRD Analysis of OSA/MK Cement Pastes

Since rCF is an inert material and does not participate in chemical reactions, the cement pastes without fibre and sand filler were analysed by X-ray after 28 and 120 days, and patterns are presented in Figure 3. As mentioned earlier, the A1 and A2 pastes were prepared with a complex binder consisting of ash, MK, and cement. As cement and OSA contain many minerals of similar origin [9], the mineralogical composition of Portland cement-MK-OSA composites differs little from that of conventional hydrated cement. Significant differences arise due to the introduction of ash minerals, such as CaO, $Ca(OH)_2$,

$CaCO_3$, and $CaSO_4$, involved in the hydration reactions. Cement substitution with ash, which possesses a higher $CaSO_4$ content (>14%) [9], influences hydration, as an increase in $CaSO_4$ content accelerates the alite (C_3S) reaction. In contrast, the incorporation of ash, characterised by a higher $CaCO_3$ and CaO content [9], may contribute to monocarboaluminate formation (Mc) if enough aluminate is present. Both ashes contain a low amount of aluminium, and most of it is present in a K-feldspar compound that is hardly soluble. Thus, it may participate in the hydration reactions only at a later stage. Therefore, pozzolanic materials, such as MK, are needed to activate the hydration reactions. Replacing PC with OSA does not change the overall mineralogical composition of the hydrated samples after 28 and 120 days (Figure 3). All cementitious phases, including AFt, Mc, CH, alite (C_3S), belite (C_2S), and calcite, together with silicon or magnesium oxides, are found in both compositions (A1 and A2 pastes). Nevertheless, the intensity of the phases, particularly AFt, Mc, and CH, differs significantly. Contrary to plain systems with ash [9], the incorporation of MK results in an increase in both AFt and Mc content, although their intensity changes throughout the hydration process. It was shown in [69–71] that the sulfate and calcite content both variably affect the hydration of clinker, leading to changes in AFt and Mc formation. It is known that the presence of calcite destabilises the AFm in favour of AFt and Mc [69], whereas a larger volume of AFt is stabilised by the presence of sulfate [69]. MK, on the other hand, consumes portlandite to produce an additional CSH phase, and Mc in the presence of calcite, if CH content is in excess [51,72,73]. Due to the presence of MK in the A1 system, which has a higher sulfate content, the content of AFt should increase, and Mc should decrease with time, as indicated in [74]. However, Figure 4 shows that the AFt content decreases in the A1 system, whereas it increases in the A2 system, which possesses a higher content of carbonates. The precipitation of ettringite requires additional calcium from the system, which may be supplied by portlandite through the hydration of C3S [74]. The lower peak of CH and increase in Mc in the A1 system after long-term hydration confirm observations presented in [74], that the alumina from metakaolin participates in the formation of ettringite until the gypsum and $CaSO_4$ are depleted (Figure 3). On the contrary, the Mc and CH peaks increase throughout the hydration in the A2 system (Figure 3), confirming the synergistic effect of MK and calcite mentioned in [71]. The low intensity of peaks at 32–33° (2θ°) suggests the higher dissolution rate of C_3S and/or C_2S, thus explaining an increase in CH content in the A2 system after long-term hydration. Consequently, an excess of CH facilitates the MK reaction with carbonates, increasing Mc content.

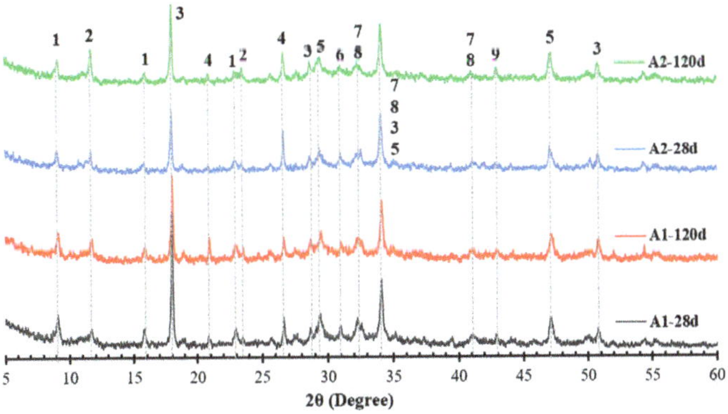

Figure 3. XRD of A1 and A2 pastes, composed of blended binder (OSA/MK/PC) after 28 and 120 hydration days. 1—AFt, 2—Mc, 3—CH, 4—SiO_2 (Quartz), 5—$CaCO_3$, 6—$CaMg(CO_3)_2$, 7—alite (C_3S), 8—belite (C_2S), and 9—MgO.

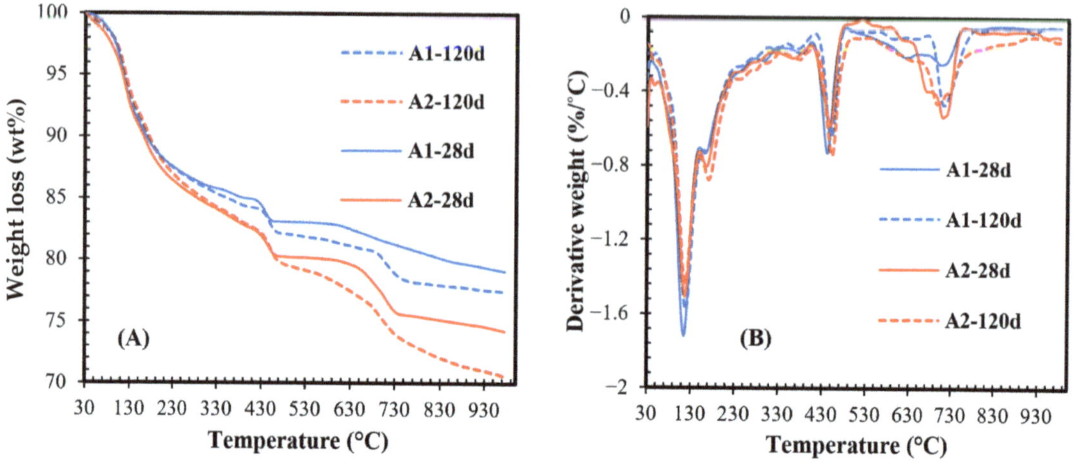

Figure 4. TG (**A**) and DTG (**B**) curves of A1 and A2 pastes, composed of blended binder (OSA/MK/PC), after 28 and 120 hydration days.

3.2. Thermogravimetric Analysis of OSA/MK Cement Pastes

Figure 4 shows the TGA and derivative thermogravimetry (DTG) curves of the A1 and A2 pastes after 28 and 120 days. Based on these measured data, the content of the hydration products and chemically bound water was calculated [51] and is summarised in Table 4. As mentioned above, both types of OSA initially contain CH [9], which is one of the main hydration products of cement paste. This means that only a relative comparison can be made when evaluating changes in the assemblage of reaction products in compositions with ash. The TGA curves (Figure 4A) show that the mass losses up to 450 °C in both samples do not change significantly with increasing hydration time, while they increase in the higher temperature range. Dehydration of CSH and AFt occurs at temperatures of up to 170 °C, while Mc and CH decompose at approximately 190 °C and 450 °C, respectively (Figure 4B). The decomposition of carbonates occurs at a broader temperature range from 700 to 790 °C, whereas the broader peak of carbonates in A2 paste suggests the partial carbonation of hydrates or decomposition of less crystalline forms of carbonates [75]. The calculated data in Table 4 shows that the bound water increases over time for both samples, indicating the higher volume of reaction products, primarily CSH, formed (Figure 4B). The CSH is an amorphous material; therefore, it is not observed in the XRD patterns (Figure 3). According to the TGA results, both compositions give nearly the same CH content, even though the intensity of CH in XRD curves differs among samples after 28 days. Nevertheless, after long-term hydration, the XRD data correlate with the TGA data, as the CH content decreases in A1 paste and increases in A2 paste. Moreover, the CH content is lowest in A1 and highest in A2 (up to 49%). The content of CC is also higher in the A2 set (up to 63%) and nearly doubles after long-term hydration. According to the findings reported by Hargis et al. (2024) [76], calcite may form rhombohedral crystals less than 5 μm in size, contributing to the formation of a more compact structure. Since the volume of hydration products is higher in A2 paste, this suggests that the pore structure in A2 was affected more than that in A1. To verify this, the nitrogen physisorption technique was used to measure the size of pores in the range of up to 50 nm, representing small capillary pores.

Table 4. Bound water, CH, and CC content of A1 and A2 pastes at different ages from TGA data.

Sample	A1 Paste		A2 Paste	
Hydration time	28 days	120 days	28 days	120 days
Bound water (%)	20.58	21.89	24.66	25.85
Portlandite (%)	12.54	10.67	12.30	15.94
Carbonate (%)	11.05	12.51	14.53	20.43

3.3. Nitrogen Gas Physisorption Analysis of OSA/MK Cement Pastes

Figure 5 shows the average pore diameter and specific surface area (SBET) of the A1 and A2 pastes. After short-term hydration (28 days), the pore diameter of the A1 paste becomes about twice that of the A2 paste. Over time, as long-term hydration (120 days) progresses, the difference between the samples decreases significantly. Notably, the average pore diameter in A2 (8.75 nm) is in the range of CSH gel pores [77], whereas for A1 it is slightly higher than 10 nm, which is characteristic of small capillary pores. Meanwhile, the specific surface area (SBET) increased by approximately 2.6 and 1.5 times for the A1 (80.4 m^2/g) and A2 (97.2 m^2/g) pastes, respectively. This increase correlates well with the TGA data, since the wb of A2 is higher than that of A1, implying the formation of a higher content of additional CSH in A2 due to chemical reactions with MK.

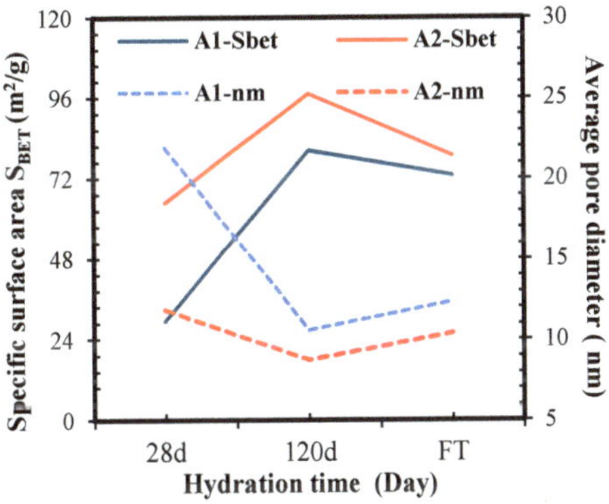

Figure 5. Effect of hydration time on pore size and specific surface area of A1 and A2 pastes, composed of blended binder (OSA/MK/PC).

3.4. Morphology of rCF-Reinforced OSA/MK/PC Mortars

The SEM observation process was performed on the fracture surface of several rCF-reinforced OSA/MK/PC mortars. These mortars had different compositions of OSA and varying amounts of rCF. The scanning process focused on the surface features of the embedded fibres in the matrices. It was noticed that the fibres had the same features in all samples, regardless of the type of OSA and rCF content. Therefore, the most characteristic morphology is presented in Figure 6. SEM cross-section images of the interface between rCF and mortar binder (OSA/MK/PC) after 120 days of hydration (Figure 6A) reveal good integration of rCF in the bulk material. The surface of the fibre is completely covered with abundant hydration products, as shown in Figure 6B. In contrast, the upper side of the fibre is almost smooth (Figure 6C) due to the detachment of the hydration products during sample preparation. However, this indicates that the rCF (Figure 6B) acts as a crystallisation centre for hydration products. The adhesion to the fibre is weak due to the absence of chemical reactions between them [78,79]. The growth of hydration products on the rCF surface may contribute to reducing the micropore size, thus obstructing water ingress [9].

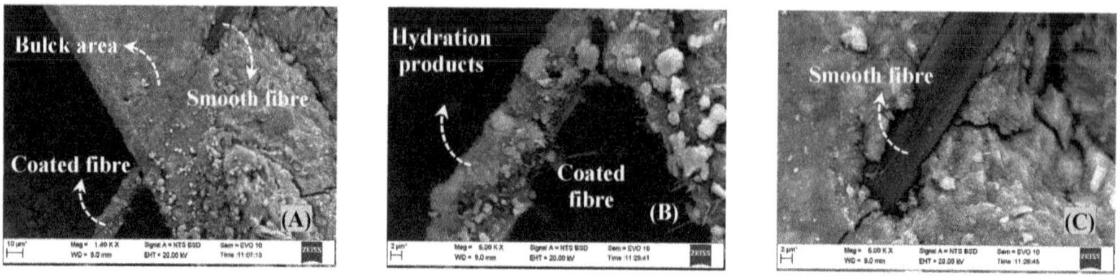

Figure 6. SEM images of the rGF interface in OSA/MK/PC mortar composites. (**A**) general view; (**B**,**C**) enlarged view of (**A**).

3.5. Compressive Properties of rCF-Reinforced OSA/MK/PC Mortars

The typical uniaxial compressive stress–strain curves of the M-F, A1-F, and A2-F mortars after 28 and 120 hydration days are presented in Figure 7. As shown, the ascending part (elastic region) of the curves was approximately the same for all mortars, indicating that the type of binder and rCF do not affect the behaviour of the composite at this stage. When the curves deviate from linearity, the propagation of internal microcracks begins. Furthermore, it is clear that the higher the rCF content, the broader the descending part of the curve. This stress reduction indicates the coalescence of multiple microcracks into macrocracks, while the enhanced ductility behaviour is shown by compositions with higher rCF content [80,81]. Conversely, samples with lower rCF content show a steeper descending branch.

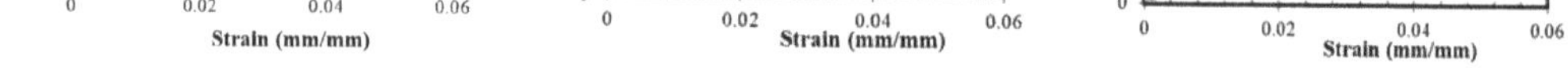

Figure 7. Typical stress–strain curves of mortars with different contents of rCF under uniaxial compression: (**A**), M-F (**B**), A1-F, and (**C**) A2-F mortars.

Based on the compressive stress–strain curves, the average compressive strengths of the M-F, A1-F, and A2-F mortars were calculated, and their values are summarised in Figure 8A, B after 28 and 120 days, respectively. As shown, the average compressive strength of the M-F, A1-F, and A2-F mortars after 28 days was estimated in the ranges of 28.5 MPa, 25.8 MPa, and 31.8 MPa, respectively. These values did not change significantly after a long-term hydration period and remained approximately in the same range, where the curing time and environment can contribute to the strength development only [57]. Additionally, the dependence of strength on fibre content is similar in systems with ash, although it is irreversible compared to the reference one (M). However, these values are lower than those of the reference paste without fibres (42.97 $\pm$ 1.42 MPa), indicating that rCF delays matrix hydration, a phenomenon also reported by Zhao et al. (2024) and Heo et al. (2020) [55,81]. Also, the A1-3F and A2-3F mortars (with higher rCF content) resulted in the highest strength (28.0 MPa and 33.8 MPa, respectively), contrary to the control mortar (M-F), reaching its highest strength with an intermediate content of rCF (30.1 MPa). Nonetheless, the A1-F group was obviously the weakest among the samples. Although the scattering of results within the groups suggests stress concentrations, possibly arising due to the nonuniform distribution of fibres [53,79], a continuous increase in strength with an increment of fibre content was observed after 120 days, as shown in Figure 8B. The increase in strength for the M group was around 15% and is consistent with results presented in [54,82,83]. The A1-F mortars showed the highest strength increment from 8 to 25% (up to 35.1 MPa), whereas the A2-F mortars demonstrated only a 5–9% increase (up to 35.3 MPa). The overall results reveal that the system with OSA, possessing a higher content of sulfates (A1), requires a longer hydration time to achieve a strength comparable to the reference mortar. Moreover, the A1 set is more sensitive to rCF content than the reference and A2 groups, thus implying that the impact of the binder's phase composition prevails. Based on the TGA data (Table 3), the A2 composition contains a higher content of carbonates. The compressive strength of pure calcium carbonate is a few times lower than that of its mixture with aragonite due to the orthorhombic shape of aragonite crystals, the interlocking and interpenetration of which mainly contribute to the strength gain [76]. This implies that the development of strength in mortars with OSA depends mainly on the amount and transformation of hydrates that contribute to the refinement of the pore structure.

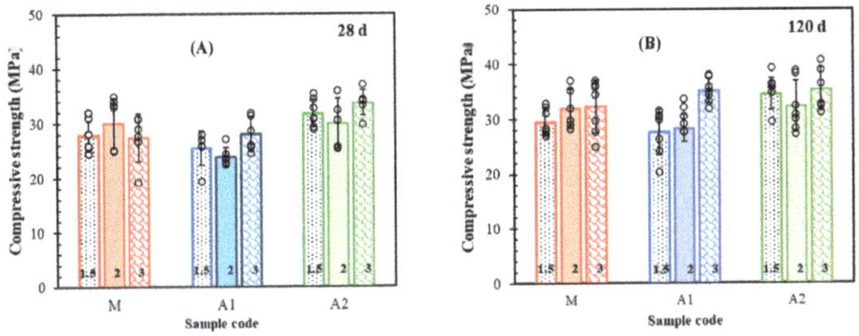

Figure 8. Impact of rCF content on compressive strength of mortars after (**A**) 28 days and (**B**) 120 days of hydration.

Finally, the elastic modulus of the M-F, A1-F, and A2-F mortars was obtained from the linear elastic phase of the compressive stress–strain curves and is displayed after 28 days (Figure 9A) and 120 days (Figure 9B) of hydration, and after freeze–thawing (F-T) (Figure 9C). The data show that both the hydration time and binder type only slightly affect elastic modulus, with values ranging from 1.5 to 1.7 GPa, likely because microfibres do not influence the elastic behaviour before crack formation. While the increased rCF content does not reveal a clear trend in the elastic modulus prior to F-T cycles, differences between samples become apparent after freeze–thawing (Figure 9C): A1-F mortars display a higher modulus than A2-F mortars. This suggests that the presence of OSA—associated with a higher sulfate content—contributes to a more rigid matrix, which may better resist material deterioration during freeze-thaw cycles.

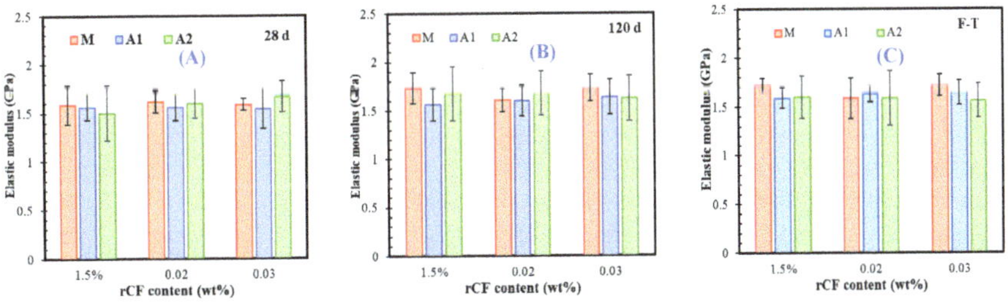

Figure 9. Effect of rCF content on elastic modulus of mortars after (**A**) 28 days, (**B**) 120 days, and (**C**) freeze–thawing process.

3.6. Flexural Properties of rCF-Reinforced Mortars

Figure 10A,B present typical load–displacement curves in flexure for the M-F, A1-F, and A2-F mortars after 28 and 120 days, respectively. Flexure curves after freeze–thawing are also included (Figure 10C). As expected, the incorporation of a higher rCF content results in a better response to flexural load, increasing the displacement. The load peak increases with an increase in hydration time and reduces after freeze–thawing. The average flexural strength results after 28 and 120 days are presented in Figure 11. It is worth mentioning that the addition of commercial carbon fibres results in 6 to 9 MPa flexural strength after 28 days [82]. In this study, the application of rCF falls within this range, as the reference composition (M) yields a value of 6.5 MPa (Figure 11A). Regardless of the binder type, the strength of all compositions increases as the fibre content rises. The increment is relatively low, around 6% for the M and A2 compositions, whereas the A1 system, being the lowest, demonstrates the highest increase of 18%. The impact of rCF content on strength development becomes stronger after long-term hydration (Figure 11B), particularly in the A1 system, resulting in a 32% increase. On the contrary, the A2 composition demonstrates only a 5% increase, although it remains the strongest among all groups. It should be noted that in the A2 sample set, a slight reduction in flexural strength was observed at intermediate fibre content, likely due to the large scattering of the results, possibly caused by nonuniform distribution or poor bonding of fibres with the matrix, which reduces the load transfer effect mentioned by others [54,55]. Nevertheless, data show that binder type is a sensitive factor at lower fibre content, while at higher fibre volumes, the strength gain depends mainly on the fibre's ability to redistribute stress and prevent crack growth [54].

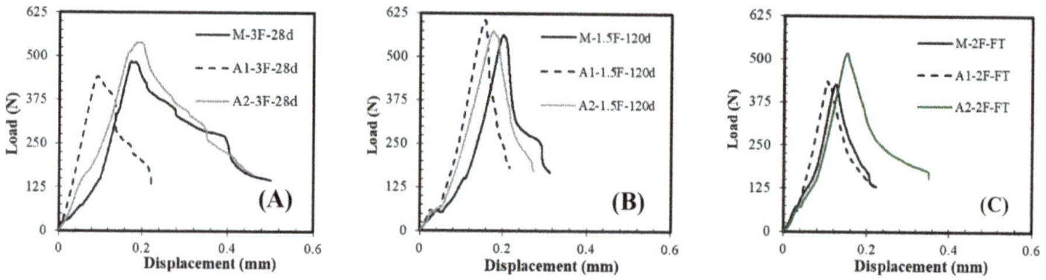

Figure 10. Typical load–displacement curves of mortars with different contents of rCF under flexure after (**A**,**B**) 28 and 120 days and (**C**) freeze–thawing.

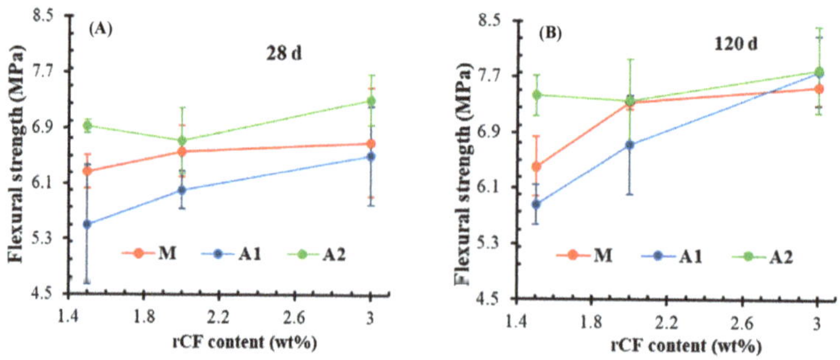

Figure 11. Impact of rCF content on flexural strength of mortars after (**A**) 28 and (**B**) 120 days.

3.7. Freeze–Thaw Cycling

After freeze–thaw cycling, the significance of the binder type becomes more apparent. Figure 12 shows flexural and compressive strength results after freeze–thawing. The results reveal that the type of binder and fibre content are crucial to the frost resistance of composites. Compared with samples hydrated for 28 days (Figure 11A), the flexural strength of all samples reduced by three to 15% after freeze–thawing (Figure 12A). Interestingly, only the A2 set, being the highest group (Figure 11A), showed a weakening effect with an increase in fibre content after freeze–thaw cycling (Figure 12A). Contrastingly, in comparison with 28-day hydrated samples (Figure 8A), the compressive strength (Figure 12B) increased (7–18%) for the sets with intermediate or the highest fibre dosages; however, a few compositions showed weakening (up to 10%) in strength. Nevertheless, the comparison of data obtained from long-term hydration (Figure 8B) reveals that the compressive strength of all sets reduced after freeze–thawing. The reduction in the A1 set is higher than in the A2 group, although the ranges are similar, at 6–15% and 9–13%, respectively. However, the A2 set is less sensitive to fibre content than other compositions. This implies that continuous hydration took place during the freeze thawing process, resulting in an increase in strength and thus inferring the binder type as a more sensitive factor considering reinforcement with rCF.

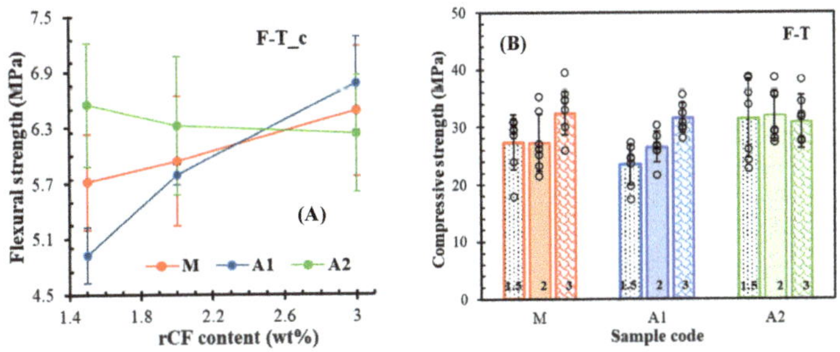

Figure 12. Impact of fibre content on (**A**) flexural and (**B**) compressive strength of mortars after freeze–thawing.

3.8. Resistance to Water Absorption and Sorptivity

The density of cement-based material depends on the compactness of hydration products, which reduces the structure's porosity and obstructs water's ingress into deeper layers [83]. Microcapillary pores (less than 50 nm) are mainly involved in capillary suction, whereas large pores or voids contribute little to capillary transport [51]. However, they may increase the average path of water migration and, thus, slow down the water ingress. On the other hand, recovered carbon fibres (rCFs) serve as micro-reinforcements, as their diameter ranges from 10 to 20 µm, while their length is less than 10–15 mm. However, it was shown in [83,84] that incorporating fibres of different origins increases the ingress rate due to poor interfacial bonding between fibre and cement matrix. Figure 13 shows the water intake rates for samples with the highest dosage of rCF. The linear increment is seen for all groups after 28 days and after freeze–thawing for A1 and A2. As the hydration time increases, water absorption gradually declines from a linear relationship. Although the water ingress rate is very similar among samples, the A2 group is the least dependent on the curing conditions. The sorptivity coefficients, calculated from the linear part of the curves obtained during the first six hours after immersion, are summarised in Figure 14. In general, the sorptivity data are irrelevant to the trend of compressive strength (Figures 8 and 12B), except for the reference group with intermediate fibre content after the freeze–thawing test. The data in Figure 14 reveals that the impact of fibre content is marginal. However, the effect of binder type is important, as the sorptivity of the A2 group remains at the same level after all curing conditions. In contrast, the A1 set, having the lowest strength, exhibits sorptivity comparable to that of the A2 set after long-term hydration and freeze–thawing. Moreover, the freeze-thawing had an insignificant effect on the sorptivity of both sets A1 and A2 compared with long-term hydration. It is worth mentioning that the sorptivity of compositions without MK was up to 40% higher [9]. This implies that the contribution of chemical reactions between ash and MK changes the content of hydration products and increases the contact area between rCF and matrix [82].

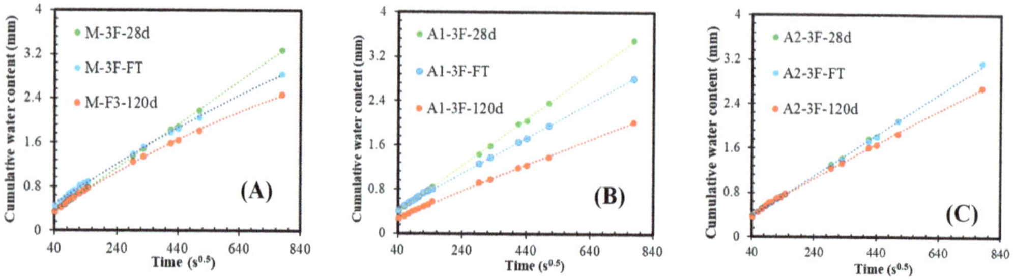

Figure 13. The impact of curing conditions and binder type on the capillary absorption rate of the (**A**) M-F, (**B**) A1-F, and (**C**) A2-F mortars.

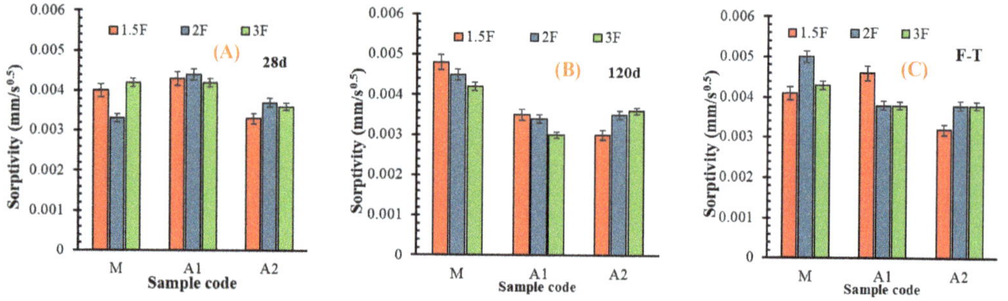

Figure 14. Sorptivity of fibre-reinforced mortars after (**A**) 28 days, (**B**) 120 days, and (**C**) freeze–thawing.

3.9. Environmental Impact Assessment

The environmental impact categories of mortar, OSA1/MK/PC, and OSA2/MK/PC + rCF composite production are summarised in Table 5. Their contributions are presented in Figure 15. A slight increase in environmental impacts was observed when mixing mortar with OSA and MK additives (MK/PC + rCF composite scenario) by up to 32%. Mineral resource scarcity increased by up to 73%. This is due to the use of plasticisers, which were absent in the case of mortar, as well as the energy consumed in producing MK (mineral source) by kaolin calcination. In the case of the OSA/MK/PC + rCF composite, very promising results were obtained in all categories compared to the mortar and MK/PC + rCF composite scenarios. These included global warming (−3749 and −4225 kg CO_2 eq), terrestrial ecotoxicity (−1651 and −1567 kg 1,4-DCB), human non-carcinogenic toxicity (−2280 and −1844 kg 1,4-DCB), and fossil resource scarcity (−387 and −422 kg oil eq). These significant reductions resulted from eliminating the sending of rCF to landfills and avoiding its emissions. Additionally, the amount of cement used in composite materials was significantly reduced. Furthermore, the total mass of the composite decreased significantly from 2233 kg (mortar) to 1830 kg (OSA2/MK/PC + rCF). This reduced the energy required to produce its components and its emissions as well. Finally, it was noted that the composition of OSA does not affect the environmental impacts of the composite. Based on these results, OSA, MK, and rCF have promising potential to enhance the environmental performance of mortar. This makes it more environmentally friendly and reduces its environmental burden.

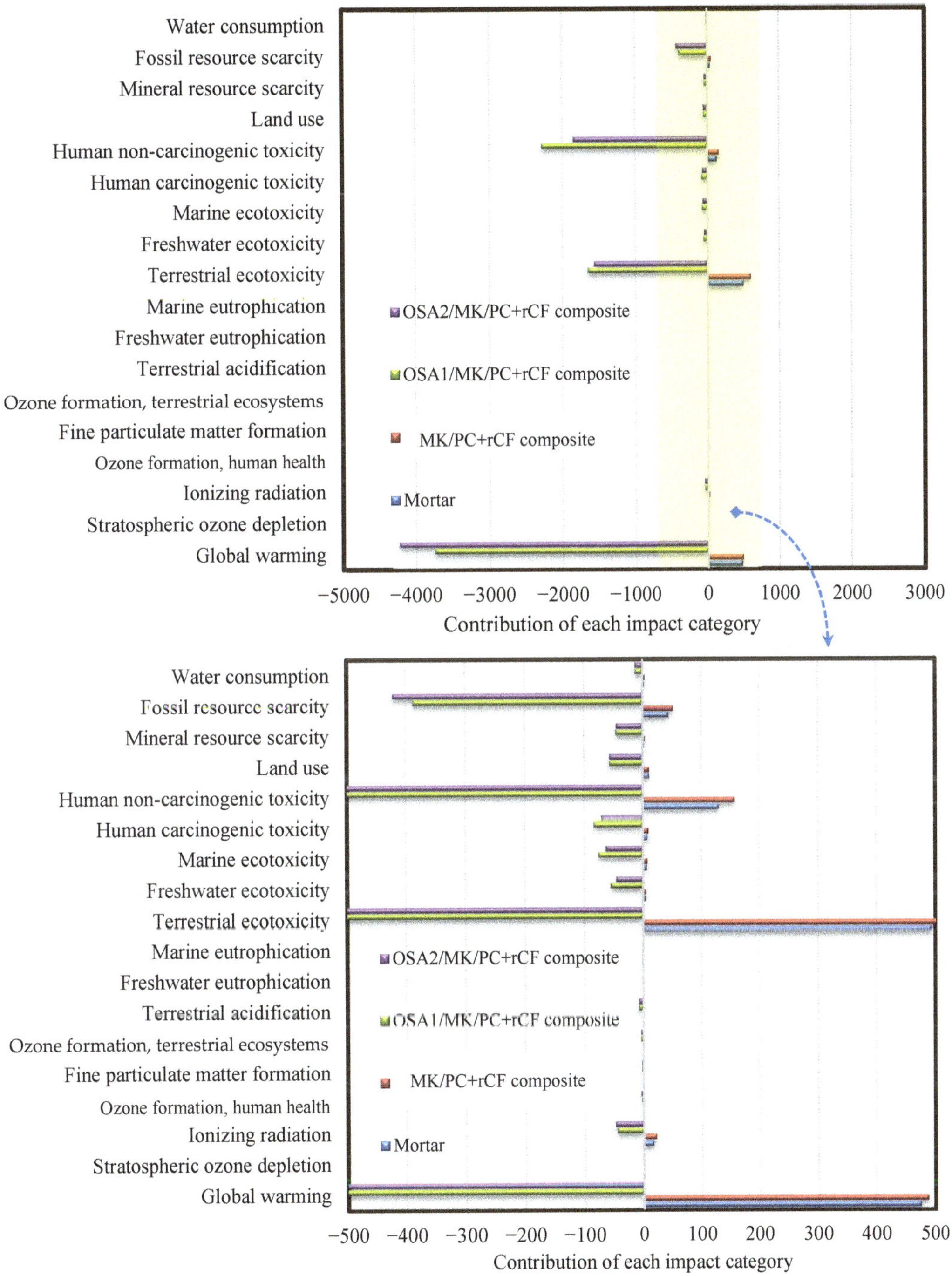

Figure 15. Distribution of impact categories for each production scenario.

Table 5. Environmental impact categories of production of mortar, OSA/MK/PC, and OSA/MK/PC + rCF composites.

Impact Category	Unit	Mortar	MK/PC + rCF Composite	OSA1/MK/PC + rCF Composite	OSA2/MK/PC + rCF Composite
Global warming	kg CO_2 eq	474.8	488.6	−3748.7	−4224.7
Stratospheric ozone depletion	kg CFC11 eq	0.0	0.0	0.0	0.0
Ionising radiation	kBq Co-60 eq	17.3	23.0	−45.5	−48.9
Ozone formation, human health	kg NOx eq	0.8	0.9	−3.7	−3.9
Fine particulate matter formation	kg PM2.5 eq	0.3	0.3	−2.5	−2.5
Ozone formation, terrestrial ecosystems	kg NOx eq	0.9	0.9	−4.0	−4.3
Terrestrial acidification	kg SO_2 eq	0.8	0.8	−7.5	−7.6
Freshwater eutrophication	kg P eq	0.1	0.1	−0.2	−0.2
Marine eutrophication	kg N eq	0.0	0.0	0.0	0.0
Terrestrial ecotoxicity	kg 1,4-DCB	493.5	600.6	−1651.1	−1567.3
Freshwater ecotoxicity	kg 1,4-DCB	4.6	5.8	−54.9	−45.9
Marine ecotoxicity	kg 1,4-DCB	6.2	7.8	−75.8	−63.5
Human carcinogenic toxicity	kg 1,4-DCB	8.3	9.4	−84.0	−70.6
Human non-carcinogenic toxicity	kg 1,4-DCB	129.9	157.4	−2279.6	−1844.4
Land use	m^2a crop eq	10.9	11.4	−57.3	−57.0
Mineral resource scarcity	kg Cu eq	2.5	4.3	−45.8	−45.3
Fossil resource scarcity	kg oil eq	45.0	52.9	−387.4	−421.7
Water consumption	m^3	3.1	3.5	−13.5	−13.0

4. Conclusions

This study examined the partial replacement of cement with 20% sulfate-rich or carbonate-rich oil shale ash (OSA), combined with 6% metakaolin (MK). Recycled short carbon fibres (rCFs) from the pyrolysis of wind turbine blades (WTBs) were added to advanced cement mortar. Their effects on compressive and flexural strength were monitored during extended hydration and after freeze–thaw cycles. The effects of MK, OSA, and rCF on cementitious phase development, hydration products, specific surface area, mechanical properties, and composite sorptivity were systematically assessed. Life cycle assessment (LCA) was also conducted. XRD analysis showed that interaction with MK increased monocarboaluminate (Mc) in mixtures where cement was replaced with carbonate-rich OSA (A2 group). This led to a refined pore structure and greater volumes of reaction products, primarily calcium silicate hydrates (C-S-H), as confirmed by nitrogen physisorption and thermogravimetric analysis (TGA). The results demonstrated that binder mineralogy, rather than fibre content, most strongly influenced the formation of a dense microstructure, improving water resistance and maintaining mechanical performance after long-term hydration and freeze–thaw cycles. However, a higher rCF content (3 wt.%) proved most effective in producing a strong, frost-resistant composite when cement was substituted by OSA and MK, due to the fibres' ability to distribute stress and inhibit crack growth. Finally, LCA results indicated that OSA, MK, and rCF have promising potential to improve the environmental performance of mortar across most impact categories.

Author Contributions: Conceptualization, R.K.-D. and S.Y.; Methodology, R.K.-D.; Software, I.S.; Validation, R.K.-D. and S.Y.; Formal analysis, R.K.-D. and A.B.; Investigation, R.K.-D. and I.S.; Resources, R.K.-D. and S.Y.; Data curation, I.S.; Writing—original draft, R.K.-D. and S.Y.; Writing—review & editing, R.K.-D. and S.Y.; Supervision, S.Y. All authors have read and agreed to the published version of the manuscript.

Funding: This project has received funding from the Research Council of Lithuania (LMTLT), agreement No. S-MIP-23-118.

Data Availability Statement: The original contributions presented in this study are included in the article. Further inquiries can be directed to the corresponding author.

Acknowledgments: The authors extend their gratitude to partners from Riga Technical University (Latvia) and the University of Tartu (Estonia) for the ash supply. J. Česnienė is acknowledged for her input with TGA and XRD analysis.

Conflicts of Interest: The authors declare no conflicts of interest.

References

1. Available online: https://www.statista.com/topics/12721/cement-industry-in-europe/#topicOverview (accessed on 20 October 2025).
2. Available online: https://www.statista.com/statistics/1488014/market-size-of-cement-industry-europe/ (accessed on 20 October 2025).
3. Available online: https://publications.jrc.ec.europa.eu/repository/handle/JRC131246 (accessed on 20 October 2025).
4. Cavalett, O.; Watanabe, M.D.B.; Voldsund, M.; Roussanaly, S.; Cherubini, F. Paving the way for sustainable decarbonization of the European cement industry. *Nat. Sustain.* **2024**, *7*, 568–580. [CrossRef]
5. Mohamad, N.; Muthusamy, K.; Embong, R.; Kusbiantoro, A.; Hashim, M.H. Environmental impact of cement production and Solutions: A review. *Mater. Today Proc.* **2022**, *48*, 741–746. [CrossRef]
6. Voicu, G.; Ciobanu, C.; Istrate, I.A.; Tudor, P. Emissions Control of Hydrochloric and Fluorhydric Acid in cement Factories from Romania. *Int. J. Environ. Res. Public Health* **2020**, *17*, 1019. [CrossRef] [PubMed]
7. Sudhakar, C.V.; Reddy, G.U. Impacts of cement industry air pollutants on the environment and satellite data applications for air quality monitoring and management. *Environ. Monit. Assess.* **2023**, *195*, 840. [CrossRef]
8. Sambataro, L.; Bre, F.; Ukrainczyk, N.; Koenders, E.A. Environmental benchmarks for the European cement industry. *Sustain. Prod. Consum.* **2024**, *45*, 429–449. [CrossRef]
9. Kalpokaitė-Dičkuvienė, R.; Pitak, I.; Baltušnikas, A.; Česnienė, J.; Kriūkienė, R.; Lukošiūtė, S.I. Functional and microstructural alterations in hydrated and freeze–thawed cement-oil shale ash composites. *Case Stud. Constr. Mater.* **2023**, *19*, e02302. [CrossRef]
10. Ashteyat, A.M.; Al Rjoub, Y.S.; Obaidat, A.T.; Kirgiz, M.; Abdel-Jaber, M.; Smadi, A. Roller Compacted Concrete with Oil Shale Ash as a Replacement of Cement: Mechanical and Durability Behavior. *Int. J. Pavement Res. Technol.* **2022**, *17*, 151–168. [CrossRef]
11. Usta, M.C.; Yörük, C.R.; Hain, T.; Paaver, P.; Snellings, R.; Rozov, E.; Gregor, A.; Kuusik, R.; Trikkel, A.; Uibu, M. Evaluation of new applications of oil shale ashes in building materials. *Minerals* **2020**, *10*, 765. [CrossRef]
12. Silva, J.V.S.; Reis, E.D.; de Azevedo, R.C.; Poggiali, F.S.J. Towards eco-friendly cement-based materials: A review on incorporating oil shale ash. *Discov. Civ. Eng.* **2024**, *1*, 23. [CrossRef]
13. Al-Thunibat, I.M.; Al-Harahsheh, A.M.; Aljbour, S.H.; Shawabkeh, A. Chemical and mechanical properties of Attarat (Jordan) Oil shale ash and its engineering viable options. *Solid Fuel Chem.* **2023**, *57*, 138–146. [CrossRef]
14. Alaloul, W.S.; Al Salaheen, M.; Malkawi, A.B.; Alzubi, K.; Al-Sabaeei, A.M.; Musarat, M.A. Utilizing of oil shale ash as a construction material: A systematic review. *Constr. Build. Mater.* **2021**, *299*, 123844. [CrossRef]
15. Alaloul, W.S.; Al Salaheen, M.; Alzubi, K.; Musarat, M.A. Utilizing calcined and raw fly oil shale ash in the carbonation process of OPC cement-paste and mortar. *Case Stud. Constr. Mater.* **2024**, *20*, e02945. [CrossRef]
16. Wei, H.; Zhang, Y.; Cui, J.; Han, L.; Li, Z. Engineering and environmental evaluation of silty clay modified by waste fly ash and oil shale ash as a road subgrade material. *Constr. Build. Mater.* **2019**, *196*, 204–213. [CrossRef]
17. Nov, S.; Barak, S.; Cohen, H.; Knop, Y. Treated oil shale ashes as cement and fine aggregates substitutes for the concrete industry. *ACS Omega* **2023**, *8*, 46608–46613. [CrossRef] [PubMed]
18. Rabab'ah, S.R.; Sharo, A.A.; Alqudah, M.M.; Ashteyat, A.M.; Saleh, H.O. Effect of using Oil Shale Ash on geotechnical properties of cement-stabilized expansive soil for pavement applications. *Case Stud. Constr. Mater.* **2023**, *19*, e02508. [CrossRef]
19. Siddique, R.; Klaus, J. Influence of metakaolin on the properties of mortar and concrete: A review. *Appl. Clay Sci.* **2009**, *43*, 392–400. [CrossRef]
20. Dong, Y.; Pei, L.; Fu, J.; Yang, Y.; Liu, T.; Liang, H.; Yang, H. Investigating the mechanical properties and durability of Metakaolin-Incorporated mortar by different curing methods. *Materials* **2022**, *15*, 2035. [CrossRef]
21. Cai, R.; Tian, Z.; Ye, H.; He, Z.; Tang, S. The role of metakaolin in pore structure evolution of Portland cement pastes revealed by an impedance approach. *Cem. Concr. Compos.* **2021**, *119*, 103999. [CrossRef]
22. Homayoonmehr, R.; Ramezanianpour, A.A.; Mirdarsoltany, M. Influence of metakaolin on fresh properties, mechanical properties and corrosion resistance of concrete and its sustainability issues: A review. *J. Build. Eng.* **2021**, *44*, 103011. [CrossRef]
23. Sun, X.; Zhao, Y.; Tian, Y.; Wu, P.; Guo, Z.; Qiu, J.; Xing, J.; Xiaowei, G. Modification of high-volume fly ash cement with metakaolin for its utilization in cemented paste backfill: The effects of metakaolin content and particle size. *Powder Technol.* **2021**, *393*, 539–549. [CrossRef]
24. Sujjavanich, S.; Suwanvitaya, P.; Chaysuwan, D.; Heness, G. Synergistic effect of metakaolin and fly ash on properties of concrete. *Constr. Build. Mater.* **2017**, *155*, 830–837. [CrossRef]

25. Barabanshikov, Y.; Pham, T.H.; Akimov, S. Effect of metakaolin and MC adhesive additives on the mechanical properties of concrete. In *Sustainable Energy Systems: Innovative Perspectives, Proceedings of the SES 2020, Saint-Petersburg, Russia, 29–30 October 2020*; Lecture Notes in Civil Engineering; Springer: Cham, Switzerland, 2020; pp. 300–309. [CrossRef]
26. Hosseinzadeh, M.; Dehestani, M.; Hosseinzadeh, A. Exploring elastic properties of fly ash recycled aggregate concrete: Insights from multiscale modeling and machine learning. *Structures* **2023**, *59*, 105720. [CrossRef]
27. Herrmann, H.; Schnell, J. Short Fibre Reinforced Cementitious Composites and Ceramics. In *Advanced Structured Materials*; Springer: Cham, Switzerland, 2019. [CrossRef]
28. Pakravan, H.R.; Ozbakkaloglu, T. Synthetic fibers for cementitious composites: A critical and in-depth review of recent advances. *Constr. Build. Mater.* **2019**, *207*, 491–518. [CrossRef]
29. Guo, P.; Bao, Y.; Meng, W. Review of using glass in high-performance fiber-reinforced cementitious composites. *Cem. Concr. Compos.* **2021**, *120*, 104032. [CrossRef]
30. Carloni, C.; Bournas, D.A.; Carozzi, F.G.; D'Antino, T.; Fava, G.; Focacci, F.; Giacomin, G.; Mantegazza, G.; Pellegrino, C.; Perinelli, C.; et al. Fiber Reinforced Composites with Cementitious (Inorganic) Matrix. In *Design Procedures for the Use of Composites in Strengthening of Reinforced Concrete Structures: State-of-the-Art Report of the RILEM Technical Committee 234-DUC*; Springer: Dordrecht, The Netherlands, 2015; pp. 349–392.
31. Abedi, M.; Hassanshahi, O.; Rashiddel, A.; Ashtari, H.; Meddah, M.S.; Dias, D.; Arjomand, M.; Choong, K.K. A sustainable cementitious composite reinforced with natural fibers: An experimental and numerical study. *Constr. Build. Mater.* **2023**, *378*, 131093. [CrossRef]
32. Kampragkou, P.; Kamperidou, V.; Stefanidou, M. Evaluation of Hydrothermally Treated Wood Fibre Performance in Cement Mortars. *Fibers* **2024**, *12*, 21. [CrossRef]
33. Aduwenye, P.; Chong, B.W.; Gujar, P.; Shi, X. Mechanical properties and durability of carbon fiber reinforced cementitious composites: A review. *Constr. Build. Mater.* **2024**, *452*, 138822. [CrossRef]
34. Yang, W.; Huang, Y.; Li, C.; Tang, Z.; Quan, W.; Xiong, X.; He, J.; Wu, W. Damage prediction and long-term cost performance analysis of glass fiber recycled concrete under freeze-thaw cycles. *Case Stud. Constr. Mater.* **2024**, *21*, e03795. [CrossRef]
35. Danish, A.; Mosaberpanah, M.A.; Salim, M.U.; Amran, M.; Fediuk, R.; Ozbakkaloglu, T.; Rashid, M.F. Utilization of recycled carbon fiber reinforced polymer in cementitious composites: A critical review. *J. Build. Eng.* **2022**, *53*, 104583. [CrossRef]
36. Lamba, N.; Raj, R.; Singh, P. Feasibility of recycled Carbon Fiber-Reinforced polymer fibers in cementitious composites: An experimental investigation. *Arab. J. Sci. Eng.* **2024**, *49*, 13577–13591. [CrossRef]
37. Wang, Y.; Zhang, S.; Luo, D.; Shi, X. Effect of chemically modified recycled carbon fiber composite on the mechanical properties of cementitious mortar. *Compos. Part B Eng.* **2019**, *173*, 106853. [CrossRef]
38. Won, J.-P.; Lee, S.-J.; Kim, Y.-J.; Jang, C.-I.; Lee, S.-W. The effect of exposure to alkaline solution and water on the strength–porosity relationship of GFRP rebar. *Compos. Part B Eng.* **2008**, *39*, 764–772. [CrossRef]
39. Liao, D.; Gu, T.; Liu, J.; Chen, S.; Zhao, F.; Len, S.; Dou, J.; Qian, X.; Wang, J. Degradation behavior and ageing mechanism of E-glass fiber reinforced epoxy resin composite pipes under accelerated thermal ageing conditions. *Compos. Part B Eng.* **2023**, *270*, 111131. [CrossRef]
40. Chen, R.; Xu, X.; Lu, S.; Zhang, Y.; Lo, S. Pyrolysis study of waste phenolic fibre-reinforced plastic by thermogravimetry/Fourier transform infrared/mass spectrometry analysis. *Energy Convers. Manag.* **2018**, *165*, 555–566. [CrossRef]
41. Ren, Y.; Xu, L.; Han, Z.; Xiao, S.; Sun, Y.; Nan, Z.; Shu, J.; Li, L.; Shen, Z. Study on recycling carbon fibers from carbon fiber reinforced polymer waste by microwave molten salt pyrolysis. *Fuel* **2024**, *377*, 132819. [CrossRef]
42. Yousef, S.; Eimontas, J.; Striūgas, N.; Abdelnaby, M.A. Pyrolysis kinetic behaviour and thermodynamic analysis of waste wind turbine blades (carbon fibres/unsaturated polyester resin). *Energy Sources Part A Recovery Util. Environ. Eff.* **2023**, *45*, 10505–10522. [CrossRef]
43. Liu, P.; Barlow, C.Y. Wind turbine blade waste in 2050. *Waste Manag.* **2017**, *62*, 229–240. [CrossRef]
44. Yousef, S.; Eimontas, J.; Zakarauskas, K.; Striūgas, N. Recovery of styrene-rich oil and glass fibres from fibres-reinforced unsaturated polyester resin end-of-life wind turbine blades using pyrolysis technology. *J. Anal. Appl. Pyrolysis* **2023**, *173*, 106100. [CrossRef]
45. Li, J.-H.; Jiang, H.; Chen, W.-W.; Wu, Y.-C.; Xu, M.-X.; Di, J.-Y.; Li, W.; Lu, Q. Selective production of phenol from the end-of-life wind turbine blade through catalytic pyrolysis. *Fuel* **2024**, *378*, 132877. [CrossRef]
46. Yousef, S.; Eimontas, J.; Striūgas, N.; Abdelnaby, M.A. Synthesis of value-added aromatic chemicals from catalytic pyrolysis of waste wind turbine blades and their kinetic analysis using artificial neural network. *J. Anal. Appl. Pyrolysis* **2023**, *177*, 106330. [CrossRef]
47. Yousef, S.; Eimontas, J.; Stasiulaitiene, I.; Zakarauskas, K.; Striūgas, N. Recovery of energy and carbon fibre from wind turbine blades waste (carbon fibre/unsaturated polyester resin) using pyrolysis process and its life-cycle assessment. *Environ. Res.* **2023**, *245*, 118016. [CrossRef]

48. Yousef, S.; Kalpokaitė-Dičkuvienė, R. Sustainable mortar reinforced with recycled glass fiber derived from pyrolysis of wind turbine blade waste. *J. Mater. Res. Technol.* **2024**, *31*, 879–887. [CrossRef]
49. Yousef, S.; Eimontas, J.; Zakarauskas, K.; Stasiulaitiene, I.; Striūgas, N.; Tuckute, S. Catalytic pyrolysis of wind turbine blades waste for plasticizers recovery and its life cycle assessment. *J. Environ. Manag.* **2025**, *395*, 127690. [CrossRef]
50. *EN 197-1:2000/A1:2004*; Cement—Part 1: Composition, Specifications and Conformity Criteria for Common Cements. CEN: Brussel, Belgium, 2000.
51. Kalpokaitė-Dičkuvienė, R.; Baltušnikas, A.; Levinskas, R.; Česnienė, J. Incinerator residual ash—Metakaolin blended cements: Effect on cement hydration and properties. *Constr. Build. Mater.* **2019**, *206*, 297–306. [CrossRef]
52. *EN 934-3:2009*; Admixtures for Concrete, Mortar and Grout—Part 3: Admixtures for Masonry Mortar—Definitions, requirements, Conformity and Marking and Labelling. CEN: Brussel, Belgium, 2009.
53. Leben, K.; Mõtlep, R.; Paaver, P.; Konist, A.; Pihu, T.; Paiste, P.; Heinmaa, I.; Nurk, G.; Anthony, E.J.; Kirsimäe, K. Long-term mineral transformation of Ca rich oil shale ash waste. *Sci. Total Environ.* **2019**, *658*, 1404–1415. [CrossRef]
54. Huang, H.; Zhang, Z.; Wu, Z.; Liu, Y. Reinforcement Effect of Recycled CFRP on Cement Based Composites: With a Comparison to Commercial Carbon Fiber Powder. *Struct. Durab. Health Monit.* **2024**, *18*, 409–423. [CrossRef]
55. Zhao, Y.; Zhang, J.; Qiang, S.; Lu, H.; Li, J. Effect of carbon fibers and graphite particles on mechanical properties and electrical conductivity of cement composite. *J. Build. Eng.* **2024**, *94*, 110036. [CrossRef]
56. Dhandapani, Y.; Joseph, S.; Bishnoi, S.; Kunther, W.; Kanavaris, F.; Kim, T.; Irassar, E.; Castel, A.; Zunino, F.; Machner, A.; et al. Durability performance of binary and ternary blended cementitious systems with calcined clay: A RILEM TC 282-CCL, review. *Mater. Struct.* **2022**, *55*, 145. [CrossRef]
57. Yousef, S.; Kalpokaitė-Dičkuvienė, R.; Baltušnikas, A.; Pitak, I.; Lukošiūtė, S.I. A new strategy for functionalization of char derived from pyrolysis of textile waste and its application as hybrid fillers (CNTs/char and graphene/char) in cement industry. *J. Clean. Prod.* **2021**, *314*, 128058. [CrossRef]
58. Huang, F.; Liu, J.; Li, X.; Li, C.; Hu, Z.; Shen, X.; Chen, B. Impact of silica fume on the long-term stability of cement-based materials with low water-to-binder ratio under different curing conditions. *Constr. Build. Mater.* **2024**, *450*, 138604. [CrossRef]
59. *EN 196-1:2005*; Methods of Testing Cement—Part 1: Determination of Strength. CEN: Brussel, Belgium, 2005.
60. *EN 1015-18:2002*; Methods of Test for Mortar for Masonry—Part 18: Determination of Water Absorption Coefficient Due to Capillary Action of Hardened Mortar. CEN: Brussel, Belgium, 2002.
61. *LST L 1413.11:2005*; Mortar. Test methods. Determination of Frost Resistance. LSD: Vilnius, Lithuania, 2005.
62. Santos, T.; Almeida, J.; Silvestre, J.; Faria, P. Life cycle assessment of mortars: A review on technical potential and drawbacks. *Constr. Build. Mater.* **2021**, *288*, 123069. [CrossRef]
63. Akbar, A.; Liew, K.M. Multicriteria performance evaluation of fiber-reinforced cement composites: An environmental perspective. *Compos. Part B Eng.* **2021**, *218*, 108937. [CrossRef]
64. Diaz-Basteris, J.; Rivero, J.C.S.; Menéndez, B. Life cycle assessment of restoration mortars and binders. *Constr. Build. Mater.* **2022**, *326*, 126863. [CrossRef]
65. Agrela, F.; Rosales, M.; Alonso, M.L.; Ordóñez, J.; Cuenca-Moyano, G.M. Life-Cycle Assessment and Environmental Costs of Cement-Based Materials Manufactured with Mixed Recycled Aggregate and Biomass Ash. *Materials* **2024**, *17*, 4357. [CrossRef] [PubMed]
66. Almeida, J.; Faria, P.; Ribeiro, A.B.; Silva, A.S. Life cycle assessment of mortars produced partially replacing cement by treated mining residues. *Appl. Sci.* **2021**, *11*, 7947. [CrossRef]
67. Onuaguluchi, O.; Mohamed, B.; Adwan, A.; Li, L.; Banthia, N. Sludge-derived biochar as an additive in cement mortar: Mechanical strength and life cycle assessment (LCA). *Constr. Build. Mater.* **2024**, *425*, 135959. [CrossRef]
68. Balea, A.; Fuente, E.; Monte, M.C.; Blanco, A.; Negro, C. Recycled Fibers for Sustainable Hybrid Fiber Cement Based Material: A Review. *Materials* **2021**, *14*, 2408. [CrossRef] [PubMed]
69. Zajac, M.; Rossberg, A.; Le Saout, G.; Lothenbach, B. Influence of limestone and anhydrite on the hydration of Portland cements. *Cem. Concr. Compos.* **2014**, *46*, 99–108. [CrossRef]
70. Adu-Amankwah, S.; Zajac, M.; Stabler, C.; Lothenbach, B.; Black, L. Influence of limestone on the hydration of ternary slag cements. *Cem. Concr. Res.* **2017**, *100*, 96–109. [CrossRef]
71. Adu-Amankwah, S.; Black, L.; Skocek, J.; Ben Haha, M.; Zajac, M. Effect of sulfate additions on hydration and performance of ternary slag-limestone composite cements. *Constr. Build. Mater.* **2018**, *164*, 451–462. [CrossRef]
72. Damidot, D.; Lothenbach, B.; Herfort, D.; Glasser, F. Thermodynamics and cement science. *Cem. Concr. Res.* **2011**, *41*, 679–695. [CrossRef]
73. Antoni, M.; Rossen, J.; Martirena, F.; Scrivener, K. Cement substitution by a combination of metakaolin and limestone. *Cem. Concr. Res.* **2012**, *42*, 1579–1589. [CrossRef]
74. Zunino, F.; Scrivener, K. The reaction between metakaolin and limestone and its effect in porosity refinement and mechanical properties. *Cem. Concr. Res.* **2021**, *140*, 106307. [CrossRef]

75. Saeki, N.; Kurihara, R.; Ohkubo, T.; Teramoto, A.; Suda, Y.; Kitagaki, R.; Maruyama, I. Semi-dry natural carbonation at different relative humidities: Degree of carbonation and reaction kinetics of calcium hydrates in cement paste. *Cem. Concr. Res.* **2024**, *189*, 107777. [CrossRef]
76. Hargis, C.W.; Chen, I.; Wang, Y.; Maraghechi, H.; Gilliam, R.J.; Monteiro, P.J. Microstructure development of calcium carbonate cement through polymorphic transformations. *Cem. Concr. Compos.* **2024**, *153*, 105715. [CrossRef]
77. Schlumberger, C.; Thommes, M. Characterization of Hierarchically Ordered Porous Materials by Physisorption and Mercury Porosimetry—A Tutorial Review. *Adv. Mater. Interfaces* **2021**, *8*, 2002181. [CrossRef]
78. Boris, R.; Antonovič, V.; Kerienė, J.; Stonys, R. The effect of carbon fiber additive on early hydration of calcium aluminate cement. *J. Therm. Anal. Calorim.* **2016**, *125*, 1061–1070. [CrossRef]
79. Akbar, A.; Kodur, V.; Liew, K. Microstructural changes and mechanical performance of cement composites reinforced with recycled carbon fibers. *Cem. Concr. Compos.* **2021**, *121*, 104069. [CrossRef]
80. Safiuddin, M.; Abdel-Sayed, G.; Hearn, N. Absorption and strength properties of short carbon fiber reinforced mortar composite. *Buildings* **2021**, *11*, 300. [CrossRef]
81. Heo, G.-H.; Park, J.-G.; Song, K.-C.; Jun, H.-M. Mechanical Properties of SiO_2-Coated Carbon Fiber-Reinforced Mortar Composites with Different Fiber Lengths and Fiber Volume Fractions. *Adv. Civ. Eng.* **2020**, *2020*, 8881273. [CrossRef]
82. Qiao, W.; Shi, M.; Ban, T.; Hou, B.; Zhang, W.; Kong, X. Using graphene oxide to enhance the bonding properties between carbon fibers and cement matrix to improve the mechanical properties of cement-based composites. *Constr. Build. Mater.* **2024**, *453*, 138992. [CrossRef]
83. Paktiawal, A.; Alam, M. Experimental evaluation of sorptivity for high strength concrete reinforced with zirconia rich glass fiber and basalt fiber. *Mater. Today Proc.* **2022**, *49*, 1132–1140. [CrossRef]
84. Vafaei, D.; Hassanli, R.; Ma, X.; Duan, J.; Zhuge, Y. Sorptivity and mechanical properties of fiber-reinforced concrete made with seawater and dredged sea-sand. *Constr. Build. Mater.* **2021**, *270*, 121436. [CrossRef]

materials

MDPI

Article

Mechanical Properties of Biochar–Sulfur Composites

Ewa Syguła [1], Monika Słupska [2], Maja Radziemska [1,*] and Andrzej Białowiec [3,4]

1 Institute of Environmental Engineering, Warsaw University of Life Sciences, Nowoursynowska 159, 02-776 Warsaw, Poland; ewa_sygula@sggw.edu.pl
2 Institute of Agricultural Engineering, Wroclaw University of Environmental and Life Sciences, 37 Chełmonskiego Str., 51-630 Wroclaw, Poland; monika.slupska@upwr.edu.pl
3 Department of Applied Bioeconomy, Wroclaw University of Environmental and Life Sciences, 37a Chełmonskiego Str., 51-630 Wroclaw, Poland; andrzej.bialowiec@upwr.edu.pl or andrzej.bialowiec@port.lukasiewicz.gov.pl
4 Lukasiewicz Research Network—PORT Polish Center for Technology Development, Stablowicka 147 Str., 54-066 Wroclaw, Poland
* Correspondence: maja_radziemska@sggw.edu.pl

Abstract

The study examines the mechanical strength of sulfur–biochar composites (SBCs), an underexplored area with potential for developing robust materials. Sulfur production, primarily from specialized extraction and waste generation in petroleum refining, yields about 70 million tons annually, necessitating efficient waste management. SBCs were produced using waste-derived biochar and elemental sulfur at varying sulfur contents (60–80%) and employing two fabrication methods: a muffle furnace and an electric burner. The mechanical performance of the composites was evaluated through strength and displacement measurements, with particular emphasis on the influence of processing method and sulfur content. The results demonstrate that both sulfur content and fabrication method significantly affect the mechanical behavior of SBCs. An increase in sulfur content led to a systematic improvement in ultimate strength for all samples. However, composites produced using the electric burner exhibited markedly higher ultimate forces and lower displacements compared to those fabricated in the muffle furnace, indicating superior strength and reduced brittleness. The enhanced performance is attributed to improved sulfur distribution and more effective infiltration of liquid sulfur into the porous biochar structure. These findings confirm the synergistic effect of combining sulfur with biochar and highlight the critical role of processing conditions in developing mechanically robust sulfur–biochar composites suitable for sustainable material applications.

Keywords: sulfur; biochar; mechanical properties; composite structure

1. Introduction

1.1. Application of Waste Sulfur in Materials

1.1.1. Sulfur Production

Sulfur is extracted through two sectors: specialized and attendant. The first sector relies on extracting native sulfur from primary deposits, primarily found in Iraq, the United States, Chile, Mexico, and Poland [1]. On the other hand, the second sector involves the production of waste sulfur during the purification and desulfurization processes of petroleum products [2]. The element is recovered through the Claus process, in which hydrogen sulfide decomposes into elemental sulfur [3]. Annual sulfur production is estimated to reach 70 million tons [4], with the first sector responsible for only 10% of extraction, resulting in the closure of an increasing number of mines [5]. The world is struggling to manage the enormous amounts of sulfur waste generated by the petrochemical industry. According to data from the U.S. Geological Survey, sulfur recovery from petrochemical processes increased by 19% from January to March 2022 compared to the previous year [6]. The United Arab Emirates is the largest sulfur exporter, accounting for over 95% of its production, representing 25% of global sulfur production, and it is predicted to double in the next 10–15 years [7]. However, the demand for sulfur will not increase at the same pace, which is why finding methods to manage this element is crucial. In the past, sulfur was primarily used as a medicine and was also utilized in the production of gunpowder and bleaching agents [6]. Currently, due to its numerous physical, chemical, and mechanical properties, sulfur offers a wide range of applications, including tire vulcanization, is an essential component in artificial fertilizers, and is increasingly used in the construction sector (sulfur-based concrete) [8], as well as replaces lithium in lithium-ion battery cathodes [4,9] and is used for water electrolysis for hydrogen production [10].

1.1.2. Mechanical Properties of Sulfur

Sulfur exhibits several allotropic forms: monoclinic, rhombic, and plastic, each varying in their state of matter and viscosity depending on prevailing temperatures [2]. Additionally, the mechanical properties of sulfur are influenced by thermal conditions and temperature. Its compressive strength is measured at 22 MPa, while its tensile strength is entirely dependent on cooling rate and temperature [6]. Faster cooling and higher target temperatures yield better mechanical strength parameters. Utilizing sulfur as an aggregate binder, concrete can achieve compressive strengths of up to 48 MPa, marking a 37% increase compared to traditional Portland cement (PC) concrete [11]. This improvement stems from the low porosity of sulfur-modified concrete in the absence of water. In traditional concrete, excessive water needed for cement hydration leads to void formation, thereby reducing mechanical strength. Moreover, studies have demonstrated that the addition of sulfur enhances tensile strength by 10% compared to PC. Numerous investigations corroborate the positive impact of sulfur addition on concrete's mechanical strength [9,11]. Furthermore, sulfur enhances concrete's resistance to acidic and saline environments, attributed to its lower calcium oxide content and higher aluminum oxide content relative to PC [10]. Sulfur properties are also harnessed in conjunction with polymers through inverse vulcanization. By adjusting sulfur and divinylbenzene and 1,3-diisopropenylbenzene ratios appropriately, materials with desired characteristics can be obtained. Samples with higher sulfur concentrations exhibit amorphous properties with low flexibility and high shape retention, while those with lower concentrations display low viscosity and high fluidity [12]. Incorporating sulfur into composites allows manipulation of their mechanical properties, leveraging sulfur's unique array of physicochemical properties [13]. In parallel with sulfur-based modification, modern concrete technology increasingly employs fiber reinforcement, including polypropylene (PP) fibers, to improve crack resistance and post-failure behavior. Recent studies have demonstrated that recycled PP fibers can effectively enhance the mechanical performance of cementitious materials, contributing to sustainability-oriented construction solutions. The potential combination of sulfur-modified concrete with polymer fiber reinforcement represents a promising, yet largely unexplored, research direction [14].

1.1.3. Usage of Waste Sulfur in Materials

Currently, sulfur finds application in various sectors, with the largest being the fertilizer industry. Biologically, sulfur serves as an essential element for plant organisms, offering direct nutritional value and enhancing the effectiveness of other necessary elements [15].

https://doi.org/10.3390/ma19030549

Studies have demonstrated that growth stimulators containing sulfur are readily absorbed by wheat seeds, hastening germination and ensuring early plant yield. Incorporating elemental sulfur into fertilizers can boost plant yield by up to 50% while also reducing soil pH and redox potential. In cases of arsenic-contaminated soil, sulfur diminishes its value by over 40% [16]. An important consideration in the use of sulfur-containing fertilizers is the form in which it is applied. Fertilizers containing sulfur in dry or solution form contribute to increased yields due to the nanostructure form of the element, facilitating its penetration into plants. Utilizing both forms of fertilizer positively impacts morphological parameters, plant height, and the vegetative period's development [17]. Another sector where sulfur finds utility is the construction industry. Sulfur serves as a binder in the production of concrete, referred to as sulfur–biochar composites (SBCs). Unlike traditional PC, SBC production does not necessitate the use of water or cement, which are major contributors to high CO_2 emissions [2]. Consequently, SBC yields material with twice the mechanical strength and reduced environmental impact. While the use of sulfur in construction materials offers numerous benefits, its application is limited due to its physical properties (melting point above 140 °C) [18]. Potential applications of SBCs include constructing dams, bridges, irrigation channels, or producing sewage pipes. Sulfur is commonly employed as a raw material for specialized polymers. The vulcanization process aims to produce durable tires. Sulfur's inclusion in this process enhances the physical properties of natural rubber. Additionally, the polymerization of sulfur and alkenes yields a chemical substance used for pollution control, monitoring, and remediation [19]. Polysulfides formed during the inverse vulcanization process using elemental sulfur and limonene compound from lemon peel can effectively remove mercury pollutants from water. Adding sulfur to polymers enhances their tensile strength and Young's modulus [20]. The increasing market demand is driving the exploration of solutions offering higher energy density. Lithium–sulfur batteries, boasting a high energy efficiency of 26 $kWh \cdot kg^{-1}$ and a large depth of discharge, present a promising avenue. However, their main drawback lies in the low electrical conductivity of the cathode and poor cycle life due to the dissolution of polysulfides in the electrolyte [21,22]. Similarly, sulfur may be used for porous nickel–sulfur electrodes, which may be applied for hydrogen generation from water electrolysis, especially with brackish water [10].

1.2. Application of Biochar in Materials

1.2.1. Biochar Production

Biochar is a product of the thermal decomposition of biomass under oxygen-free conditions. Mostly, it is obtained through pyrolysis, which can be categorized into slow or fast pyrolysis depending on specified parameters of time and temperature. Torrefaction, gasification, or hydrothermal carbonization (HTC) may also be a source of biochar. Pyrolysis occurs at temperatures ranging from 200 to 900 °C, during which cellulose, hemicellulose, and lignin undergo depolymerization and fragmentation reactions [23]. By limiting oxygen access, carbon combustion is prevented, leading to the production of pyrolytic gas, pyrolytic oil, and biochar [24]. The quantity and quality of the products obtained depend on the appropriate selection of process parameters and the feedstock used. Gasification of biomass primarily yields syngas, with biochar accounting for 10% of the feedstock [25]. Studies indicate that the amount of biochar produced decreases with increasing temperature [26]. However, the ash content, pH, porosity, and specific surface area of the resulting carbonization product increase [27,28]. Biochar produced at 750 °C exhibited a specific surface area of 540.36 $m^2 \cdot g^{-1}$, while biochar from a process set at 350 °C had a specific surface area of 1.44 $m^2 \cdot g^{-1}$ [29]. Both temperature and the duration of the

https://doi.org/10.3390/ma19030549

biomass process influence the characteristics of biochar. Increasing the process time has been found to decrease porosity and specific surface area [30].

1.2.2. Mechanical Properties of Biochar

The type of biomass used in the pyrolysis process significantly influences the physicochemical and mechanical properties of the resulting biochar. Researchers have demonstrated that biochar produced from sugarcane bagasse or rice husk exhibits a high nutrient content and electrical conductivity [31]. Utilizing carrot peels as raw material in the pyrolysis process yields carbonization with high adsorption potential. It has been shown that biochar derived from plant biomass has a high carbon content [32]. The reactor employed in the process can also impact the mechanical properties of biochar. Comparing the carbonization results from an HTC reactor (300 °C), a fixed bed horizontal tube furnace reactor (300 °C and 900 °C), and a fixed bed vertical batch reactor (300 °C and 900 °C), differences in Young's modulus hardness were observed. All biochars set at a temperature of 300 °C in different reactors displayed significant differences in Young's modulus hardness results. Biochar obtained from an HTC reactor exhibited the lowest result, at 143.1 MPa. The presence of volatile tars and trapped water after the process enhances the plasticity of biochar. In the vertical batch reactor, the temperature increase resulted in higher modulus yields, ranging from 1452 MPa to 3019 MPa. This phenomenon is because, at lower temperatures, the breakdown of cellulose results in a viscoplastic biochar. Interestingly, the trend observed in the horizontal reactor was the opposite. The carbonization process set at a higher temperature of 900 °C yielded a hard result of 1419 MPa, while the process set at 300 °C resulted in 3867 MPa. The type of reactor, with different N_2 flow rates, influenced this parameter [33]. Therefore, it is crucial to select appropriate process parameters and the reactor used. In other studies, it has been found that the addition of biochar to polylactic acid (PLA) increases the Young's modulus hardness of the resulting composite. This is attributed to the strong intermolecular interaction between biochar and PLA, where biochar molecules restrict the mobility of the PLA chain [34]. Additionally, it has been observed that the addition of biochar decreases the tensile strength of the composite, indicating an inability to transfer stress as the cause [35].

1.2.3. Usage of Waste Biochar in Materials

New applications of biochar are the subject of extensive research. Scientists unanimously agree on one aspect: that the feedstock and process methodology significantly impact the quality of the final product. Particularly, the mechanical and thermal properties, along with the sorption capabilities for various pollutants, are modified by the process [36–38]. Biochar is an environmentally friendly product and relatively inexpensive to obtain, with exceptional characteristics such as high adsorption capacity, specific surface area, microporosity, and ion exchange capacity. Manipulating the pyrolysis process allows the production of biochar with desired properties for various applications. Studies have demonstrated that using biochar as a soil amendment improves soil microbial activity by providing essential nutrients, neutralizing soil acidity, and positively influencing physicochemical properties such as pH, CEC, and water retention capacity [39,40]. Biochar can effectively absorb heavy metal ions present in soil and organic pollutants [41]. Due to its large specific surface area, hydrophobic properties, and porosity, biochar is successfully applied as an adsorbent for many pollutants in aquatic environments [42]. Heavy metal pollutants such as lead, zinc, copper, and cadmium can be efficiently adsorbed from water, and by selecting the appropriate biomass, their precipitation is also possible [43]. Research has also been conducted on the removal of nanoplastics from contaminated water. Biochar produced at higher temperatures with a specific surface area of $540.36\ m^2 \cdot g^{-1}$ removed up

 https://doi.org/10.3390/ma19030549

to 99% of pollutants [29]. The main research directions for biochar are its application in soil and water remediation, as mentioned earlier. However, more studies explore the possibility of utilizing biochar as a filler in composites and plastics. Scientists, as pioneers in investigating this idea, have proposed applying biochar in wood polymer composites (WPCs) and studying its properties. The hydrophobic nature of biochar combined with WPC reduces moisture absorption, thereby enhancing the composite's resistance to decay. Once again, it has been confirmed that the porous structure of biochar translates into improved properties. The more pores, the better the polymer can infiltrate, creating a mechanical interlock/scaffold [44]. Biochars characterized by a large specific surface area and high carbon content (up to 82%) exhibit high flexural and tensile strength, primarily due to the high process temperature. A sample with high ash content but low carbon content (up to 50%) and small specific surface area showed the highest strength modulus and the highest tensile strength. High electrical conductivity of the biochar enables its application for electrode production for water electrolysis and hydrogen production [45]. This publication is one of many supporting the thesis that biochar positively influences the mechanical properties of composites [46]. However, there is a study that presents the opposite findings. It compared pure PLA with PLA incorporating biochar, resulting in decreased flexural strength and impact strength. Although the authors confirmed the antithesis, they noted that sample preparation procedures may have influenced the results [47]. However, numerous similar applications of sulfur and biochar: agriculture, construction, batteries, water electrolysis, and many more, induce the concept of the synergistic combination of these two materials, leading to the strengthening of the desired properties and the mitigation of the weaknesses of both materials.

1.3. Current Research on Biochar–Sulfur Composite

Biochar–sulfur composites can also find application as soil amendments. The impacts of synthetic fertilizer and sulfur-enriched biochar (SulfaChar) on soybean and maize cultivation were compared. The biochar composite was produced by exposing it to biogas emissions from a landfill, where it adsorbed significant amounts of hydrogen sulfide present in the biogas. Improved nutrient uptake by maize plants was observed when SBC was applied. The addition of the composite increased the uptake of macronutrients such as nitrogen, phosphorus, potassium, and calcium and micronutrients, including zinc, manganese, and boron, by several percentage points compared to synthetic fertilizer. However, for soybeans, the addition of SulfaChar did not result in better results compared to fertilizer [48]. The authors were the first to confirm that hydrogen sulfide present in the composite contributed to plant growth, suggesting the use of SBC as a fertilizer. Biochar functionalized by sulfur has been successfully applied for soil and water remediation for efficient removal of metal ions, antibiotics, and dyes [49]. BSC may also be used as a bacteriocide [50] and for biogas production enhancement [51].

Previous studies on SBC have primarily focused on lithium–sulfur batteries. Lithium–sulfur batteries (Li-S) are relatively cheap compared to traditional lithium-ion batteries due to the abundance and surplus of sulfur, and they are more environmentally friendly as they do not contain toxic and heavy metals. Li-S batteries possess a high theoretical specific capacity and theoretical specific energy, but they exhibit low electrical conductivity [52]. For this reason, sulfur is often combined with materials that have high electrical conductivity, such as biochar [53,54]. Porous biochar, characterized by its large specific surface area, provides the necessary volume space for sulfur conversion to Li_2S_2 and Li_2S [46]. Studies have been conducted using a core–shell composite material consisting of a sulfonated polyaniline shell and a bamboo-derived porous carbon core loaded with sulfur as the cathode in Li-S batteries. The use of this composite material resulted in a high discharge

https://doi.org/10.3390/ma19030549

capacity of 1484 mAh·g^{-1} at 0.1 °C, high efficiency, and cyclic stability with a capacity of 853 mAh·g^{-1} maintained after 100 cycles at 0.1 °C [46,48]. Also, new trends in the application of sulfur-doped biochars for water electrolysis are under development [13].

The use of biochar and sulfur as fillers/binders in composites is a practice under development. Biochar, primarily derived from waste materials, is an environmentally friendly product, while sulfur is readily available and inexpensive due to its role in the petrochemical industry. Considering the development of new SBCs and their various emerging applications, an extremely important aspect is the durability of the composites produced.

1.4. Aim of Study

This study investigates the mechanical strength of SBC, an underexplored research area with potential for developing durable and sustainable materials. SBCs were produced using different sulfur contents and fabrication methods to evaluate the influence of processing conditions on strength and deformation behavior. Mechanical strength testing was conducted to identify the key factors governing composite performance and to assess potential application areas for the developed materials. It was hypothesized that both sulfur content and fabrication method significantly influence the mechanical strength and deformation behavior of sulfur–biochar composites.

2. Materials and Methods

2.1. Materials

2.1.1. Biochar

Apple wood chips (Browin, Lodz, Poland) were utilized to produce SBC. The preparation of apple chip biochar was conducted in a muffle furnace (SNOL, 8.1/1100, Utena, Lithuania) at a temperature of 500 °C with a holding time of 1 h. The temperature ramp-up was 10 °C·min^{-1}. The process was performed in the absence of oxygen due to the application of CO_2, with the flow rate 0.6 mL·min^{-1}. Subsequently, the biochar was ground to a mesh size of 1 mm to match the form of the sulfur for later use [8].

2.1.2. Sulfur

The crystalline sulfur was supplied by the ORLEN Poland SA group (Płock, Poland). Initially, the sulfur existed as a single agglomerate, which was subsequently crushed and ground in a laboratory mortar. The resulting sulfur was in powder form, with a mesh size of 1.0 mm, facilitating its mixing with biochar while ensuring high homogeneity [10].

2.2. Methods

2.2.1. Production of SBC

The ground test materials underwent additional sifting through a 1 mm gradation sieve to remove any larger particles. While composite materials are typically produced using high-pressure injection molding machines, the authors of this work tested whether the desired effect could be achieved using atmospheric pressure melting. The production of two types of SBCs in a muffle furnace and on an electric burner was proposed. Two different heating approaches were applied to produce SBCs to evaluate the influence of fabrication conditions on mechanical performance. The muffle furnace provides uniform, indirect heating with relatively slow heat transfer, while the electric burner enables direct and localized heating, resulting in faster sulfur melting and different solidification conditions. Four mixtures of biochar and sulfur were proposed: SBC_60 (40% biochar, 60% sulfur), SBC_65 (35% biochar, 65% sulfur), SBC_75 (25% biochar, 75% sulfur), and SBC_80 (20% biochar, 80% sulfur) as our previous study revealed that the SBCs with higher contents of biochar were not durable [51].

https://doi.org/10.3390/ma19030549

Initially, the four proposed mixtures were prepared in a muffle furnace (SNOL, 8.1/1100, Utena, Lithuania). The test material was placed in silicone molds, with the side wall of the mold at right angles to its base, to create samples with a base-to-height ratio of 1 to 1.5. Making the SBCs in the silicone mold facilitated easy removal of the samples. The premeasured proportions of mixed biochar and sulfur were then placed in the silicone mold and placed inside the muffle furnace. The process temperature was set at 140 °C with a process time of 2 h.

The second type of SBC was produced with the application of a laboratory electric burner (CAT H3.1, Ballrechten-Dottingen, Germany). Initially, the sulfur was melted at a controlled temperature of 140 °C in a thermal test chamber (KBC-100WL WAMED, Warsaw, Poland) until liquefaction. Then, the biochar was heated on an electric burner to 140 °C, with temperature control conducted using a pyrometer (Hendi 400, Rhenen, The Netherlands). Liquid sulfur was added to the heated vessel until it reached a temperature of 140 °C. The biochar was heated to homogenize the temperature of the two materials to maintain a constant mixture temperature of 140 °C throughout the process. Failure to heat the biochar would result in a lower sulfur temperature during mixing. After mixing the materials on the burner, the SBC sample was placed into the thermal test chamber for 10 min to solidify the material. The procedure for producing SBC on the burner was conducted based on the methodology for forming sulfur concrete according to Gulzar et al. [8]. The scheme of the tests conducted is shown in Figure 1, and eight mixtures of biochar and sulfur were analyzed for strength analysis.

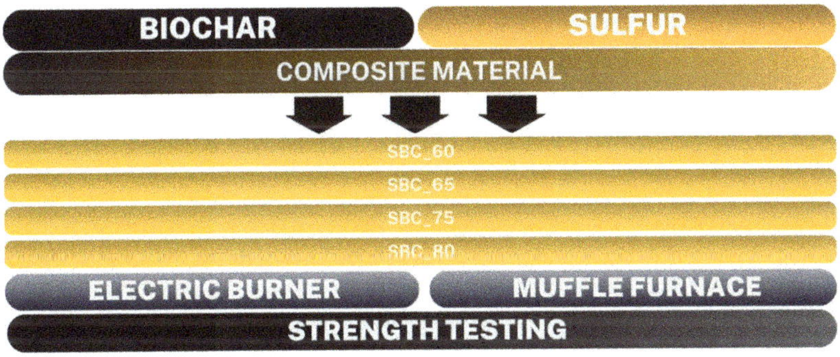

Figure 1. Diagram of individual SBC mixtures of the appropriate proportion with demonstration of analyses.

2.2.2. Structure Observation and Image Analysis

The structure of the produced SBC was examined using a Delta Optical Smart 5.0 optical microscope (Delta Optical, Mińsk Mazowiecki, Poland) at a magnification of ×100. The homogeneity and surface features of the composites were documented using a 5 MP integrated digital camera (Delta Optical DLT-Cam 5MP, Delta Optical, Mińsk Mazowiecki, Poland). The acquired images of SBC surfaces were saved in *.jpg format and further analyzed using ImageJ software (Version 1.54p).

Image processing and fractal analysis were performed using ImageJ software (Version 1.54p, NIH, Bethesda, MD, USA). Prior to analysis, the images were converted to 8-bit grayscale format and binarized using a global threshold to obtain binary images with pixel intensity values of 0 and 255. This preprocessing step was required to meet the input criteria of the fractal analysis tool. The fractal dimension was determined using the box-counting method implemented in the Fractal Box Counting plugin available in ImageJ [55]. The plugin calculates the fractal dimension by counting the number of boxes of decreasing size required to cover the binary image. All analyses were performed on the entire image area

https://doi.org/10.3390/ma19030549

using identical processing parameters to ensure consistency and comparability between samples [56].

2.2.3. Material Strength Testing

The strength properties of the biochar samples were tested in the compression test using a testing machine (Instron, Norwood, MA, USA). The samples were prepared in the shape of cylinders with a diameter of 57 mm and a height of 30 mm. In total, 24 samples were tested, 3 repetitions for each sulfur content ratio and method of SBC production [49]. During the compression test, data on force and displacement were collected. Based on the data, the ultimate strength of each sample was determined.

3. Results and Discussion

SBCs were produced in triplicate batches, utilizing both a muffle furnace and an electric burner. Four distinct ratios of sulfur and biochar were selected, yielding a total of twenty-four samples. In Figure 2, a representative test sample from each group is displayed. The initial observation reveals significant structural disparities between the samples fabricated in the muffle furnace and those in the electric heater. Before conducting the strength test, each sample was meticulously prepared to ensure that the ratio between the height and the diameter of the base fell within the range of 1 to 1.5. Additionally, sulfur is much more visible in the case of the SBC produced with the burner application, showing the higher level of distribution (Figure 2). Moreover, the determined particle size of sulfur showed that that of the SBC produced in the furnace was in a higher range than that in SBC from the burner (Table 1). This may be attributed to the liquid sulfur better penetrating the voids between biochar particles.

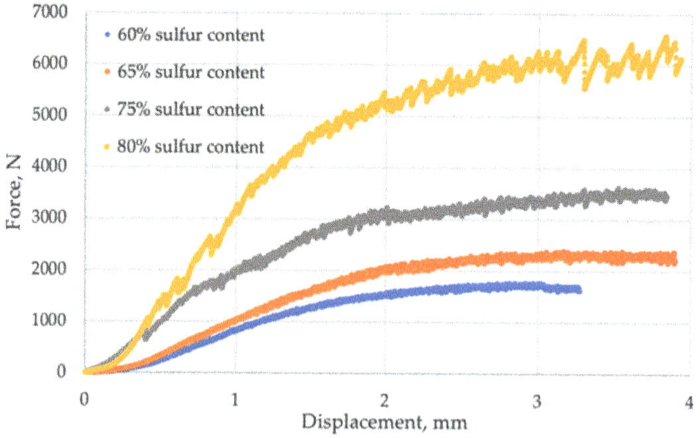

Figure 2. Strength properties of SBC after thermal treatment in a muffle furnace.

The fractal dimension (Df) values obtained for SBC produced using different sulfur contents and fabrication methods are presented in Table 2. For both production routes, an increase in sulfur content resulted in a systematic increase in fractal dimension, indicating higher surface structural complexity. For samples produced in the muffle furnace, the fractal dimension increased gradually from 1.3741 for SBC_60 to 1.5253 for SBC_80. This trend suggests a progressive increase in surface heterogeneity with increasing sulfur content. However, the absolute Df values remained relatively moderate, reflecting a less complex surface structure. In contrast, samples produced using the electric burner exhibited generally higher fractal dimension values, particularly at higher sulfur contents. The Df increased from 1.3408 for SBC_60 to 1.6198 for SBC_80. The most pronounced increase

https://doi.org/10.3390/ma19030549

was observed between SBC_65 and SBC_75, indicating a significant change in surface structural complexity associated with higher sulfur incorporation under electric burner processing conditions. The higher fractal dimension values observed for electric-burner-treated samples suggest a more complex and irregular surface morphology compared to samples produced in the muffle furnace. This difference may be attributed to variations in sulfur melting, distribution, and solidification behavior resulting from the distinct thermal conditions of the two fabrication methods. Overall, the fractal dimension analysis confirms that both sulfur content and production method influence the image-based structural complexity of SBC surfaces. Similar relationships between processing-induced structural features of biochar-containing materials and their mechanical performance have been reported in previous studies, where image-based or microstructural observations were linked to compressive strength without direct microvoid quantification [57,58]. In this context, the fractal dimension applied in the present study serves as a qualitative descriptor of surface structural complexity rather than a direct measure of porosity.

Table 1. Sulfur particles' size concerning the SBC production method and the content of sulfur.

Thermal Treatment	Sulfur Content, %	Average Size of Sulfur Particles, Pixels
Muffle Furnace	60	22.3
	65	26.1
	75	36.2
	80	66.1
Electric Burner	60	20.9
	65	24.4
	75	47.1
	80	54.8

Table 2. Surface images of sulfur–biochar composites (SBCs) produced using a muffle furnace and an electric burner at different sulfur contents, together with the corresponding fractal dimension (Df) values.

	SBC_60	SBC_65	SBC_75	SBC_80
D_f	1.3741	1.3918	1.4899	1.5253
Muffle Furnace				
D_f	1.3408	1.4273	1.6189	1.6198
Electric Burner				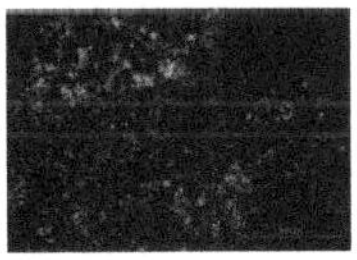

All images were acquired at the same magnification. The scale bar in each image corresponds to 1000 μm. D_f denotes the fractal dimension calculated from the binarized images.

The average results of the strength properties of the SBC samples are presented in Figures 2 and 3. It was observed that the method of composite preparation highly influenced the quality of the biochar. The mixtures prepared in the muffle furnace were characterized by lower strength properties values than SBC made using the electric burner. It can be explained by the density of the material influenced by its porosity, resulting from the air content and bond strengths between carbon molecules. The study showed that,

https://doi.org/10.3390/ma19030549

with the increase in the sulfur content in biochar, the ultimate strength also increases. For furnace treatment, the ultimate strength of the sample with a sulfur content of 60% scored about 2000 N, while the sample with 80% sulfur content achieved an about 3.5 times higher value of ultimate strength. For electric burner treatment, the ultimate strength of SBC, with a share of sulfur from 60 to 80%, increased from 8000 to 11,000 N, respectively.

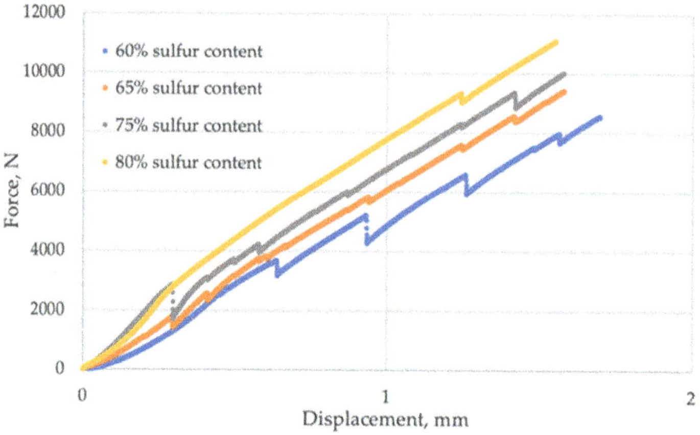

Figure 3. Strength properties of SBC after thermal treatment on an electric burner.

The samples subjected to furnace treatment exhibited increased brittleness, with the propagation of surface cracking initiating at 500 N and remaining relatively constant thereafter. This phenomenon can be attributed to higher porosity and weaker bond strengths between carbon molecules in the furnace-treated samples.

Table 2 provides data on the ultimate force and displacement for SBC samples treated under different conditions: in a muffle furnace and on an electric burner, varying the sulfur content. For samples treated in the muffle furnace, at 60% sulfur content, the ultimate force was 1770 N with a displacement of 3.27 mm. At 65% sulfur content, the ultimate force increased to 2393 N while the displacement increased to 3.90 mm. With 75% sulfur content, the ultimate force further increased to 3623 N with a displacement of 3.84 mm. At the highest sulfur content of 80%, the ultimate force substantially rose to 6577 N with a displacement of 3.93 mm. For samples treated on the electric burner, at 60% sulfur content, the ultimate force significantly increased to 8570.36 N, while the displacement decreased to 1.70 mm. With 65% sulfur content, the ultimate force further rose to 9417 N, with a slight decrease in displacement to 1.58 mm. At 75% sulfur content, the ultimate force continued to increase to 10,002 N, with displacement remaining at 1.58 mm. Finally, at 80% sulfur content, the ultimate force peaked at 11,093 N, while the displacement decreased slightly to 1.55 mm. These results suggest that samples treated on the electric burner exhibit significantly higher ultimate forces and lower displacements compared to those treated in the muffle furnace, indicating superior strength characteristics. Additionally, as the sulfur content increases, the ultimate force generally increases across both treatment methods, albeit at different rates.

Additional correlative analyses revealed that both SBC types behave differently during the performance of the force. Furnace SBC displacement increases with the increase in the force; however, in the case of the electric burner SBC, this phenomenon is the opposite (Figure 4). This may indicate that the method of SBC production significantly impacts the mechanical properties of the produced composites. When the liquid sulfur is mixed with

https://doi.org/10.3390/ma19030549

the hot biochar, it better fills the voids and porous structure of the biochar, which is in alignment with the results of sulfur particle size (Table 3).

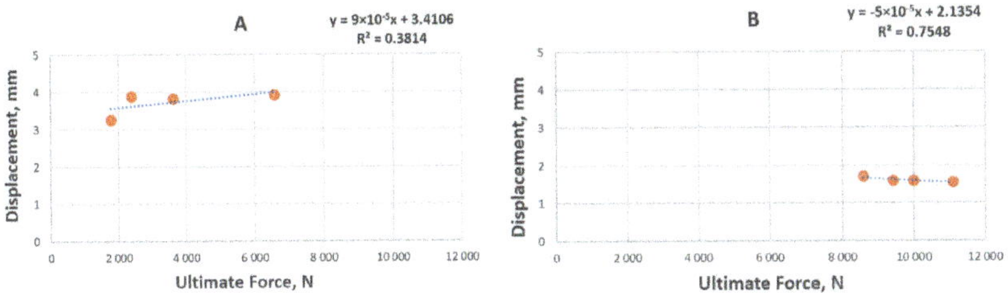

Figure 4. Relationship between ultimate force and displacement of the SBC for the furnace (**A**) and electric burner (**B**). The dotted lines represent linear regression fits; the corresponding regression equations and coefficients of determination (R^2) are shown in the plots.

Table 3. Summary of ultimate strength results.

Thermal Treatment	Sulfur Content, %	Ultimate Force, N	Displacement, mm
Muffle Furnace	60	1770.02	3.27
	65	2393.62	3.90
	75	3623.10	3.84
	80	6577.88	3.93
Electric Burner	60	8570.36	1.70
	65	9417.26	1.58
	75	10,002.45	1.58
	80	11,093.49	1.55

4. Conclusions

The preliminary tests on sulfur–biochar composites indicate a promising avenue for developing materials with commendable mechanical properties, sans the need for high-pressure injection molding machinery. An escalation in sulfur content leads to improved composite strength, highlighting the efficacy of blending sulfur with biochar to counteract the typical brittleness associated with such composites.

Significant differences emerged between samples treated with an electric burner versus those from a muffle furnace. Notably, samples treated with the electric burner exhibited a wider range of strength, with the SBC_80 variant reaching an impressive 11 kN. The correlation between biochar strength and sulfur content is apparent from the data, with biochars treated in an electric burner demonstrating notably higher durability (lower displacement) compared to those from muffle furnace treatment. Analysis of the ultimate force and displacement data suggests that samples treated in the muffle furnace experienced a notable strength increase with rising sulfur content. Conversely, samples treated with the electric burner displayed even greater increases in ultimate force, alongside lower displacements, further highlighting the superior strength characteristics of the electric-burner-treated samples. This may be related to better sulfur distribution and filling of the pores and voids between biochar particles.

The study further reveals a direct correlation between sulfur content in biochar and ultimate strength, with consistent increases observed across both furnace and electric burner treatments.

Author Contributions: Conceptualization, A.B., M.S., M.R., and E.S.; methodology, A.B. and E.S.; software, M.S. and E.S.; validation, E.S., M.R., and M.S.; formal analysis, M.S.; investigation M.S. and

https://doi.org/10.3390/ma19030549

E.S.; resources, A.B.; data curation E.S.; writing—original draft preparation, E.S.; writing—review and editing, E.S., M.S., M.R., and A.B.; visualization, E.S., M.S., and A.B.; supervision, A.B. and M.R.; project administration, E.S., M.R., and A.B.; funding acquisition, A.B. All authors have read and agreed to the published version of the manuscript.

Funding: The publication was financed by science development fund of the Warsaw University of Life Sciences—SGGW.

Informed Consent Statement: Not applicable.

Data Availability Statement: The original contributions presented in this study are included in the article. Further inquiries can be directed to the corresponding author.

Conflicts of Interest: The authors declare no conflicts of interest.

References

1. Robinson, P.R. Sulfur Removal and Recovery. In *Springer Handbook of Petroleum Technology*; Springer Handbooks; Hsu, C.S., Robinson, P.R., Eds.; Springer: Cham, Switzerland, 2017. [CrossRef]
2. Davis, M.; Okunlola, A.; Di Lullo, G.; Giwa, T.; Kumar, A. Greenhouse gas reduction potential and cost-effectiveness of economy-wide hydrogen-natural gas blending for energy end uses. *Renew. Sustain. Energy Rev.* **2023**, *171*, 112962. [CrossRef]
3. Alzahrani, A.; Petri, I.; Rezgui, Y.; Ghoroghi, A. Decarbonisation of seaports: A review and directions for future research. *Energy Strategy Rev.* **2021**, *38*, 100727. [CrossRef]
4. Pires, A.; Martinho, G. Waste hierarchy index for circular economy in waste management. *Waste Manag.* **2019**, *95*, 298–305. [CrossRef]
5. Wang, D.; Tang, Y.T.; Long, G.; Higgitt, D.; He, J.; Robinson, D. Future improvements on performance of an EU landfill directive driven municipal solid waste management for a city in England. *Waste Manag.* **2020**, *102*, 452–463. [CrossRef]
6. Hasan, M.M.; Rasul, M.G.; Khan, M.M.K.; Ashwath, N.; Jahirul, M.I. Energy recovery from municipal solid waste using pyrolysis technology: A review on current status and developments. *Renew. Sustain. Energy Rev.* **2021**, *145*, 111073. [CrossRef]
7. United States Geological Survey (USGS). *Minerals Yearbook: Area Reports—International. Africa and the Middle East, Volume III*; U.S. Department of the Interior, U.S. Geological Survey: Washington, DC, USA, 2004.
8. Gulzar, M.A.; Rahim, A.; Ali, B.; Khan, A.H. An investigation on recycling potential of sulfur concrete. *J. Build. Eng.* **2021**, *38*, 102175. [CrossRef]
9. Frilingou, N.; Xexakis, G.; Koasidis, K.; Nikas, A.; Campagnolo, L.; Delpiazzo, E.; Chiodi, A.; Gargiulo, M.; McWilliams, B.; Koutsellis, T.; et al. Navigating through an energy crisis: Challenges and progress towards electricity decarbonisation, reliability, and affordability in Italy. *Energy Res. Soc. Sci.* **2023**, *96*, 102934. [CrossRef]
10. Amanov, N.; Turaev, K.; Shadhar, M.H.; Tadjixodjayeva, U.; Jumaeva, Z.; Berdimurodov, E.; Eliboev, I.; Hosseini-Bandegharaei, A. Sulfur-based concrete: Modifications, advancements, and future prospects. *Constr. Build. Mater.* **2024**, *435*, 136765. [CrossRef]
11. Ghasemalizadeh, S. Application of Low-Portland Cement Content Binders for Potential High-Strength Concrete Masonry Unit Manufacturing. Doctoral Thesis, University of Calgary, Calgary, AB, Canada, 2025.
12. Sobczyk, W.; Sobczyk, E.J. Varying the energy mix in the EU-28 and in Poland as a step towards sustainable development. *Energies* **2021**, *14*, 1502. [CrossRef]
13. Syguła, E.; Gałęzowska, M.; Białowiec, A. Enhanced production of biogas using biochar–sulfur composite in the methane fermentation process. *Materials* **2022**, *15*, 4517. [CrossRef] [PubMed]
14. Xia, C.; Surendran, S.; Ji, S.; Kim, D.; Chae, Y.; Kim, J.; Je, M.; Han, M.-K.; Choe, W.-S.; Choi, C.H.; et al. A sulfur self-doped multifunctional biochar catalyst for overall water splitting and a supercapacitor from *Camellia japonica* flowers. *Carbon Energy* **2022**, *4*, 491–505. [CrossRef]
15. Wasielewski, R.; Nowak, M. Alternative fuel market in Poland. *Polityka Energetyczna* **2019**, *22*, 81–96. [CrossRef]
16. Bień, J. Production and use of waste-derived fuels in Poland: Current status and perspectives. *Prod. Eng. Arch.* **2021**, *27*, 36–41. [CrossRef]
17. Paszkowski, J.; Domański, M.; Caban, J.; Zarajczyk, J.; Pristavka, M.; Findura, P. The Use of Refuse Derived Fuel (RDF) in the Power Industry. *Agric. Eng.* **2020**, *24*, 83–90. [CrossRef]
18. Nam-Chol, O.; Hyo-Song, P.; Yong-Chol, S.; Yong-Hyok, R.; Yong-Nam, K. A feasibility study of energy recovery of RDF from municipal solid waste. *Energy Sources Part A Recovery Util. Environ. Eff.* **2018**, *40*, 2914–2922.
19. Dianda, P.; Mahidin, M.; Munawar, E. Production and characterization refuse derived fuel (RDF) from high organic and moisture contents of municipal solid waste (MSW). *IOP Conf. Ser. Mater. Sci. Eng.* **2018**, *334*, 012035. [CrossRef]
20. Materazzi, M.; Lettieri, P.; Taylor, R.; Chapman, C. Performance analysis of RDF gasification in a two stage fluidized bed-plasma process. *Waste Manag.* **2016**, *47*, 256–266. [CrossRef]

https://doi.org/10.3390/ma19030549

21. Świechowski, K.; Syguła, E.; Koziel, J.A.; Stępień, P.; Kugler, S.; Manczarski, P.; Białowiec, A. Low-Temperature Pyrolysis of Municipal Solid Waste Components and Refuse-Derived Fuel—Process Efficiency and Fuel Properties of Carbonized Solid Fuel. *Data* **2020**, *5*, 48. [CrossRef]
22. Zajemska, M.; Magdziarz, A.; Iwaszko, J.; Skrzyniarz, M.; Poskart, A. Numerical and experimental analysis of pyrolysis process of RDF containing a high percentage of plastic waste. *Fuel* **2022**, *320*, 123981. [CrossRef] [PubMed]
23. Czajczyńska, D.; Nannou, T.; Anguilano, L.; Krzyżyńska, R.; Ghazal, H.; Spencer, N.; Jouhara, H. Potential of pyrolysis processes in the waste management sector. *Therm. Sci. Eng. Prog.* **2017**, *3*, 171–197. [CrossRef]
24. Chatterjee, R.; Sajjadi, B.; Chen, W.Y.; Mattern, D.L.; Hammer, N.; Raman, V.; Dorris, A. Effect of Pyrolysis Temperature on PhysicoChemical Properties and Acoustic-Based Amination of Biochar for Efficient CO_2 Adsorption. *Front. Energy Res.* **2020**, *8*, 85. [CrossRef]
25. Yaashikaa, P.R.; Kumar, P.S.; Varjani, S.; Saravanan, A. A critical review on the biochar production techniques, characterization, stability and applications for circular bioeconomy. *Biotechnol. Rep.* **2020**, *28*, e00570. [CrossRef] [PubMed]
26. Mukherjee, A.; Das, P.; Minu, K. Thermogravimetric analysis and kinetic modelling studies of selected agro-residues and biodiesel industry wastes for pyrolytic conversion to bio-oil. *Biomass Convers. Biorefinery* **2014**, *4*, 259–268. [CrossRef]
27. Cai, J.; Xu, D.; Dong, Z.; Yu, X.; Yang, Y.; Banks, S.W.; Bridgwater, A.V. Processing thermogravimetric analysis data for isoconversional kinetic analysis of lignocellulosic biomass pyrolysis: Case study of corn stalk. *Renew. Sustain. Energy Rev.* **2018**, *82*, 2705–2715. [CrossRef]
28. Chrissafis, K. Kinetics of thermal degradation of polymers: Complementary use of isoconversional and model-fitting methods. *J. Therm. Anal. Calorim.* **2009**, *95*, 273–283. [CrossRef]
29. Meng, F.; Zhou, Y.; Liu, J.; Wu, J.; Wang, G.; Li, R.; Zhang, Y. Thermal decomposition behaviors and kinetics of carrageenan–poly vinyl alcohol bio-composite film. *Carbohydr. Polym.* **2018**, *201*, 96–104. [CrossRef]
30. Stępień, P.; Pulka, J.; Serowik, M.; Białowiec, A. Thermogravimetric and Calorimetric Characteristics of Alternative Fuel in Terms of Its Use in Low-Temperature Pyrolysis. *Waste Biomass Valorization* **2019**, *10*, 1669–1677. [CrossRef]
31. Vasudev, V.; Ku, X.; Lin, J. Kinetic study and pyrolysis characteristics of algal and lignocellulosic biomasses. *Bioresour. Technol.* **2019**, *288*, 121496. [CrossRef]
32. Çepelioğullar, Ö; Haykiri-Açma, H.; Yaman, S. Kinetic modelling of RDF pyrolysis: Model-fitting and model-free approaches. *Waste Manag.* **2016**, *48*, 275–284. [CrossRef]
33. Mallick, D.; Bora, B.J.; Barbhuiya, S.A.; Banik, R.; Garg, J.; Sarma, R.; Gogoi, A.K. Detailed study of pyrolysis kinetics of biomass using thermogravimetric analysis. *AIP Conf. Proc.* **2019**, *2091*, 020014. [CrossRef]
34. Chen, C.; Bu, X.; Huang, D.; Huang, Y.; Huang, H. Thermal Decomposition and Kinetics Analysis of Microwave Pyrolysis of *Dunaliella salina* Using Composite Additives. *Bioenergy Res.* **2020**, *13*, 1205–1220. [CrossRef]
35. Friedman, H.L. Kinetics of thermal degradation of char-forming plastics from thermogravimetry. Application to a phenolic plastic. *J. Polym. Sci. Part C Polym. Symp.* **1964**, *6*, 183–195. [CrossRef]
36. Lawner, B.J.; Mattu, A. Cardiac Arrest. *Cardiovasc. Probl. Emerg. Med. Discuss.-Based Rev.* **2012**, *57*, 123–137.
37. Zhang, Z.; Duan, H.; Zhang, Y.; Guo, X.; Yu, X.; Zhang, X.; Rahman, M.M.; Cai, J. Investigation of kinetic compensation effect in lignocellulosic biomass torrefaction: Kinetic and thermodynamic analyses. *Energy* **2020**, *207*, 118290. [CrossRef]
38. Escalante, J.; Chen, W.H.; Tabatabaei, M.; Hoang, A.T.; Kwon, E.E.; Lin, K.Y.; Saravanakumar, A. Pyrolysis of lignocellulosic, algal, plastic, and other biomass wastes for biofuel production and circular bioeconomy: A review of thermogravimetric analysis (TGA) approach. *Renew. Sustain. Energy Rev.* **2022**, *169*, 112914. [CrossRef]
39. Mishra, R.K.; Mohanty, K. Pyrolysis kinetics and thermal behavior of waste sawdust biomass using thermogravimetric analysis. *Bioresour. Technol.* **2018**, *251*, 63–74. [CrossRef]
40. Zhu, Z.; Cong, R.; Zhou, L.; Zheng, H.; Tu, Y.; Wu, Z. Effects of chemical properties and inherent mineral matters on pyrolysis kinetics of low-rank coals. *Processes* **2021**, *9*, 2111. [CrossRef]
41. Fan, H.; Gu, J.; Wang, Y.; Yuan, H.; Chen, Y.; Luo, B. Effect of potassium on the pyrolysis of biomass components: Pyrolysis behaviors, product distribution and kinetic characteristics. *Waste Manag.* **2021**, *121*, 255–264. [CrossRef] [PubMed]
42. Seo, M.W.; Kim, S.D.; Lee, S.H.; Lee, J.G. Pyrolysis characteristics of coal and RDF blends in non-isothermal and isothermal conditions. *J. Anal. Appl. Pyrolysis* **2010**, *88*, 160–167. [CrossRef]
43. Kumar, M.; Srivastava, N.; Upadhyay, S.N.; Mishra, P.K. Thermal degradation of dry kitchen waste: Kinetics and pyrolysis products. *Biomass Convers. Biorefinery* **2023**, *13*, 2779–2796. [CrossRef]
44. Ding, Y.; Ezekoye, O.A.; Lu, S.; Wang, C.; Zhou, R. Comparative pyrolysis behaviors and reaction mechanisms of hardwood and softwood. *Energy Convers. Manag.* **2017**, *132*, 102–109. [CrossRef]
45. Zhang, K.; Xiao, W.; Li, J.; Liu, J.; Yan, C. Two-step preparation of porous nickel–sulfur electrode for hydrogen evolution in alkaline water electrolysis. *Electrochim. Acta* **2017**, *228*, 422–427. [CrossRef]
46. Yao, Z.; Yu, S.; Su, W.; Wu, W.; Tang, J.; Qi, W. Kinetic studies on the pyrolysis of plastic waste using a combination of model-fitting and model-free methods. *Waste Manag. Res.* **2020**, *38*, 77–85. [CrossRef]

 https://doi.org/10.3390/ma19030549

47. Azam, M.; Jahromy, S.S.; Raza, W.; Jordan, C.; Harasek, M.; Winter, F. Comparison of the combustion characteristics and kinetic study of coal, municipal solid waste, and refuse-derived fuel: Model-fitting methods. *Energy Sci. Eng.* **2019**, *7*, 2646–2657. [CrossRef]
48. Kuspangaliyeva, B.; Suleimenova, B.; Shah, D.; Sarbassov, Y. Thermogravimetric study of refuse derived fuel produced from municipal solid waste of Kazakhstan. *Appl. Sci.* **2021**, *11*, 1219. [CrossRef]
49. Du, Y.; Ying, Z.; Zheng, X.; Dou, B.; Cui, G. Correlating electrochemical biochar oxidation with electrolytes during biochar-assisted water electrolysis for hydrogen production. *Fuel* **2023**, *339*, 126957. [CrossRef]
50. Rahim, H.U.; Allevato, E.; Stazi, S.R. Sulfur-functionalized biochar: Synthesis, characterization, and utilization for contaminated soil and water remediation—A review. *J. Environ. Manag.* **2024**, *370*, 122670. [CrossRef] [PubMed]
51. Peng, Y.; Su, L.; Liu, M.; Zeng, C.; Xiang, B.; Xie, Z.; Hu, Z.; Zhou, N. Biochar-enhanced sulfur: Mechanistic insights into a novel and effective bactericide. *Nanomaterials* **2025**, *15*, 697. [CrossRef]
52. Lyon, R.E. A Physical Basis for Kinetic Compensation. *J. Phys. Chem.* **2023**, *127*, 2399–2406. [CrossRef]
53. Cozzani, V.; Petarca, L.; Tognotti, L. Devolatilization and pyrolysis of refuse derived fuels: Characterization and kinetic modelling by a thermogravimetric and calorimetric approach. *Fuel* **1995**, *74*, 903–912. [CrossRef]
54. Bosmans, A.; De Dobbelaere, C.; Helsen, L. Pyrolysis characteristics of excavated waste material processed into refuse derived fuel. *Fuel* **2014**, *122*, 198–205. [CrossRef]
55. De Stefano, A.A.; Musone, L.; Horodynski, M.; Vernucci, R.A.; Galluccio, G. Fractal Dimension of the Condylar Bone Structure in Patients with Unilateral Condylar Hyperplasia: Cross-Sectional Retrospective Study. *Appl. Sci.* **2025**, *15*, 4063. [CrossRef]
56. Shambilova, G.; Korshunov, A.; Vinogradov, M.; Kadasheva, Z.; Iskakov, R.; Kalauova, A.; Makarov, G.; Kalimanova, D.; Legkov, S. Recycling of Polypropylene with Vitamin E Additives: Rheological Properties and Mechanical Characteristics. *Processes* **2025**, *13*, 3923. [CrossRef]
57. Vafaei, B.; Ghahremaninezhad, A. Effects of Biochar as a Green Additive on the Self-Healing and Properties of Sustainable Cementitious Materials. *ACS Sustain. Chem. Eng.* **2024**, *12*, 8261–8275. [CrossRef]
58. Prabahar, J.; Manimaran, A.; Manoharan, P.D.; Karthikeyan, J. The Effect of Biochar on the Properties of Alkali-Activated Slag Pastes. *Constr. Mater.* **2021**, *2*, 1–14. [CrossRef]

 https://doi.org/10.3390/ma19030549

materials

MDPI

Article

Waste-Glass-Derived Porous Silica: Synthesis and Structural Characterization

Antônia Millena de Oliveira Lima [1,2], Manuel Pedro Fernandes Graça [3,*] Imen Hammami [3,*] and Ana Angélica Mathias Macêdo [1]

1 Instituto Federal de Educação Ciência e Tecnologia do Maranhão, Campus Imperatriz, São Luís 65919-050, Brazil; amillenalima@hotmail.com (A.M.d.O.L.); anaangellica@yahoo.com.br (A.A.M.M.)
2 Foundation for Research and Scientific and Technological Development of Maranhão (FAPEMA), São Luís 65075-340, Brazil
3 i3N and Physics Department, University of Aveiro, 3810-193 Aveiro, Portugal
* Correspondence: mpfg@ua.pt (M.P.F.G.); imenhammami@ua.pt (I.H.); Tel.: +351-964236731 (M.P.F.G.); +351-932616958 (I.H.)

Abstract

This work reports the reuse of waste glass as a sustainable silica source for the synthesis of mesoporous silica, as an alternative to conventional silica precursors. Silica gel was produced through alkaline dissolution of glass powder using sodium hydroxide and subsequently employed as a precursor for the synthesis of porous silica. The waste-derived glass powder and the synthesized silica-based materials were characterized to assess their structural, morphological, surface, and textural properties. XRD analysis confirmed the amorphous nature of all samples, while FTIR spectra indicated successful silica network formation with modifications in bond connectivity. SEM imaging revealed spherical particles with average diameters of approximately 0.19 ± 0.02 μm for silica gel and 0.15 ± 0.03 μm for the mesoporous silica. Zeta potential measurements indicated a negative surface charge and good colloidal stability in aqueous media. Nitrogen sorption analysis revealed that the specific surface area was limited by the low solubility of silica gel in acidic media, which prevents ideal condensation on the surface of surfactant micelles. The results demonstrate that waste glass-derived silica gel is a promising precursor, although the synthesis conditions did not yield a highly ordered mesostructure, highlighting the need for further control of precursor solubility and pH.

Keywords: waste glass; sodium hydroxide; silica gel; mesoporous

1. Introduction

Soda–lime–silica glass, commonly referred to as soda–lime glass, is the most prevalent type of glass found in urban solid waste. Its structure is primarily composed of tetrahedral SiO_4 units, with a typical composition of approximately of 73% SiO_2, 15% Na_2O, 7% CaO, 4% MgO, and 1% Al_2O_3 (wt%) [1]. The high silica (SiO_2) content in these compositions favors the reuse of soda–lime glass as a silicon source in the synthesis of amorphous silica. Liu et al. (2019) compiled several compositions for discarded glass and identified their main sources, highlighting glass bottles as one of the largest contributors to glass waste streams [2].

This dominance of bottle-derived soda–lime glass has been further intensified by recent changes in consumption patterns. During the COVID-19 pandemic, beer consumption shifted predominantly toward disposable bottles as domestic consumption replaced social gatherings in bars. Reports indicate that approximately 70% of beer sales in Brazil relied on disposable bottles, in contrast to pre-pandemic trends. This shift led to an increase in demand for glass production in Brazil, as the beer industry traditionally relied on returnable bottle logistics. Consequently, a shortage of glass emerged, affecting large industrial sectors, particularly the beverage industry [3,4].

Given that a large fraction of these bottles are non-reusable and end up discarded, there is an urgent socio-environmental need to develop high-value upcycling strategies. A promising solution lies in transforming the discarded soda–lime glass into silica precursors for the synthesis of silica-based materials. Conventionally, the synthesis of high-order mesoporous silica relies on alkoxide precursors such as tetraethyl orthosilicate (TEOS) and tetramethyl orthosilicate (TMOS) [5,6]. However, the production of these traditional precursors is characterized by high energy consumption, high costs, and the use of toxic organic solvents, which results in a significant environmental footprint. Consequently, there is growing interest in cost-effective, biocompatible, and eco-friendly alternatives, such as waste-derived soda–lime glass, as sustainable silicon sources [7].

Mesoporous silica consists of a three-dimensional network of SiO_4 tetrahedra linked by siloxane bonds, forming an inorganic polymeric structure with well-defined pore sizes in the mesoporous range (2–50 nm), where variations in network connectivity and the presence of modifier species strongly influence the structural, dielectric, and transport properties of silicate-based amorphous materials [8]. The synthesis of such materials typically involves the use of structure-directing agents, such as surfactants, which template the formation of pores. After removal of the surfactant, the silica surface is predominantly covered by silanol (Si–OH) groups, although residual siloxane groups may remain. The surface properties of silica are governed by the density and distribution of these functional groups and can be further tailored through surface modification or functionalization, enabling the design of materials with application-specific properties [9–13]. Previous studies have demonstrated that glass-waste streams can be effectively converted into soluble silicate precursors to produce high-value porous and mesoporous silicas. For example, recycled glass cullet has been upcycled into highly porous silica microspheres (porosity $\approx$ 69%), which showed excellent performance in the adsorption of organic dyes for wastewater treatment [14]. In addition, waste-derived siliceous materials have been reported to yield high-purity amorphous silica with specific surface areas of several hundred m^2/g [15,16], exhibiting catalytic and adsorption performances comparable to those obtained using commercial silicon precursors. More recently, glass waste has been successfully employed as a silicon source to obtain high-purity silica (up to 99.4%) for the synthesis of 45S5 bioglass intended for biomedical applications [17]. These results highlight the versatility and effectiveness of glass-waste recycling strategies in both environmental remediation and biomedical engineering.

Among the various types of mesoporous silicas, SBA-15 (Santa Barbara Amorphous-15) is one of the most extensively studied structures due to its ordered two-dimensional hexagonal pore arrangement, relatively large pore diameters, and thick pore walls that confer enhanced mechanical and hydrothermal stability [18,19]. Typically synthesized using the triblock copolymer Pluronic P123 as a template in acidic media, SBA-15 features large mesopores (5–30 nm) and a complementary micropore network [20,21]. These characteristics make it exceptionally suitable for the immobilization of large biomolecules and drug delivery systems [22,23]. However, the high cost of traditional silicon sources for SBA-15 synthesis remains a barrier to large-scale application, further justifying the investigation of waste-derived alternatives.

In this context, the present work investigates the reuse of "long-neck" glass bottles to produce silicon precursors for the synthesis of SBA-15-type mesoporous silica, serv-

https://doi.org/10.3390/ma19030601

ing as a sustainable alternative to traditional precursors. Long-neck glass bottles are typically manufactured from soda–lime–silica glass, whose composition is broadly consistent across beverage containers. According to literature, these glasses generally contain 70–74 wt% SiO_2 as the network former, 12–15 wt% Na_2O as the main flux, 9–12 wt% CaO as the principal stabilizer, and small amounts of MgO and Al_2O_3 [24]. Transition-metal oxides, such as iron or chromium, may be present in green or amber bottles to provide coloration and light protection. Several characterization techniques were used to evaluate the effectiveness of these waste-derived precursors. X-ray diffraction (XRD) was used to confirm the preservation of the amorphous state, while Fourier transform infrared spectroscopy (FTIR) investigated the evolution of Si–O–Si and Si–OH bonding environments. Zeta potential measurements assessed the surface charge and colloidal stability, which are critical for dispersion behavior in biomedical applications. Finally, nitrogen (N_2) adsorption analysis was used to evaluate the textural properties and the impact of silica gel solubility on the development of the mesoporous framework. While challenges regarding solubility in acidic media were identified, this method represents a significant step toward the sustainable valorization of post-consumer glass waste.

2. Materials and Methods

2.1. Preparation of Silica Gel and SBA-15-Type Mesoporous Silica from Waste Glass

Glass bottles were collected from local businesses in the city of Imperatriz (MA, Brazil). The bottles were initially cleaned with running water and detergent to remove surface contaminants and adhesive residues, followed by rinsing with distilled water and drying at room temperature. The dried bottles were manually crushed using a hammer and subsequently pulverized with a mortar and pestle to obtain a fine glass powder. While this approach is suitable for laboratory-scale preparation, ball milling represents a scalable and industrially relevant alternative for producing glass powder in larger quantities. This method is easily replaced by using a ball mill for large-scale production. The experimental procedure for converting these waste glass precursors into silica gel and mesoporous SBA-15 is schematically illustrated in Figure 1.

The amount of 5 g of glass powder was added to 30 mL of an aqueous sodium hydroxide solution (6 mol·L^{-1} NaOH), promoting the alkaline dissolution of silica according to reaction (1):

$$SiO_2(s) + 2NaOH(aq) \longrightarrow Na_2SiO_3(aq) + H_2O(l) \quad (1)$$

The suspension was maintained under constant stirring at 75 °C for 24 h. Following the reaction, the mixture was filtered to obtain a sodium silicate solution. Silica gel was then precipitated by adding concentrated hydrochloric acid (HCl). The resulting white silica gel was aged overnight at room temperature. As summarized in Figure 1, the obtained silica gel was subsequently used as the silicon source for SBA-15-type mesoporous silica preparation.

Mesoporous silica SBA-15 was synthesized using a hydrothermal method adapted from Rechotnek et al. [25]. Briefly, 4 g of the template Pluronic P123 (CAS No. 9003-11-6; $EO_{20}PO_{70}EO_{20}$) were dissolved in a mixture of 20 mL of HCl and 130 mL of distilled water under stirring for 1 h. The temperature of the solution was then increased to 45 °C. Subsequently, 8.5 g of the waste-derived silica gel was added to the reaction medium, and the mixture was stirred for 20 h. The resulting suspension was then kept in a preheated oven at 80 °C for 24 h under static conditions. The solid product was collected by vacuum filtration, washed with 1 L of distilled water, and dried at 55 °C for 12 h. Finally, the white solid was calcined at 550 °C for 6 h to remove the organic template. The resulting mesoporous material was designated as MS-GP, corresponding to a P123-templated SBA-15-type mesoporous silica derived from waste glass, without long-range hexagonal ordering.

https://doi.org/10.3390/ma19030601

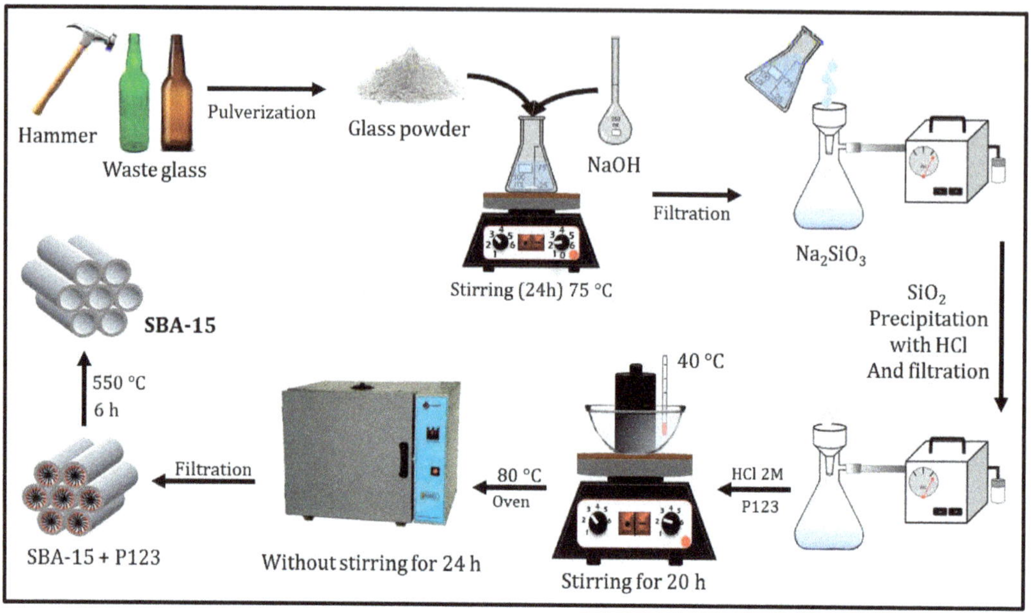

Figure 1. Schematic representation of the synthesis route used to convert post-consumer waste glass into silica gel and mesoporous SBA-15-type mesoporous silica, including glass pulverization, alkaline dissolution in NaOH, silica gel precipitation by acidification, and subsequent hydrothermal synthesis and calcination.

The materials synthesized and characterized in this study, along with their respective identification codes, are summarized in Table 1. In this study, the term SBA-15-type mesoporous silica refers to materials synthesized using the conventional P123-templated acidic route, although long-range hexagonal ordering was not achieved.

Table 1. Sample identification and descriptions of the materials prepared in this study.

Sample Code	Description	Structural Characteristic
GP	Glass powder	Amorphous, non-porous
SG	Silica gel derived from glass powder	Amorphous, texturally porous silica
MS-GP	Mesoporous silica synthesized with P123	Disordered mesoporous network

2.2. Characterization Methods

X-ray diffraction (XRD) patterns were recorded using a Rigaku Miniflex II diffractometer (Rigaku Corporation, Tokyo, Japan) equipped with Cu Kα radiation (λ = 1.5406 Å). Data were collected in the 2θ range of 10–80° with a step size of 0.02° and a counting time of 2 s per step.

Fourier-transform infrared (FTIR) spectra were recorded in transmission mode in the range of 400–4000 cm^{-1} using an Alpha II FTIR spectrometer (Bruker Optics, Ettlingen, Germany). Samples were prepared as KBr pellets by mixing the powder with KBr at a mass ratio of 1:200 mg.

The surface morphology of the prepared samples was investigated by scanning electron microscopy (SEM) using a TESCAN Vega 3 microscope (TESCAN, Brno, Czech Republic). Before imaging, the samples were coated with a thin layer of gold (around 10 nm) using a sputter coater to ensure surface conductivity and prevent charging during analysis. The elemental composition of the samples was analyzed semi-quantitatively by

https://doi.org/10.3390/ma19030601

energy-dispersive X-ray spectroscopy (EDS) using a Bruker QUANTAX EDS (Bruker Nano GmbH, Berlin, Germany) system coupled to the SEM, operated at an accelerating voltage of 20 kV. Particle size distribution was determined from SEM images by measuring the diameters of particles using ImageJ software 1.8.0.

Zeta potential (ζ) measurements were performed using a Malvern Zetasizer Nano ZL instrument (Malvern Panalytical Ltd., Malvern, UK). For each measurement, 2.0 mg of sample was suspended in 10.0 mL of phosphate-buffered saline (PBS) at pH 7.4 and sonicated for 5 min to ensure homogeneous dispersion before analysis.

Surface area and pore size distributions were determined using N_2 adsorption–desorption isotherms obtained at −196 °C on a Micromeritics Tristar II 3020 system (Micromeritics Instrument Corp., Norcross, GA, USA). The specific surface area was calculated using the Brunauer–Emmett–Teller (BET) method [26], while the pore size distribution and volume were derived from the adsorption branch of the isotherms using the Barrett–Joyner–Halenda (BJH) method [27].

3. Results and Discussion

3.1. SEM and EDS Characterization

The surface morphology and elemental composition of the samples were evaluated using SEM and EDS analysis (Figure 2).

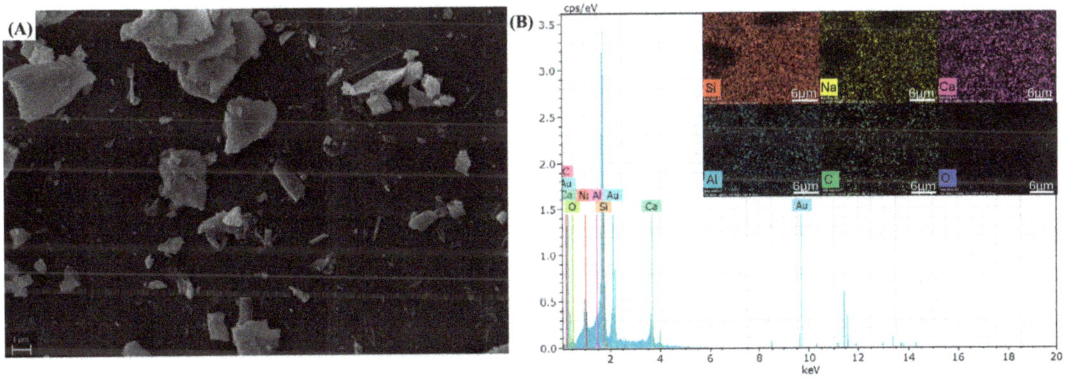

Figure 2. (**A**) SEM images and (**B**) EDS spectra and elemental mapping for the glass powder (GP) sample.

As shown in Figure 2A, the glass powder (GP) exhibits an inhomogeneous morphology composed of angular and irregular flaky particles. This irregular morphology is typical for mechanically crushed waste glass, as previously reported in the literature [28]. EDS analysis, presented in Figure 2B, confirms the presence of Si, O, Na, Ca, and Al, consistent with the typical composition of soda–lime–silica glass.

The silica gel (SG), obtained after alkaline dissolution and acid precipitation, shows a significant morphological transformation (Figure 3A). The SG sample consists of well-distributed, spherical particles with a mean diameter of approximately 0.19 ± 0.02 µm, as determined from the particle size distribution (Figure 3E). This morphology is commonly observed for silica gels prepared using sol–gel routes [29–31]. The EDS results, illustrated in Figure 3D, indicate that while the Al and Ca present in the GP precursors were successfully removed during the purification process, Na atoms persist within the structure despite the HCl precipitation.

https://doi.org/10.3390/ma19030601

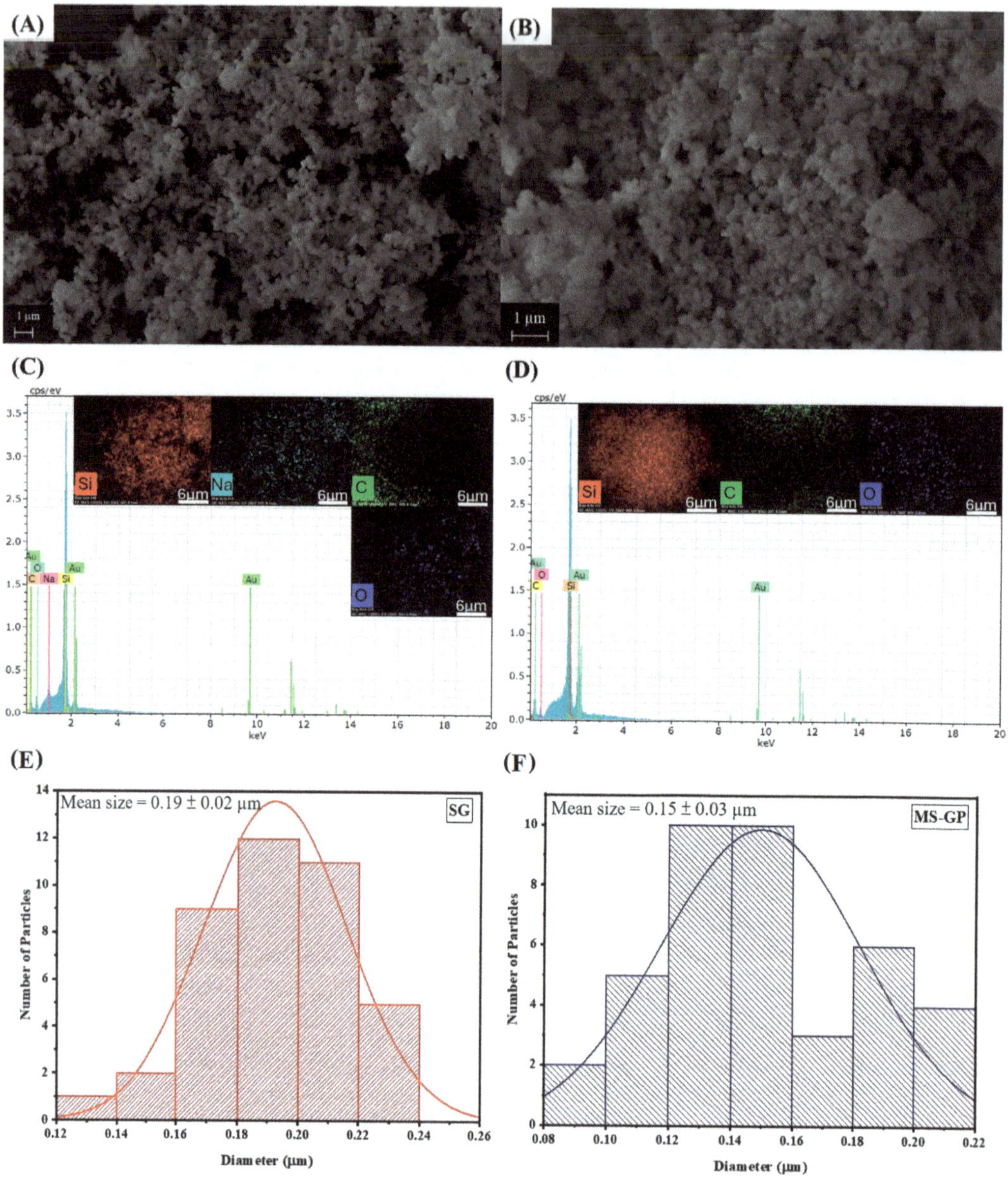

Figure 3. SEM images, EDS mapping, and particle size distributions for (**A**,**C**,**E**) SG and (**B**,**D**,**F**) MS-GP samples.

For MS-GP (Figure 3B), a spheric morphology similar to that of the SG sample was observed, with a slightly reduced mean particle size of 0.15 µm ± 0.03 (Figure 3F). Although SBA-15 materials are reported to exhibit a rod-like structure [32], the retention of a spherical form in this study is attributed to the pH conditions maintained during the synthesis involving the P123 template. It is noteworthy that sphere-like mesoporous silicas are highly desirable for applications in high-performance liquid chromatography (HPLC) as this geometry facilitates uniform column packing and enhances the separation efficiency of

https://doi.org/10.3390/ma19030601

diverse molecular species [33,34]. The EDS mapping and spectra for MS-GP (Figure 3D) show a high-purity framework composed predominantly of Si and O atoms. The absence of Na in the final MS-GP structure indicates that the residual sodium ions found in the SG sample were effectively leached out during the hydrothermal treatment and final washing steps of the SBA-15-type mesoporous silica synthesis. Additionally, the Au peaks observed in all EDS spectra are attributed to the gold-sputtering process required for sample conductivity during analysis.

3.2. XRD Analysis

Figure 4 presents the XRD patterns of the GP, SG, and MS-GP samples. All samples exhibit a characteristically amorphous structure, evidenced by the absence of sharp diffraction peaks. Each pattern displays a prominent broad hump centered at approximately $2\theta = 25°$, which is attributed to the short-range structural order within the disordered silica network [13]. This observed behavior is consistent with the literature for silica-based materials, confirming that the glassy state was maintained throughout the chemical transformation from glass powder to silica gel and mesoporous materials [35,36].

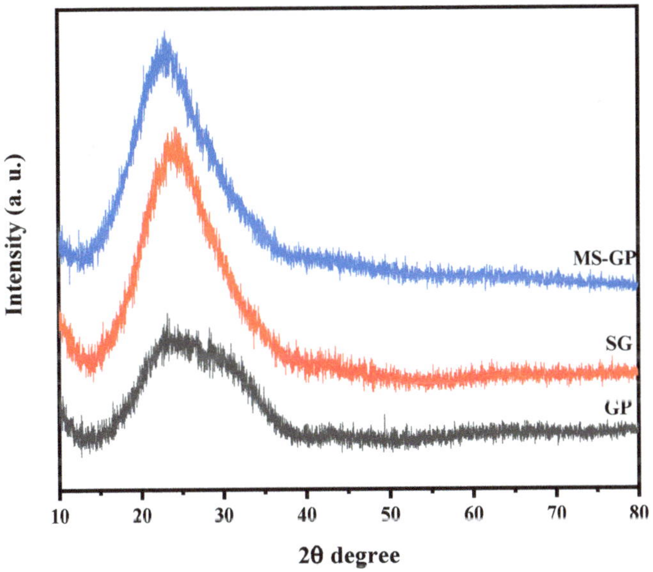

Figure 4. XRD patterns for GP, SG, and MS-GP samples.

To evaluate the structural ordering of the mesoporous framework, small-angle XRD patterns were recorded and are shown in Figure 5. In general, highly ordered mesoporous silicas, such as SBA-15, exhibit three well-resolved peaks at low angles, corresponding to the (100), (110), and (200) reflection planes of a 2D hexagonal (p6 mm) symmetry [37,38]. However, these characteristic reflections are absent in all analyzed samples. While the lack of ordered reflections was expected for the GP and SG samples, the absence of these peaks in the MS-GP sample suggests a disordered mesostructure. This result can be attributed to the limited solubility of the waste-derived silica gel in the acidic reaction medium during the synthesis. This incomplete dissolution hindered the effective condensation and organization of silica species around the surfactant micelles, preventing the formation of a highly periodic hexagonal arrangement.

https://doi.org/10.3390/ma19030601

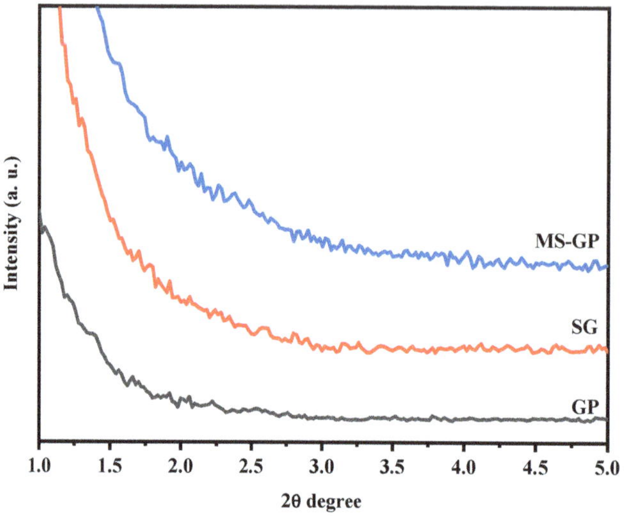

Figure 5. Small-angle XRD patterns for GP, SG, and MS-GP samples.

3.3. FTIR Analysis

Figure 6 presents the FTIR spectra of the GP, SG, and MS-GP samples. All spectra exhibit characteristic bands associated with the silica network. The bands centered at approximately 1090 cm^{-1}, 806 cm^{-1}, and 465 cm^{-1} are attributed to the asymmetric stretching, symmetric stretching, and bending vibrations of Si–O–Si bonds in siloxane groups, respectively [39]. The band located at around 965 cm^{-1} is assigned to the stretching vibration of Si–OH groups. In addition, the broad band at around 3445 cm^{-1} and the band at 1630 cm^{-1} correspond to the stretching and bending vibrations of O–H bonds, arising from both surface silanol groups and adsorbed water molecules [40,41].

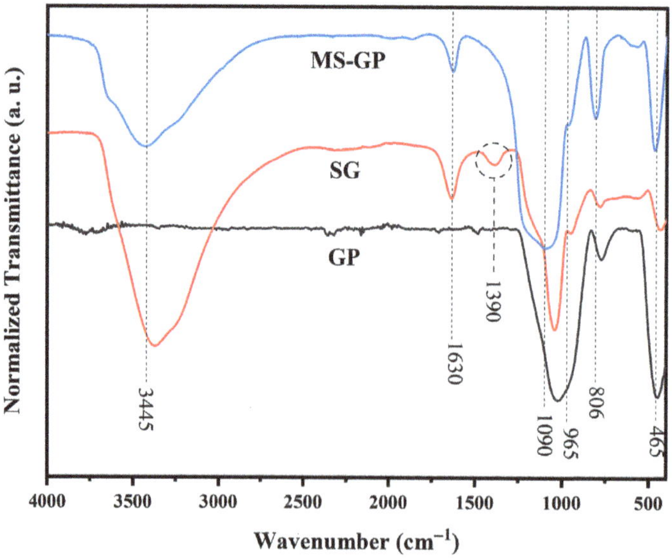

Figure 6. FTIR spectra for GP, SG, and MS-GP samples.

https://doi.org/10.3390/ma19030601

Differences between the GP and SG spectra confirm the formation of silica gel following alkaline dissolution and acid precipitation. The increased absorbance of the band at around 3400 cm^{-1} in the SG sample indicates a higher concentration of silanol groups within its structure. Additionally, the SG sample exhibits a distinct band at 1390 cm^{-1}, which can be attributed to residual carbonates (CO_3^{2-}) resulting from the interaction of the alkaline silicate solution with atmospheric CO_2 during processing and drying [42,43].

Significant shifts observed in the siloxane-related bands (1090, 806, and 465 cm^{-1}) for the SG sample compared to the GP precursor are attributed to the structural reorganization of the silicate network during the sol–gel transition. While EDS identified Al in the GP sample (Figure 2B), its absence in the SG sample (Figure 3D) indicates a successful chemical purification during the acid precipitation and washing steps, where soluble Al species were removed. Consequently, the spectral displacements in the SG spectrum reflect the transition from the multicomponent oxide network of the soda–lime glass to a high-purity, hydrated silica gel framework.

For the MS-GP sample, noticeable shifts in these characteristic bands are observed compared to the SG sample. These shifts are primarily attributed to structural rearrangements of the silica network induced by the calcination step used to remove the Pluronic P123 template [44]. Thermal treatment at elevated temperatures promotes changes in Si–O–Si bond angles and local network densification, which are known to result in shifts of vibrational bands toward higher wavenumbers [45,46]. Similar structural rearrangements and variations in Si–O–Si connectivity have been previously reported for silica-based glasses and glass-ceramics subjected to thermal and electrical characterization studies [47]. In this context, the calcination at 550 °C, applied to the MS-GP, increased Si–O–Si bond angles and induced partial condensation of surface silanol groups, explaining the observed spectral displacements.

3.4. *Zeta Potential Analysis (ξ)*

The surface charge and colloidal stability of the GP, SG, and MS-GP samples were evaluated via zeta potential (ξ) measurements in phosphate-buffered saline (PBS) at pH 7.4. The results, presented in Figure 7, show that all samples exhibit a negative surface charge, with values of −36.0 mV, −36.4 mV, and −30.6 mV for GP, SG, and MS-GP, respectively. The negative surface charge is attributed to the presence of silanol groups (Si–OH) on the silica surface, which partially deprotonate to form negatively charged silanolate groups ($Si–O^-$) under near-neutral or slightly basic conditions [48].

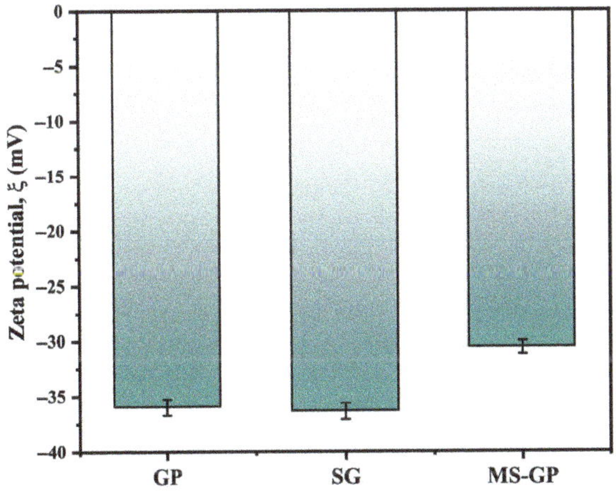

Figure 7. Zeta potential of GP, SG, and MS-GP samples.

https://doi.org/10.3390/ma19030601

Furthermore, the magnitude of the zeta potential provides important insight into the colloidal stability of particles in aqueous suspensions. In general, absolute zeta potential values greater than approximately 30 mV, either positive or negative, are associated with moderate to good electrostatic stabilization due to interparticle repulsion, which limits aggregation. The values obtained in this study indicate that the waste-derived silica particles exhibit good dispersion stability in aqueous media. This characteristic, combined with their negative surface charge, makes these materials promising candidates for biomedical applications, such as targeted drug delivery systems and cancer therapy, where colloidal stability under physiological conditions is essential [49–51].

3.5. Nitrogen Sorption Analysis

The N_2 adsorption–desorption isotherm of the GP sample is shown in Figure 8. The GP sample exhibits negligible nitrogen uptake, resulting in a very low specific surface area of 0.1 m^2/g. This behavior is typical of dense, non-porous glass materials and confirms the absence of accessible micro or mesoporosity in the glass powder.

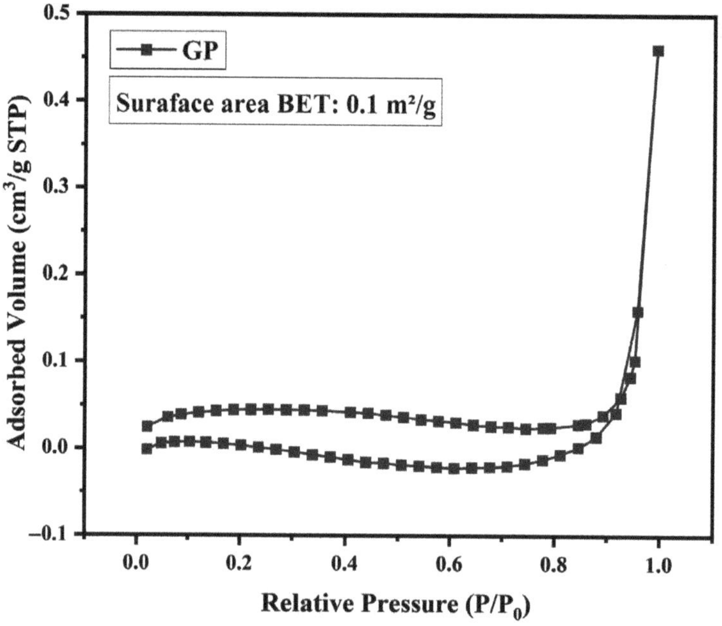

Figure 8. N_2 sorption isotherms for the GP sample.

Figure 9 shows the N_2 adsorption–desorption isotherms and pore size distributions of the SG and MS-GP samples. Both samples exhibit type II isotherms according to IUPAC classification, which are characteristic of non-porous or macroporous solids and materials with limited mesoporosity. A weak and narrow hysteresis loop is observed at high relative pressures ($P/P0 \gtrsim 0.8$), indicating the presence of limited capillary effects. This narrow hysteresis behavior is commonly associated with open, heterogeneous, and weakly connected pore systems and is attributed to textural porosity rather than to a well-developed, ordered mesoporous network. The main textural parameters derived from the adsorption–desorption data are summarized in Table 2.

https://doi.org/10.3390/ma19030601

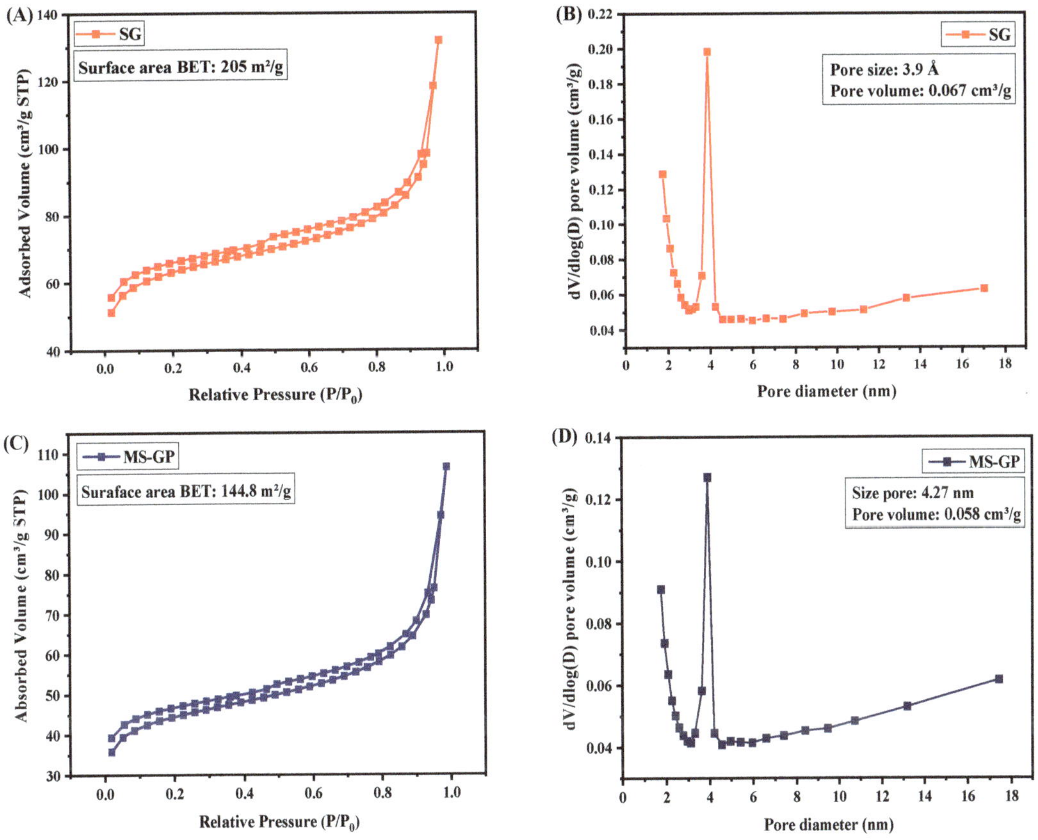

Figure 9. N_2 sorption isotherms for (**A**) SG and (**C**) MS-GP. Size distribution for (**B**) GS and (**D**) MS-GP.

Table 2. Textural properties of glass-derived materials obtained from N_2 adsorption–desorption analysis.

Sample	BET Surface Area (m^2/g)	Pore Size (Å)	Pore Volume (cm^3/g)
GP	0.1	—	—
SG	205	3.9	0.067
MS-GP	144.8	4.27	0.058

As shown in Figure 9A,C, the BET surface areas of SG and MS-GP are approximately 205 m^2/g and 145 m^2/g, respectively. These values are significantly lower than those typically reported for highly ordered SBA-15 materials synthesized from alkoxide precursors such as TEOS or TMOS, which generally exhibit Type IV isotherms and specific surface areas in the range of 500–1300 m^2/g [52,53].

The reduced surface area observed for the MS-GP sample can be attributed to the limited solubilization of the silica gel under acidic synthesis conditions, which hindered effective interaction between silica species and the P123 surfactant micelles. Moreover, the calcination step at 550 °C used to remove the organic template can promote further condensation of silanol groups and partial densification of the silica framework, leading to a reduction in accessible surface area [54,55].

https://doi.org/10.3390/ma19030601

The pore size distribution curves of SG and MS-GP samples, presented in Figure 9B,D, possess average pore diameters in the mesoporous range, with the MS-GP sample exhibiting larger pores (Table 2). The gradual increase in pore volume observed at pore diameters above approximately 8 nm suggests the presence of larger mesopore-sized cavities associated with textural porosity or poorly connected pore domains rather than a uniform, well-ordered mesoporous channel system. Such features are commonly associated with pore coalescence, partial pore blocking, or mesopore disconnection, particularly when silica condensation is incomplete or heterogeneous [56]. The presence of larger pores reduces the overall surface area by decreasing the contribution of smaller, high-surface-area pores, which explains the reduced BET values despite measurable pore volumes [57].

Overall, the N_2 sorption results indicate that while the SG and MS-GP samples exhibit some degree of mesoporosity, the formation of a highly ordered SBA-15-type mesoporous structure is limited. These results are consistent with the weak hysteresis observed in the adsorption–desorption isotherms and with the absence of long-range mesostructural ordering in small-angle XRD, as well as the particle aggregation observed by SEM. The results highlight the critical influence of precursor solubility and synthesis conditions on mesostructure formation when waste-derived silica sources are employed. Several strategies may be considered to improve silica solubility, including enhanced alkaline activation, controlled pH adjustment during dissolution, or the use of alternative dissolution agents.

4. Conclusions

In this work, waste soda–lime glass was successfully valorized as a silicon source through alkaline dissolution using NaOH (6 mol·L^{-1}), leading to the formation of silica gel with uniformly distributed, spherical particles. Structural characterization confirmed that all samples retained an amorphous silica structure, while FTIR analysis revealed shifts in Si–O–Si vibrational bands after sol–gel processing and calcination, indicating modifications in bond angles and silica network connectivity. N_2 sorption analysis revealed that the specific surface areas of silica gel and the mesoporous silica samples were comparable. This behavior was attributed to limitations in silica gel solubilization under acidic synthesis conditions, which constrained effective interaction with the surfactant template and hindered the development of a highly ordered mesostructure. Overall, this study demonstrates a proof of concept for the use of waste glass-derived silica gel as a precursor for mesoporous silica synthesis, while explicitly highlighting the structural limitations associated with precursor solubility and synthesis conditions. The results emphasize that careful control of precursor chemistry, solubility, and pH is essential to achieve improved mesostructural ordering when waste-derived silica sources are employed.

Author Contributions: Conceptualization, A.M.d.O.L. and A.A.M.M.; methodology, A.M.d.O.L. and A.A.M.M.; software, A.M.d.O.L., I.H. and M.P.F.G.; validation, A.M.d.O.L., I.H., M.P.F.G. and A.A.M.M.; formal analysis, A.M.d.O.L. and A.A.M.M.; investigation, A.M.d.O.L. and A.A.M.M.; resources, A.A.M.M.; data curation, I.H., M.P.F.G. and A.A.M.M.; writing—original draft preparation, A.M.d.O.L.; writing—review and editing, I.H., M.P.F.G. and A.A.M.M.; visualization, I.H., M.P.F.G. and A.A.M.M.; supervision, A.A.M.M. All authors have read and agreed to the published version of the manuscript.

Funding: This research was funded by Fundação de Amparo à Pesquisa e ao Desenvolvimento Científico e Tecnológico do Maranhão—FAPEMA (BPV-00342/22). The authors are also thankful to the Brazilian agency CNPq and Portuguese Foundation for Science and Technology (FCT), I.P., within the projects Number UIDB/04564/2020, UIDP/04564/2020, for the financial support.

Institutional Review Board Statement: Not applicable.

Informed Consent Statement: Not applicable.

https://doi.org/10.3390/ma19030601

Data Availability Statement: The original contributions presented in this study are included in the article. Further inquiries can be directed to the corresponding authors.

Conflicts of Interest: The authors declare no conflicts of interest.

References

1. Karazi, S.M.; Ahad, I.U.; Benyounis, K.Y. Laser Micromachining for Transparent Materials. In *Reference Module in Materials Science and Materials Engineering*; Elsevier: Amsterdam, The Netherlands, 2017.
2. Liu, Y.; Shi, C.; Zhang, Z.; Li, N. An Overview on the Reuse of Waste Glasses in Alkali-Activated Materials. *Resour. Conserv. Recycl.* **2019**, *144*, 297–309. [CrossRef]
3. Pandemia e Auxílio Emergencial Mudaram o Jeito de o Brasileiro Beber Cerveja. Available online: https://www.cnnbrasil.com.br/economia/macroeconomia/com-pandemia-brasileiro-troca-cerveja-grande-por-long-neck-e-falta-vidro/ (accessed on 31 December 2025).
4. Falta de Vidro Complica Entregas de Cerveja Long Neck e Preocupa Os Bares Na Chegada Do Verão | G1. Available online: https://g1.globo.com/economia/noticia/2021/12/14/falta-de-vidro-complica-entregas-de-cerveja-long-neck-e-preocupa-os-bares-na-chegada-do-verao.ghtml (accessed on 31 December 2025).
5. Laurinavičius, G.; Poškus, V. Mesoporous Silica Synthesis: Different Precursors, Catalysts and Structure Directing Agents. *Silicon* **2025**, *17*, 2381–2391. [CrossRef]
6. Borówka, A. Effects of Twin Methyl Groups Insertion on the Structure of Templated Mesoporous Silica Materials. *Ceram. Int.* **2019**, *45*, 4631–4636. [CrossRef]
7. Owoeye, S.S.; Jegede, F.I.; Borisade, S.G. Preparation and Characterization of Nano-Sized Silica Xerogel Particles Using Sodium Silicate Solution Extracted from Waste Container Glasses. *Mater. Chem. Phys.* **2020**, *248*, 122915. [CrossRef]
8. Babu, N.C.R.; Valente, M.A.; Rao, N.N.; Graça, M.P.F.; Raju, G.N.; Piasecki, M.; Kityk, I.V.; Veeraiah, N. Low Temperature Dielectric Dispersion and Electrical Conductivity Studies on Fe_2O_3 Mixed Lithium Yttrium Silicate Glasses. *J. Non-Cryst. Solids* **2012**, *358*, 3175–3186. [CrossRef]
9. Bahrami, Z.; Badiei, A.; Atyabi, F. Surface Functionalization of SBA-15 Nanorods for Anticancer Drug Delivery. *Chem. Eng. Res. Des.* **2014**, *92*, 1296–1303. [CrossRef]
10. Szegedi, A.; Popova, M.; Goshev, I.; Mihály, J. Effect of Amine Functionalization of Spherical MCM-41 and SBA-15 on Controlled Drug Release. *J. Solid State Chem.* **2011**, *184*, 1201–1207. [CrossRef]
11. Yang, X.; Li, Z.; Liu, Y.; Xing, Y.; Wei, J.; Yang, B.; Zhang, C.; Yang, R.T.; Tsai, C.-J. Research Progress of Gaseous Polycyclic Aromatic Hydrocarbons Purification by Adsorption. *Aerosol Air Qual. Res.* **2019**, *19*, 911–924. [CrossRef]
12. Sevimli, F.; Yılmaz, A. Surface Functionalization of SBA-15 Particles for Amoxicillin Delivery. *Microporous Mesoporous Mater.* **2012**, *158*, 281–291. [CrossRef]
13. Edeler, D.; Drača, D.; Petković, V.; Natalio, F.; Maksimović-Ivanić, D.; Mijatović, S.; Schmidt, H.; Kaluđerović, G.N. Impact of the Mesoporous Silica SBA-15 Functionalization on the Mode of Action of $Ph_3Sn(CH_2)_6OH$. *Mater. Sci. Eng. C* **2019**, *100*, 315–322. [CrossRef]
14. Samad, S.A.; Arafat, A.; Lester, E.; Ahmed, I. Upcycling Glass Waste into Porous Microspheres for Wastewater Treatment Applications: Efficacy of Dye Removal. *Materials* **2022**, *15*, 5809. [CrossRef]
15. Lin, L. Y.; Bai, H. Efficient Method for Recycling Silica Materials from Waste Powder of the Photonic Industry. *Environ. Sci. Technol.* **2013**, *47*, 4636–4643. [CrossRef]
16. Montini, D.; Cara, C.; D'Arienzo, M.; Di Credico, B.; Mostoni, S.; Nisticò, R.; Pala, L.; Scotti, R. Recent Advances on Porous Siliceous Materials Derived from Waste. *Materials* **2023**, *16*, 5578. [CrossRef] [PubMed]
17. Hammal, A.; Nabgali, N. Using Glass Waste to Prepare a Type of Bioceramic. *Baghdad Sci. J.* **2024**, *22*, 938–945. [CrossRef]
18. Ravikovitch, P.I.; Neimark, A.V. Characterization of Micro- and Mesoporosity in SBA-15 Materials from Adsorption Data by the NLDFT Method. *J. Phys. Chem. B* **2001**, *105*, 6817–6823. [CrossRef]
19. Karczmarska, A.; Laskowska, W.; Stróż, D.; Pawlik, K. Inside the Framework: Structural Exploration of Mesoporous Silicas MCM-41, SBA-15, and SBA-16. *Materials* **2025**, *18*, 3597. [CrossRef]
20. Medeiros de Paula, G.; do Nascimento Rocha de Paula, L.; Freire Rodrigues, M.G. Production of MCM-41 and SBA-15 Hybrid Silicas from Industrial Waste. *Silicon* **2022**, *14*, 439–447. [CrossRef]
21. Ryoo, R.; Ko, C.H.; Kruk, M.; Antochshuk, V.; Jaroniec, M. Block-Copolymer-Templated Ordered Mesoporous Silica: Array of Uniform Mesopores or Mesopore–Micropore Network? *J. Phys. Chem. B* **2000**, *104*, 11465–11471. [CrossRef]
22. Yang, P.; Huang, S.; Kong, D.; Lin, J.; Fu, H. Luminescence Functionalization of SBA-15 by $YVO_4:Eu^{3+}$ as a Novel Drug Delivery System. *Inorg. Chem.* **2007**, *46*, 3203–3211. [CrossRef]
23. Juárez, J.M.; Téllez, J.F.; Bercero, M.Á.L.; Moyano, E.L.; Gómez Costa, M.B. Synthesis of MCM-41 and SBA-15 from Rice Husk Silica and Their Carbon Replicas for Hydrogen Adsorption. *J. Chem. Tech. Biotech.* **2025**, *100*, 1609–1617. [CrossRef]

https://doi.org/10.3390/ma19030601

24. de Oliveira, I.R.B.; Sorce, A.R.; Gaglieri, M.V.V.; Cassanjes, F.C.; Maestrelli, S.C. Influence of the Addition of Glass from Long Neck Bottles in the Properties of the Reactive Powder Concrete. *Res. Soc. Dev.* **2022**, *11*, e519111335853. [CrossRef]
25. Rechotnek, F.; Follmann, H.D.; Silva, R. Mesoporous Silica Decorated with L-Cysteine as Active Hybrid Materials for Electrochemical Sensing of Heavy Metals. *J. Environ. Chem. Eng.* **2021**, *9*, 106492. [CrossRef]
26. Brunauer, S.; Emmett, P.H.; Teller, E. Adsorption of Gases in Multimolecular Layers. *J. Am. Chem. Soc.* **1938**, *60*, 309–319. [CrossRef]
27. Barrett, E.P.; Joyner, L.G.; Halenda, P.P. The Determination of Pore Volume and Area Distributions in Porous Substances. I. Computations from Nitrogen Isotherms. *J. Am. Chem. Soc.* **1951**, *73*, 373–380. [CrossRef]
28. Ortega, J.M.; Letelier, V.; Solas, C.; Miró, M.; Moriconi, G.; Climent, M.Á.; Sánchez, I. Influence of Waste Glass Powder Addition on the Pore Structure and Service Properties of Cement Mortars. *Sustainability* **2018**, *10*, 842. [CrossRef]
29. Suprapto, S.; Azizah, P.A.N.; Ni'mah, Y.L. Silica Gel from Chemical Glass Bottle Waste as Adsorbent for Methylene Blue: Optimization Using BBD. *J. Renew. Mater.* **2023**, *11*, 4007. [CrossRef]
30. Wang, C.; Tang, L.; Cui, L.; Chen, S.; Liu, J.; Dong, Y. SO_2 Adsorption and Kinetics Analysis during Supported Triethanolamine Acetate Ionic Liquid Desulfurization. *ACS Omega* **2022**, *7*, 41107–41119. [CrossRef]
31. Nakamura, Y.K.; Murakami, T.N.; Iwata, N.Y.; Hirano, Y.; Tokuoka, Y.; Kawashima, N. Preparation of Porous Diopside Microspheres from Spherical Silica Gels Impregnated with $Ca(NO_3)_2$ and $MgCl_2$. *J. Ceram. Soc. Jpn.* **2004**, *112*, 133–137. [CrossRef]
32. Wang, Z.; Sun, R.; Wang, P.; Wang, W. Unit-Cell Wide SBA-15 Type Mesoporous Silica Nanoparticles. *Microporous Mesoporous Mater.* **2021**, *328*, 111491. [CrossRef]
33. Ma, Y.; Qi, L.; Ma, J.; Wu, Y.; Liu, O.; Cheng, H. Large-Pore Mesoporous Silica Spheres: Synthesis and Application in HPLC. *Colloids Surf. A Physicochem. Eng. Asp.* **2003**, *229*, 1–8. [CrossRef]
34. Wan, H.; Liu, L.; Li, C.; Xue, X.; Liang, X. Facile Synthesis of Mesoporous SBA-15 Silica Spheres and Its Application for High-Performance Liquid Chromatography. *J. Colloid Interface Sci.* **2009**, *337*, 420–426. [CrossRef]
35. Abou Hussein, E.M.; Rammah, Y.S. Fabrication of Zirconium Borosilicate Glasses Using Municipal Waste Ash for Radiation Shielding Applications; Optical and Structural Investigations. *J. Phys. Chem. Solids* **2023**, *183*, 111633. [CrossRef]
36. Sangwan, P.; Barala, P.; Hooda, V. Enhancing Melatonin Photostability and Controlled Release Using pH-Responsive Mesoporous Silica Nanoparticles for Agricultural Applications. *Biocatal. Agric. Biotechnol.* **2025**, *64*, 103497. [CrossRef]
37. Kruk, M.; Jaroniec, M.; Ko, C.H.; Ryoo, R. Characterization of the Porous Structure of SBA-15. *Chem. Mater.* **2000**, *12*, 1961–1968. [CrossRef]
38. Ferreira Soares, D.C.; De Sousa Andrada, A.; Andrade Ramaldes, G. Silica Nanoparticles Containing Gadolinium Complex as Potential Alternative to Anticancer Radiotherapy. *Part. Sci. Technol.* **2015**, *33*, 331–338. [CrossRef]
39. Samadi-Maybodi, A.; Sedighi-Pashaki, E. Comprehensive Study of Loading and Release of Sodium Valproate Drug Molecule from Functionalized SBA-15 with Aminopropyl Groups Through Co-Condensation Modification Method. *Mater. Chem. Phys.* **2021**, *257*, 123622. [CrossRef]
40. Bhuyan, D.; Selvaraj, K.; Saikia, L. Pd@SBA-15 Nanocomposite Catalyst: Synthesis and Efficient Solvent-Free Semihydrogenation of Phenylacetylene under Mild Conditions. *Microporous Mesoporous Mater.* **2017**, *241*, 266–273. [CrossRef]
41. Betiha, M.A.; Moustafa, Y.M.; El-Shahat, M.F.; Rafik, E. Polyvinylpyrrolidone-Aminopropyl-SBA-15 Schiff Base Hybrid for Efficient Removal of Divalent Heavy Metal Cations from Wastewater. *J. Hazard. Mater.* **2020**, *397*, 122675. [CrossRef]
42. Torres-Carrasco, M.; Palomo, J.G.; Puertas, F. Sodium Silicate Solutions from Dissolution of Glasswastes. Statistical Analysis. *Mater. Construcción* **2014**, *64*, e014. [CrossRef]
43. Konduru, H.; Karthiyaini, S.; Shanmugasundaram, M. Comparative Study of Silica Fume and Sodium Silicate as Replacement of Active Reactive Silica in Bauxite Residue Based Geopolymer Mortar. *Case Stud. Constr. Mater.* **2025**, *22*, e04302. [CrossRef]
44. Pisciella, P.; Pelino, M. FTIR Spectroscopy Investigation of the Crystallisation Process in an Iron Rich Glass. *J. Eur. Ceram. Soc.* **2005**, *25*, 1855–1861. [CrossRef]
45. Agarwal, A.; Tomozawa, M. Correlation of Silica Glass Properties with the Infrared Spectra. *J. Non-Cryst. Solids* **1997**, *209*, 166–174. [CrossRef]
46. Yuan, X.; Cormack, A.N. Si–O–Si Bond Angle and Torsion Angle Distribution in Vitreous Silica and Sodium Silicate Glasses. *J. Non-Cryst. Solids* **2003**, *319*, 31–43. [CrossRef]
47. Graça, M.P.F.; da Silva, M.F.; Sombra, A.S.B.; Valente, M.A. Electrical Characterization of SiO_2: $LiNbO_3$ Glass and Glass–Ceramics Using Dc Conductivity, TSDC Measurements and Dielectric Spectroscopy. *J. Non-Cryst. Solids* **2007**, *353*, 4390–4394. [CrossRef]
48. Rosenholm, J.M.; Lindén, M. Wet-Chemical Analysis of Surface Concentration of Accessible Groups on Different Amino-Functionalized Mesoporous SBA-15 Silicas. *Chem. Mater.* **2007**, *19*, 5023–5034. [CrossRef]
49. Pochapski, D.J.; Carvalho Dos Santos, C.; Leite, G.W.; Pulcinelli, S.H.; Santilli, C.V. Zeta Potential and Colloidal Stability Predictions for Inorganic Nanoparticle Dispersions: Effects of Experimental Conditions and Electrokinetic Models on the Interpretation of Results. *Langmuir* **2021**, *37*, 13379–13389. [CrossRef]

50. Onugwu, A.L.; Nwagwu, C.S.; Onugwu, O.S.; Echezona, A.C.; Agbo, C.P.; Ihim, S.A.; Emeh, P.; Nnamani, P.O.; Attama, A.A.; Khutoryanskiy, V.V. Nanotechnology Based Drug Delivery Systems for the Treatment of Anterior Segment Eye Diseases. *J. Control. Release* **2023**, *354*, 465–488. [CrossRef]
51. Heurtault, B.; Saulnier, P.; Pech, B.; Proust, J.-E.; Benoit, J.-P. Physico-Chemical Stability of Colloidal Lipid Particles. *Biomaterials* **2003**, *24*, 4283–4300. [CrossRef]
52. Khanh Nguyen, Q.N.; Yen, N.T.; Hau, N.D.; Tran, H.L. Synthesis and Characterization of Mesoporous Silica SBA-15 and ZnO/SBA-15 Photocatalytic Materials from the Ash of Brickyards. *J. Chem.* **2020**, *2020*, 456194. [CrossRef]
53. Hernández Morales, R.; Pacheco Sosa, J.G.; Torres Torres, J.G.; Pérez Vidal, H.; Lunagómez Rocha, M.A.; Cruz Romero, D.; Escobar Aguilar, J.; Barrera, M.C. Pt/Ga-SBA-15 Composites Synthesis and Characterization. *Superf. Y Vacío* **2020**, *33*. [CrossRef]
54. Barczak, M. Template Removal from Mesoporous Silicas Using Different Methods as a Tool for Adjusting Their Properties. *New J. Chem.* **2018**, *42*, 4182–4191. [CrossRef]
55. Aumond, T.; Pinard, L.; Batiot-Dupeyrat, C.; Sachse, A. Non-Thermal Plasma: A Fast and Efficient Template Removal Approach Allowing for New Insights to the SBA-15 Structure. *Microporous Mesoporous Mater.* **2020**, *296*, 110015. [CrossRef]
56. Ali, U.; Meinusch, R.; Turke, K.; Schreiner, P.R.; Smarsly, B.M. Tuning the Mesopore Network in Meso-Macroporous Silica Monoliths by Hydrothermal Treatment—A Physisorption Study. *Langmuir* **2025**, *41*, 13845–13859. [CrossRef] [PubMed]
57. Angotzi, M.S.; Mameli, V.; Cara, C.; Borchert, K.B.; Steinbach, C.; Boldt, R.; Schwarz, D.; Cannas, C. Meso-and Macroporous Silica-Based Arsenic Adsorbents: Effect of Pore Size, Nature of the Active Phase, and Silicon Release. *Nanoscale Adv.* **2021**, *3*, 6100–6113. [CrossRef] [PubMed]

https://doi.org/10.3390/ma19030601

materials

MDPI

Article

Study and Reuse of Silicone Implants After Radiotherapy

Aleksandra Kozłowska [1], Marta Worzakowska [2], Bogdan Tarasiuk [2], Maria Piziorska [1] and Beata Podkościelna [2,*]

[1] Department of Radiotherapy, University Clinical Hospital No. 1, 13 Radziwiłłowska Street, 20-080 Lublin, Poland; olla.kozlowska@gmail.com (A.K.); maria.piziorska@gmail.com (M.P.)
[2] Department of Polymer Chemistry, Faculty of Chemistry, Institute of Chemical Sciences, Maria Curie-Sklodowska University, Gliniana St. 33, 20-614 Lublin, Poland; marta.worzakowska@mail.umcs.pl (M.W.); bogtar2020@wp.pl (B.T.)
* Correspondence: beata.podkoscielna@mail.umcs.pl

Abstract

Studies on the influence of radiotherapy on the structure, thermal properties, and thermo-oxidative decomposition of breast silicone implants were conducted. Additionally, the potential use of breast silicone implant waste as a component in layered composites was investigated. ATR-FTIR, DSC, and TG/DTG/FTIR analyses confirmed that radiation does not affect the structure, thermal properties, or oxidative decomposition behavior of the shell and gel layers of breast silicone implants. The conducted tests demonstrated the successful fabrication of composite materials using a PUM matrix and breast silicone implant waste. The presence of the PUM matrix in the manufactured composites influenced the crystallization and melting behavior of the silicone phase. Moreover, the incorporation of a silicone implant waste layer into the composites increased their thermal stability while decreasing the glass transition temperature, storage modulus and hardness compared to neat PUM. The type of implant waste layer used (shell or gel) in the preparation of the PUM composites did not significantly affect the melting and glass transition temperatures, thermal stability, or oxidative decomposition behavior of the newly developed materials. As demonstrated, new layered composite materials based on silicone implant waste (shell and gel), with properties valuable for practical applications, were successfully developed.

Keywords: breast silicone implants; post-mastectomy radiotherapy; reuse of breast implants; TG/DTG/FTIR

1. Introduction

Immediate implant-based reconstruction after mastectomy is now a common clinical practice. Indications for post-mastectomy radiation therapy (PMRT) remain unchanged by the presence of implants and include high-risk factors such as tumor size > 5 cm, chest wall or skin involvement, positive margins, and ≥4 positive lymph nodes. Decisions in cases with 1–3 positive lymph nodes are increasingly individualized. Managing PMRT alongside immediate reconstruction requires balancing oncological safety with esthetic outcomes [1–3]. While conventional fractionation schedules (50 Gy in 25 fractions) are still used, moderate hypofractionation (40–42.5 Gy in 15–16 fractions) has become the standard of care, demonstrating comparable efficacy without increasing implant-related complications [3,4]. In prepectoral reconstruction, the implant is placed above the pectoral muscle, offering more natural breast contours, reduced postoperative pain, and faster recovery compared to the submuscular approach. However, careful patient selection is essential due to thinner soft-tissue coverage and an increased risk of post-radiation rippling and ischemia. An ideal candidate for the prepectoral technique is a non-smoker with a moderate body mass index (BMI). Acellular dermal matrices (ADMs), used as biological scaffolds, help reduce animation deformity (involuntary breast movement during muscle contraction) associated with radiation-induced fibrosis, improve esthetic outcomes, and enhance tissue integration [5,6]. In clinical practice, complications associated with PMRT in patients undergoing prepectoral reconstruction are classified as acute or late effects. Early complications occur during treatment or within six months after radiotherapy and include dermatitis, implant exposure, seroma, and fat necrosis. Late complications, occurring after six months, include capsular contracture, telangiectasia, and skin atrophy or fibrosis [7]. To minimize treatment-related morbidity, advanced radiation delivery techniques are essential. Volumetric modulated arc therapy (VMAT) enables highly conformal dose distribution, shaping isodose curves to minimize irradiation of the heart and lungs while maintaining dose homogeneity within the clinical target volume. To address the risk of skin recurrence, which is often thin and stretched, bolus materials (tissue-equivalent layers) are used to increase the surface dose. Furthermore, deep inspiration breath hold (DIBH) is routinely applied in left-sided cases to increase the distance between the heart and the radiation field, thereby reducing long-term cardiotoxicity. Patients should be informed about the importance of gentle daily skin care, prophylactic moisturizing, loose clothing, and sun protection during and after radiotherapy [2,3].

Modern silicone breast implants are extensively tested and considered safe; however, they are not lifetime devices. Radiotherapy after mastectomy clearly increases the risk of implant-related complications. Implants are also removed or replaced for esthetic and personal reasons, including changes in patient preference or dissatisfaction with appearance. In addition, replacement is often recommended based on duration of use, as implants typically have a lifespan of approximately 10–15 years. Prolonged contact with biological tissues may lead to material aging and changes in properties, necessitating replacement or removal [6,7]. Implants removed from one patient are not reused in another, due to sterility concerns, risk of complications, and safety, hygiene, and ethical considerations. They are certified single-use medical devices. Silicone breast implants removed for medical or esthetic reasons are treated as medical waste and disposed of, most commonly via thermal treatment. The most widely used method is high-temperature incineration in specialized facilities, which leads to the complete decomposition of the silicone material.

Surgical procedures involving breast implants generate significant amounts of medical waste, particularly in the form of single-use plastics. Approximately 4 kg of plastic waste is produced during an implant replacement [8]. Meanwhile, attention is drawn to the significant carbon footprint of breast reconstruction procedures involving implants, mainly due to the production and transportation of medical equipment, and the use of single-use materials [9]. Therefore, there is a growing emphasis on the need to implement strategies that reduce waste and improve the sustainable management of resources in plastic surgery [10]. Despite growing interest in sustainability issues, there are relatively few reports in the scientific literature regarding the recyclability of used medical implants. While these materials cannot be reused for medical applications due to biological safety concerns, they could potentially be used as a source of raw materials for producing new materials for non-medical applications [11]. In this context, this article presents a new approach involving using used implants as a component in producing new composite materials.

In the present work, we investigate the effects of ionizing radiation on silicone breast implants and propose the reuse of silicone implant waste obtained after radiotherapy or explantation for personal reasons (in cases not previously exposed to radiotherapy) as a component of layered composites. In our study, both layers of silicone implant waste, the outer shell and inner gel, were used to fabricate novel composite materials. The structure of

https://doi.org/10.3390/ma19091798

the implants and the prepared composites was confirmed using attenuated total reflection Fourier transform infrared spectroscopy (ATR-FTIR). The effects of ionizing radiation on glass transition temperature, melting and crystallization behavior, thermal stability in inert and oxidative atmospheres, and thermal degradation pathways under heating in synthetic air were evaluated. Furthermore, the influence of the silicone layer type (shell or gel) on the thermal properties, thermal stability, viscoelastic properties, hardness and thermo-oxidative degradation behavior of the prepared composites was investigated using differential scanning calorimetry (DSC) and thermogravimetric/derivative thermogravimetric analysis coupled with Fourier transform infrared spectroscopy (TG/DTG/FTIR), dynamic mechanical analysis (DMA) and Brinell hardness.

2. Materials and Methods

2.1. Materials

Commercially available gel-filled breast implants, consisting of silicone gel (polysiloxane) enclosed in a silicone elastomer shell (Mentor, OH, USA), were used. 2,4-diisocyanato-1-methylbenzene, poly(oxypropylene)diol (Mw = 2000 g/mol), 2-hydroxyethyl methacrylate, 3-(trimethoxysilyl)propyl methacrylate, 2-ethylhexyl acrylate, 2,2-dimethoxy-2-phenylacetophenone (photoinitiator), and dibutyltin dilaurate were purchased from Sigma-Aldrich (Darmstadt, Germany). The liquid composition was cured using TL-K 40W/10-R UV-A lamps (Philips, Amsterdam, the Netherlands), emitting radiation in the wavelength range of 315–380 nm, with a UV-A irradiance of 8.0 W (0 h, IEC-International Electrotechnical Commission).

2.2. Preparation of the Composites

2.2.1. Synthesis of Oligomeric Urethane-Methacrylate Resin (OUM)

In the first step, an oligomeric urethane-methacrylate resin (OUM) was synthesized by reacting 0.05 mol of 2,4-diisocyanato-1-methylbenzene with 0.025 mol of poly(propylene)diol. The reaction was carried out at 80 °C for 5 h. The resulting urethane prepolymer was then cooled to 50 °C, followed by the addition of 0.05 mol of 2-hydroxyethyl methacrylate and 22.0 g of 2-ethylhexyl acrylate, used as a reactive diluent. The reaction was continued at 55 °C for 2 h. Subsequently, the mixture was again cooled to 50 °C, and 0.20 g of dibutyltin dilaurate was added as a catalyst. The system was stirred for an additional 2 h to promote and complete the formation of the urethane–methacrylate oligomer. Detailed physicochemical data are provided in Ref. [12]. Figure 1 presents the chemical structure of the obtained oligo(urethane-methacrylate) resin.

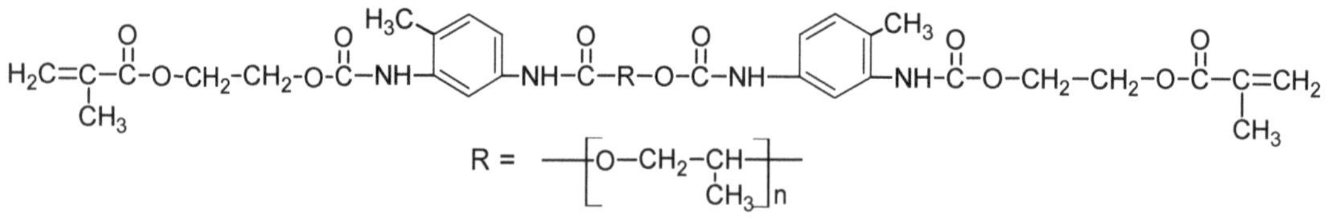

Figure 1. Chemical structure of resin oligo(urethane–methacrylate) (OUM).

2.2.2. Synthesis of New Compositions Based on Waste Breast Implants

In the second step, a urethane–methacrylate composition (PUM) was prepared by mixing the following components: oligo(urethane–methacrylate) resin (OUM), 3-(trimethoxysilyl)propyl methacrylate (10 wt%), and the photopolymerization initiator 2,2-dimethoxy-2-phenylacetophenone (4 wt%). Two types of composites were prepared depending on the type of silicone from the breast implants used, namely the shell or the

 https://doi.org/10.3390/ma19091798

gel fraction. The procedure was as follows: two thin PUM films were initially cured under UV irradiation for 2 min. Subsequently, a layer consisting of PUM, gel, or silicone shell containing liquid methacrylates was placed between two glass plates and exposed to UV irradiation for 10 min. After curing, the solid samples were subjected to further testing. A schematic diagram of the mold polymerization process is shown in Figure 2. The chemical structure of the functional groups at the PUM polymer–silicone interface is presented in Figure 3.

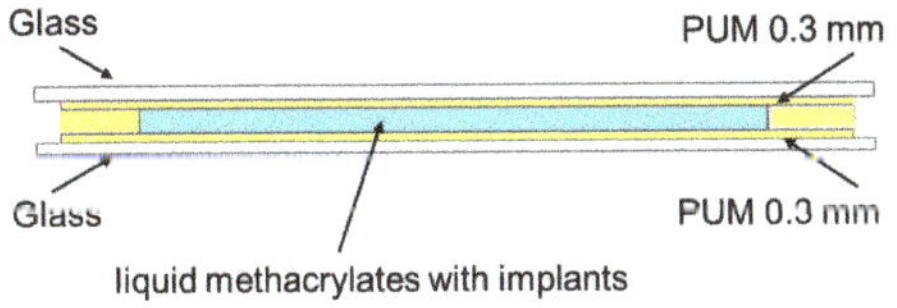

Figure 2. Schematic diagram of the mold for preparing the silicone recycling composite.

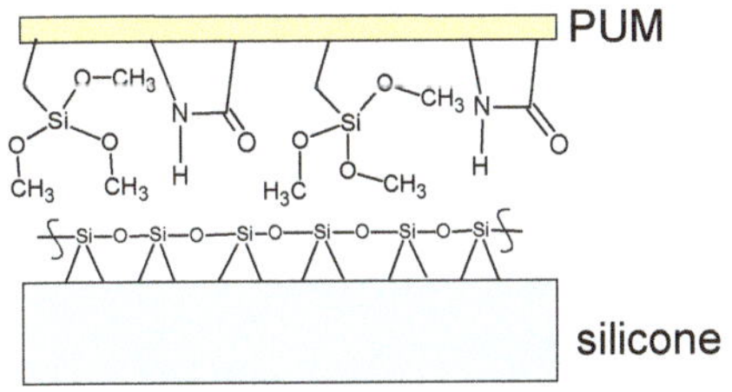

Figure 3. Chemical structure of functional groups at the PUM polymer–silicone interface.

2.2.3. Methodology of Irradiation of Silicone Breast Implants

The implant was irradiated using a standard protocol commonly applied in clinical practice. The treatment plan was developed using the MONACO treatment planning system. A total dose of 50 Gy was delivered in 25 fractions of 2 Gy using 6 MV photon beams generated by a Versa HD linear accelerator (Elekta, Stockholm, Sweden). The volumetric modulated arc therapy (VMAT) technique was employed.

2.3. *Methods*

2.3.1. Attenuated Total Reflection Fourier Transform Infrared Spectroscopy (ATR-FTIR) Measurements

ATR-FTIR spectra of the investigated silicone breast implants (before and after radiotherapy) and the prepared layered composites were recorded using a Tensor 27 FTIR spectrometer equipped with a diamond ATR crystal (Bruker, Bremen, Germany). Each spectrum was collected over the range of 600–4000 cm^{-1} with a resolution of 4 cm^{-1} and 64 scans.

2.3.2. Differential Scanning Calorimetry (DSC) Measurements

DSC measurements were performed using a DSC 204 Phoenix calorimeter (Netzsch, Selb, Germany). Samples with a mass of approximately 10 mg were placed in aluminum crucibles with pierced lids and analyzed under an inert atmosphere (argon, flow rate: 30 mL/min). The heating and cooling rate was 10 °C min^{-1}. The DSC procedure was as follows: the sample was first cooled to −150 °C, then heated to 150 °C, cooled again to −150 °C, and finally reheated to 150 °C. Each measurement was repeated three times, and the values represent the arithmetic mean of these measurements. Characteristic temperatures, including the glass transition temperature (T_g), crystallization temperature (T_c), and

https://doi.org/10.3390/ma19091798

melting temperature (T_m), as well as the corresponding enthalpies of crystallization (ΔH_c) and melting (ΔH_m), were determined based on the DSC curves.

2.3.3. Thermogravimetric/Derivative Thermogravimetric (TG/DTG) Measurements

Thermogravimetric (TG/DTG) analysis was performed to evaluate the thermal properties of the tested silicone samples and the prepared composites using an STA 449 Jupiter F1 instrument (Netzsch, Selb, Germany). In a typical experiment, approximately 10 mg of sample was placed in an open corundum crucible and heated from 40 °C to 1000 °C at a heating rate of 10 °C min^{-1}. All analyses were carried out under two different furnace atmospheres: inert (helium, flow rate: 40 mL/min) and oxidative (synthetic air, flow rate: 100 mL/min). Each measurement was repeated three times, and the results are the arithmetic mean of these measurements. Based on the TG/DTG curves, the following parameters were determined: the initial decomposition temperature corresponding to a 5% mass loss ($T_{5\%}$), the temperatures of the maximum decomposition rate (T_{max}), mass losses at individual decomposition stages (Δm), and the residual mass at 1000 °C (m_r).

2.3.4. Simultaneous Thermogravimetric/Derivative Thermogravimetric Analysis Coupled with Fourier Transform Infrared Spectroscopy (TG/DTG/FTIR) Measurements

Simultaneous TG/DTG/FTIR analysis was carried out using an STA 449 Jupiter F1 instrument (Netzsch, Selb, Germany) coupled online with a TGA 585 FTIR spectrometer (Bruker, Selb, Germany). FTIR spectra of gaseous products evolved during the oxidative decomposition of the tested silicone implants and the prepared composites were collected to identify the emitted compounds and assess their potential toxicity during disposal via combustion. The spectra were recorded in the wavenumber range of 600–4000 cm^{-1} with a resolution of 4 cm^{-1} and 32 scans per spectrum.

2.3.5. Dynamic Mechanical Analysis (DMA) Measurements

DMA was performed on a DMA Q 800 (TA Instruments, New Castle, DE, USA). Tests were con-ducted using a tension clamp. Measurements for all samples were carried out from −150 °C up to the temperature at which the samples became too soft to be tested. A constant heating rate of 3 °C min^{-1} and an oscillation frequency of 1 Hz were applied. Rectangular samples with dimensions of 10 mm in length, 6 mm in width, and 1 mm in thickness were used. The storage modulus ($E_{20°C}$) and maximum tangent delta (tg δ) were evaluated.

2.3.6. Brinell Hardness Measurements

The Brinell hardness (HK) was determined using an HPK hardness tester (Schifferstadt, Germany). In this procedure, a steel ball indenter with a diameter of 5 mm was hydraulically pressed into the material surface under a load (F) of 13.5 kgf, applied for 60 s. The resulting indentation depth (h), expressed in centimeters, was measured. Ten measurements were performed for each sample. The HK values are the arithmetic mean of these measurements with the calculated standard deviation. The Brinell hardness (HK), expressed in megapascals (MPa), was then calculated using the following equation:

$$\mathrm{HK}\,[\mathrm{MPa}] = \frac{F}{1.57 \cdot h \cdot 0.098066} = F_1 \cdot 0.098066$$

where F is the applied load (13.5 kgf) and h is the indentation depth (cm), with $F_1 = F/(1.57 \times h)$.

https://doi.org/10.3390/ma19091798

2.3.7. Surface Morphology Characterization

Morphological images and surface roughness were determined for each sample using a Contour GT-K1 optical profilometer (Veeco Instruments, Inc., Plainview, NT, USA) which has a very high level of accuracy in the size range from subnanometers to 10 mm. The experiments were repeated three times for each sample. The average roughness (Ra) parameter was determined.

3. Results

The investigated silicone breast implants, both after radiotherapy and those not previously subjected to radiotherapy (referred to as "before radiotherapy"), consist of two distinct layers. The outer layer (shell) is organoleptically harder, yet elastic, and less transparent. In contrast, the inner layer (gel) is softer, more fluid, and more transparent. Due to the layered structure of the implants, both the shell and gel were analyzed before and after radiotherapy. The shell before radiotherapy is denoted as (a), and the gel before radiotherapy as (b). The corresponding layers after radiotherapy are denoted as (a′) and (b′). Furthermore, both implant layers were used as components in composites with PUM, denoted as (c).

3.1. *ATR-FTIR Results*

3.1.1. ATR-FTIR Spectra of Silicone Breast Implants

Figure 4 presents the ATR-FTIR spectra of the investigated silicone breast implants before (a, b) and after (a′, b′) radiotherapy. The spectra include both the shell (a, a′) and gel (b, b′) layers of the implants. All FTIR spectra exhibit characteristic absorption bands corresponding to asymmetric stretching vibrations of $-CH_3$ in $Si-CH_3$ groups (2905 and 2962 cm^{-1}), asymmetric deformation vibrations of $-CH_3$ (1412 cm^{-1}), symmetric deformation vibrations of $-CH_3$ in $Si-CH_3$ (1258 cm^{-1}), and asymmetric stretching vibrations of Si–O–Si in the silicone backbone (1010 and 1080 cm^{-1}). Additionally, rocking vibrations of $Si-CH_3$ (863 cm^{-1}) and coupled vibrations involving Si–C stretching in $Si-CH_3$ and $-CH_3$ rocking (787 cm^{-1}) are observed. Importantly, both the shell and gel spectra, recorded before and after radiotherapy, show absorption bands in the same wavenumber regions. This indicates that radiotherapy exposure does not induce detectable structural changes in the investigated implants. Furthermore, the ATR-FTIR analysis confirms that the main component of the tested silicone breast implants, in both shell and gel layers, is polydimethylsiloxane (PDMS) [13–18].

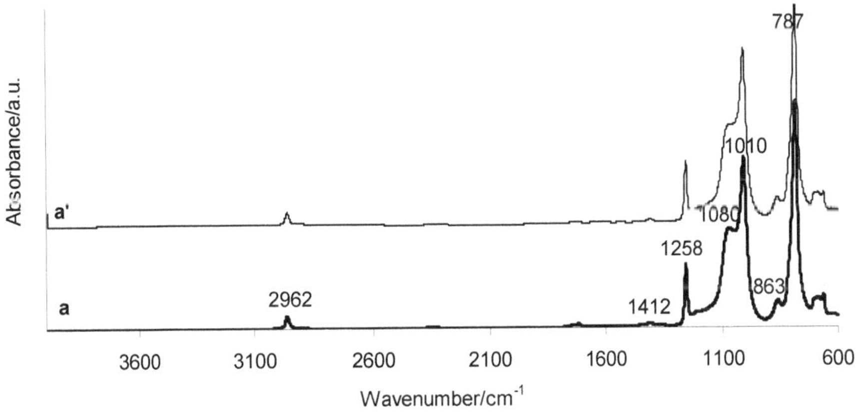

Figure 4. *Cont.*

https://doi.org/10.3390/ma19091798

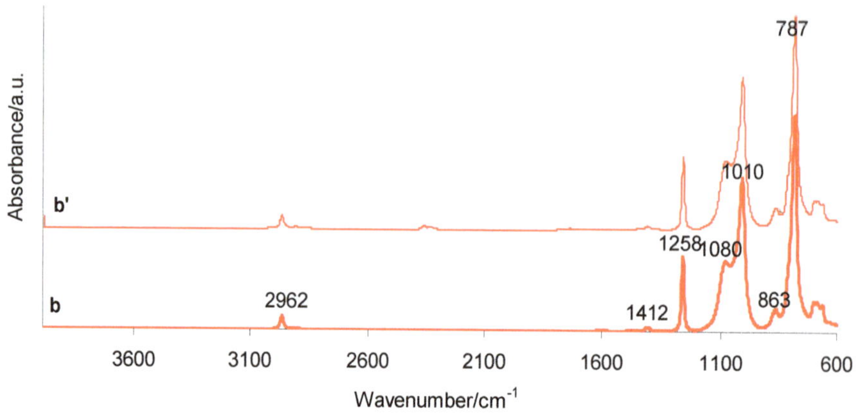

Figure 4. ATR-FTIR spectra of silicone breast implants: shell before (a) and after radiotherapy (a′), and gel before (b) and after radiotherapy (b′).

3.1.2. ATR-FTIR Spectra of the Prepared Composites

Due to the lack of structural changes in silicone breast implants after radiotherapy, it can be assumed that these materials are not subjected to degradation under ionizing radiation. Therefore, both pre- and post-radiotherapy silicone breast implants can be used for composite preparation without significant changes in their properties. Figure 5 presents the ATR-FTIR spectra of the prepared composites containing the shell fraction of breast implant waste (a + c) and those containing the gel fraction (b + c). For comparison, the spectra of both implant layers (a and b) and the PUM matrix (c) are also shown.

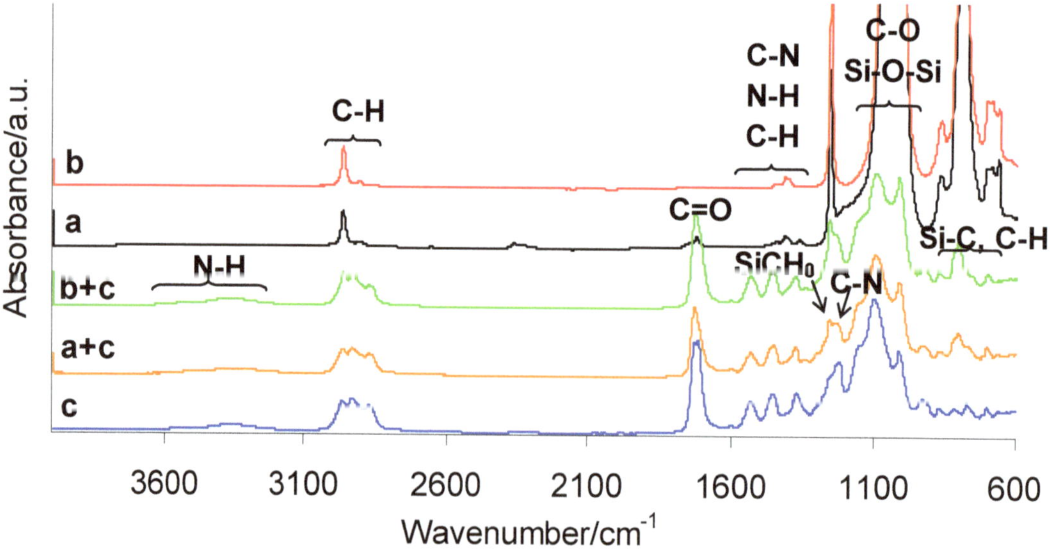

Figure 5. ATR-FTIR spectra of silicone breast implants: shell (a), gel (b), PUM (c), and prepared composites: shell–PUM (a + c) and gel–PUM (b + c).

As it is clearly observed, the ATR-FTIR spectra of both PUM and the prepared composites exhibit a broad absorption band in the range of 3500–3100 cm^{-1}, which is attributed to N–H stretching vibrations. In addition, a weaker band at 3269–3280 cm^{-1} corresponds

https://doi.org/10.3390/ma19091798

to hydrogen-bonded N–H stretching vibrations in both the composites and PUM. The C–H stretching vibrations associated with PUM, silicone implants (C–H in Si–CH_3), and the composites (C–H in both Si–CH_3 and PUM) appear in the range of 2968–2864 cm^{-1}. The absorption band at 1722 cm^{-1} indicates the presence of C=O stretching vibrations in both the composites and the neat PUM. Bands observed at 1523–1570 cm^{-1} are attributed to C–N stretching vibrations and N–H bending vibrations in the structure of PUM and the composites. Bending vibrations of C–H groups appear in the region of 1360–1440 cm^{-1}. The FTIR spectra of the composites also show two characteristic maxima at 1228 cm^{-1} and 1255 cm^{-1}, corresponding to C–N stretching in urethane groups and deformation vibrations of –CH_3 in Si–CH_3, respectively. Intense absorption bands in the range of 1010–1140 cm^{-1} are associated with C–O stretching vibrations of ether groups and Si–O–Si stretching vibrations of the silicone backbone. Additionally, rocking vibrations of Si–CH_3 (863 cm^{-1}) and coupled vibrations involving Si–C stretching in Si–CH_3 and –CH_3 rocking (787 cm^{-1}) are clearly observed [17–19]. The presence of all these characteristic bands confirms the expected chemical structure of the prepared materials and the successful fabrication of composites based on PUM and silicone breast implant waste.

3.2. DSC Results

DSC analysis of the investigated silicone breast implants before and after radiotherapy, as well as of the prepared composites, enabled the determination of the glass transition temperature (T_g), melting temperature (T_m), crystallization temperature (T_c), enthalpy of fusion (ΔH_f), and degree of crystallinity (χ). The obtained DSC curves for the shell and gel fractions of the silicone implant waste, PUM, and the prepared composites are shown in Figure 6. The numerical data derived from the DSC curves, including results for implants before and after radiotherapy, are presented in Table 1. The glass transition temperature (T_g) characterizes the transition of the amorphous phase from a rubbery to a glassy state. At T_g, long-range translational and rotational motions of polymer chains become frozen. T_g is directly related to polymer chain mobility; higher mobility corresponds to lower T_g values [20]. Based on the DSC results, the polymer chain mobility in the investigated implants can be considered high. The T_g values for the shell and gel before radiotherapy are comparable, at −123.9 °C (shell) and −121.3 °C (gel), respectively. These values are consistent with literature data [21–24]. After radiotherapy, no significant changes in T_g are observed for either layer; any slight variations fall within the experimental error. Therefore, it can be concluded that radiotherapy does not affect the T_g of the shell and gel layers of silicone implants and does not induce detectable changes in their chemical structure. For the prepared composites, two glass transition regions are observed, corresponding to two T_g values. The first T_g appears at −119.4 °C for the shell–PUM composite and −117.6 °C for the gel–PUM composite. The second T_g occurs at approximately −61.0 °C for both systems. In comparison, the neat PUM matrix exhibits a T_g of −50.8 °C. The decrease in the second T_g by approximately 10 °C in the composites compared to pure PUM is most likely due to the plasticizing effect of the siloxane chains present in PDMS silicone implant waste on the polymer matrix, resulting from their high flexibility (T_g values below at −121 °C). Upon heating, the investigated implants undergo melting at low temperatures, approximately −36.6 to −39.5 °C (shell) and −39.5 to −42.3 °C (gel), regardless of radiotherapy exposure. The corresponding enthalpy of fusion (ΔH_m), determined from the area of the endothermic peaks, is presented in Table 1. Noticeable differences in ΔH_m between the shell and gel layers are observed. In both irradiated and non-irradiated samples, the shell exhibits lower ΔH_m values than the gel. Since ΔH_m reflects the energy required to melt a material, these differences are directly related to structural differences between the two layers [25–30]. According to the literature, the shell of silicone implants consists mainly of a highly cross-

https://doi.org/10.3390/ma19091798

linked silicone elastomer formed via vulcanization of silicone rubber [31,32]. Its primary component is polydimethylsiloxane (PDMS), as confirmed by ATR-FTIR analysis. In contrast, the gel layer consists of a lightly cross-linked silicone gel (PDMS) with a cohesive consistency and contains relatively mobile polymer chains [31,32]. Silicone implants are designed as amorphous materials at room temperature and at physiological conditions, which ensures their biocompatibility, thermal stability, and flexibility. However, under specific conditions such as low temperature (freezing) and cooling below the melting point, PDMS can undergo processes in which polymer chains organize into regular structures (lamellae), leading to the formation of a crystalline phase. The DSC melting peaks of the silicone implant waste are asymmetric and broadened over a wide temperature range, which is characteristic of macromolecular amorphous systems. Despite their cross-linked structure, reheating after melting induces partial ordering of PDMS chains during cooling at a controlled rate. As a result, a crystallization peak (exothermic recrystallization) is observed below −70 °C. This ordering process depends on the degree of cross-linking, molecular weight, and the presence of fillers. It indicates the solidification of the amorphous phase, suggesting semi-crystalline behavior at low temperatures. Based on the enthalpy of fusion, calculated from the area of the crystallization peak, the degree of crystallinity (χ) can be determined. In this context, crystallinity refers to the ability of the material to undergo partial crystallization after melting, despite the cross-linked structure of PDMS-based silicone implants. The enthalpy of fusion is proportional to the degree of crystallinity. According to literature data, the enthalpy of fusion for fully crystalline PDMS (ΔH_f°) is 61.19 J/g [33–36]. The degree of crystallinity (χ) can be calculated using the equation below:

$$\chi = (\Delta H_f / \Delta H_f^{\circ}) \cdot 100\% \quad (1)$$

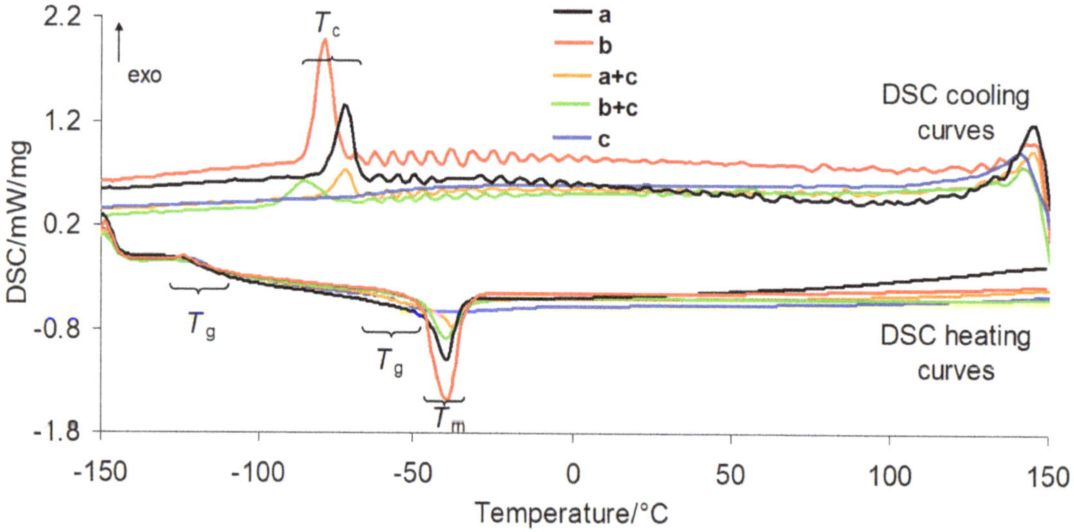

Figure 6. DSC curves for silicone implants: shell (a), gel (b), PUM (c) and the prepared composites: shell–PUM (a + c) and gel–PUM (b + c).

https://doi.org/10.3390/ma19091798

Table 1. DSC data.

Sample	T_g/°C	T_m/°C	ΔH_m/J/g	T_c/°C	ΔH_f/J/g	χ/%
a	−123.9	−36.6	30.92	−71.6	30.55	49.9
a′	−120.8	−39.5	30.53	−72.1	30.30	49.5
b	−121.3	−42.3	55.56	−78.9	53.54	87.5
b′	−120.3	−39.5	55.33	−78.2	53.29	87.1
a + c	−119.4/−60.9	−37.7	13.70	−72.2	13.55	22.1
b + c	−117.6/−61.1	−39.5	20.90	−84.7	20.10	33.4
c	−50.8	-	-	-	-	-

The calculated χ values indicate lower crystallinity of the implant shell compared to the gel fraction. This directly confirms that the shell and gel exhibit different polymer network structures associated with varying cross-link densities. A higher cross-link density significantly restricts the mobility of PDMS polymer chains and their ability to organize into ordered structures. In general, polymers with lower molecular weight and a lower degree of cross-linking show a greater tendency to crystallize, as shorter chains can move and arrange more easily. This explains the higher crystallization ability of the gel fraction of the silicone implants (Table 1). Based on the results presented in Table 1, it can also be concluded that radiotherapy has no significant influence on the melting and crystallization behavior of the investigated implants, nor on their degree of crystallinity. In contrast, neat PUM does not exhibit melting or crystallization upon heating and cooling; only a glass transition region is observed in the DSC curves. For the composites, however, both melting and recrystallization peaks are observed. The melting of silicone implant phases within the composites occurs at approximately −37.7 °C (shell) and −39.5 °C (gel). These values are comparable to those observed for the neat silicone implant layers. However, within the composites, the silicone phases exhibit significantly lower melting enthalpy (ΔH_m) compared to the pure PDMS layers. Similarly, the enthalpy of fusion (ΔH_f) of the silicone waste is markedly reduced in the composites. This reduction is attributed to the presence of the PUM matrix, which decreases the crystallization ability of PDMS. PUM matrix in the composites acts as a spatial constraint (steric hindrance effect). It limits the mobility of PDMS chains by interfacial interactions, thereby hindering the formation of crystalline siloxane domains and reducing the relative content of PDMS in the overall composite mass [36].

3.3. TG/DTG in Inert Atmosphere

The TG/DTG curves of the investigated silicone implants before and after radiotherapy under heating in an inert atmosphere are presented in Figure 7. The TG/DTG curves for PUM and the prepared composites are shown in Figure 8. The corresponding data extracted from the TG/DTG curves are summarized in Table 2. As clearly observed, the thermal decomposition behavior of the silicone implant shell before and after radiotherapy is very similar. The initial decomposition temperature ($T_{5\%}$) is approximately 330 °C, which is consistent with literature data [37–39]. The decomposition occurs in one main stage, consisting of at least three steps, identified as T_{max1}, T_{max2}, and T_{max3} (Table 2, Figure 7). Upon heating to 1000 °C, the silicone shell does not fully decompose, leaving a residual mass of approximately 36%. This is attributed to its high cross-link density, which promotes the formation of thermally stable intermediates and residues. In contrast, the thermal stability of the gel fraction is not affected by radiotherapy. The $T_{5\%}$ values are also approximately 330 °C. The decomposition of the gel occurs in one main stage consisting of at least two steps, denoted as T_{max1} and T_{max2}. These decomposition steps occur at lower temperatures compared to those observed for the shell, which is directly related to the lower cross-link density of the gel. As a result, the gel undergoes complete decomposition at approximately 560 °C (Table 2, Figure 7).

https://doi.org/10.3390/ma19091798

The thermal stability of the prepared shell and gel-based composites is approximately 104–108 °C lower than that of the neat silicone implant layers, reaching values of 222.5–226.8 °C (Table 2 and Figure 8). However, it remains approximately 27–31 °C higher than that of the pure PUM matrix. This behavior is directly related to the presence of highly thermally stable silicone implant waste and to interfacial interactions between the shell/gel phases and the PUM matrix.

In addition, the thermal decomposition of the composites proceeds in several poorly resolved stages with T_{max} values summarized in Table 2. The first (T_{max1}) and second (T_{max2}) stages occur in a temperature range similar to that of neat PUM, indicating that these steps are associated with the degradation of chemical bonds in the PUM matrix. Above 380 °C, further degradation of the composites is primarily governed by pyrolysis of bonds within the silicone shell or gel phases, as well as decomposition of PUM residues. This process leads to the release of volatile products and a significant reduction in residual mass at 1000 °C, particularly for the shell-based composites.

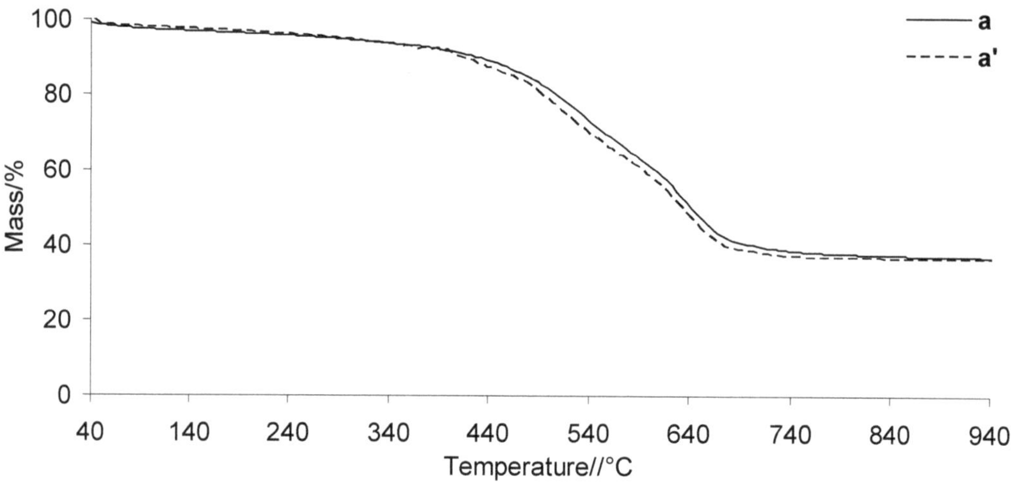

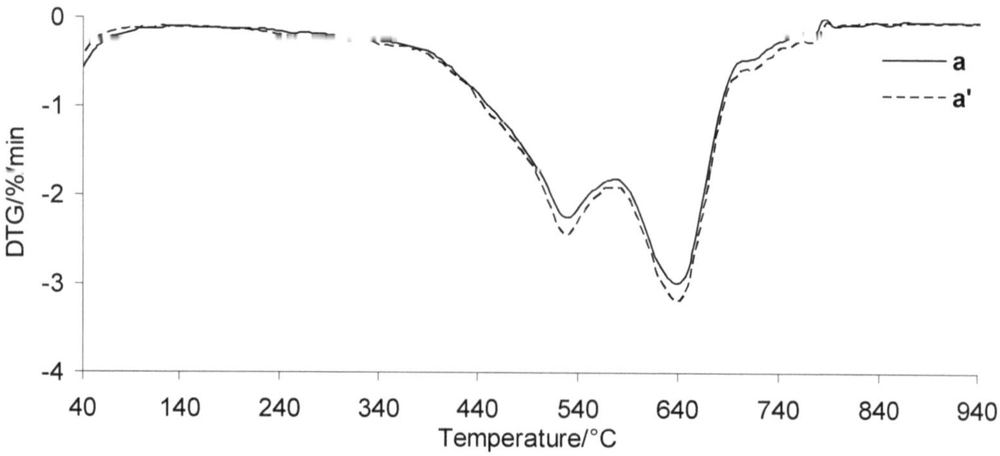

Figure 7. *Cont.*

https://doi.org/10.3390/ma19091798

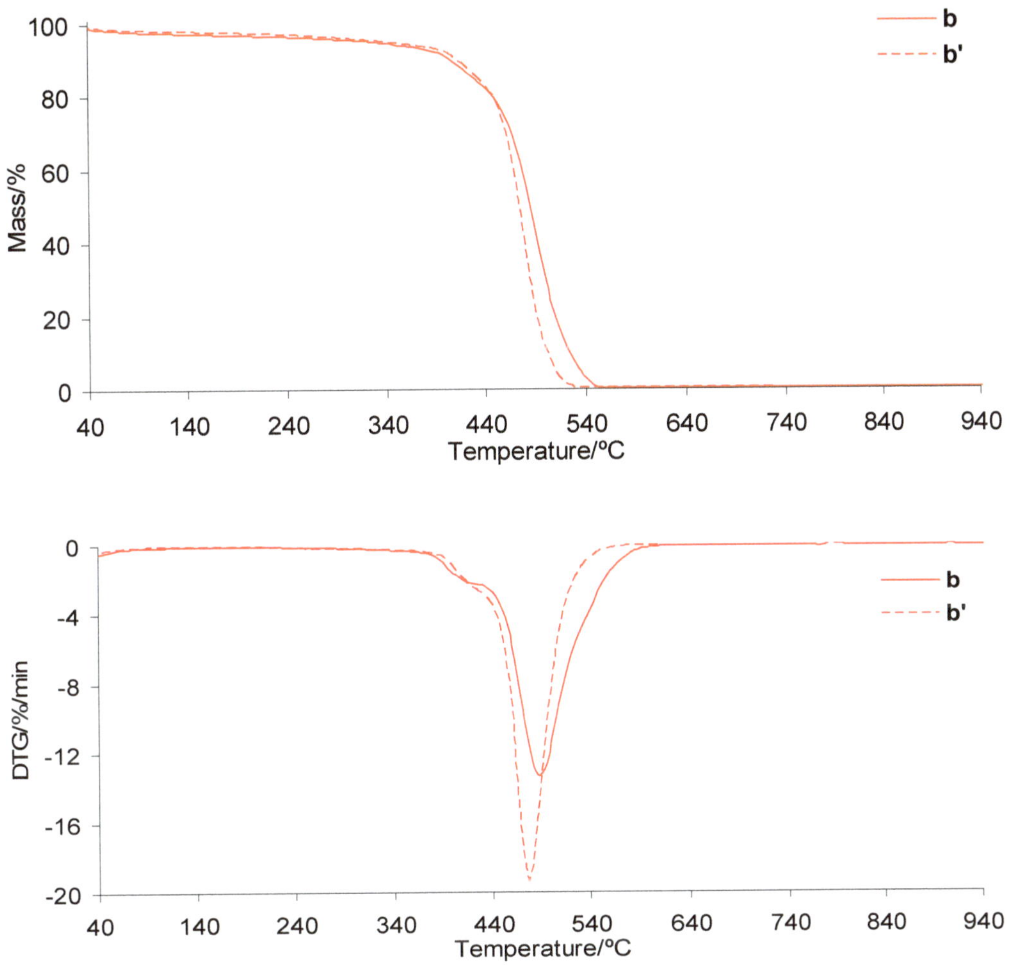

Figure 7. TG/DTG curves of silicone implants: shell before (a) and after radiotherapy (a′), and gel before (b) and after radiotherapy (b′).

Table 2. TG/DTG data in inert atmosphere.

Sample	$T_{5\%}$/°C	T_{max1}/T_{max2}/T_{max3}/T_{max4}/T_{max5}/°C	Δm_1/%	m_r/%
a	330.8	529.2/639.5/717.2	63.2	36.8
a′	330.9	530.5/639.8/721.1	63.8	36.2
b	330.1	411.1/487.7	100	0
b′	330.5	415.4/477.9	100	0
a + c	226.8	221.5/360.9/405.6/514.8/665.3	92.9	7.1
b + c	222.5	220.6/345.5/419.4/500.3/661.8	100	0
c	195.3	222.4/381.6/541.5	100	0

m_r—residual mass at 1000 °C.

https://doi.org/10.3390/ma19091798

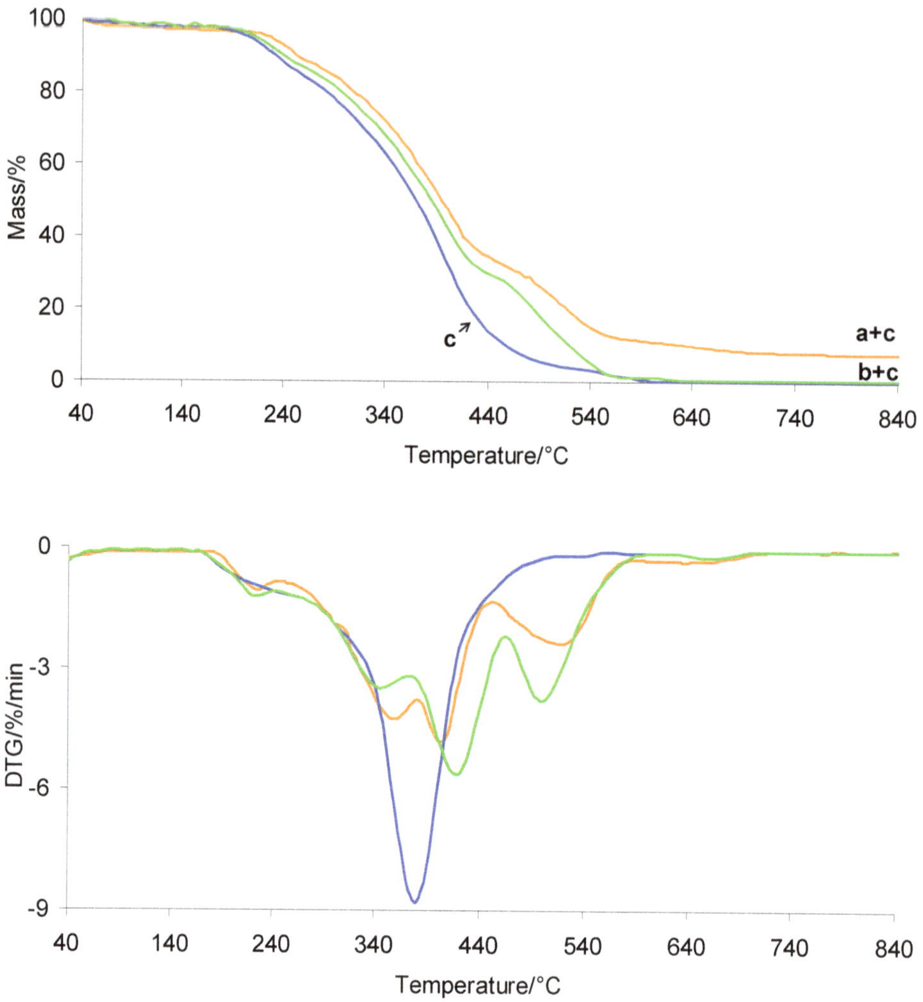

Figure 8. TG/DTG curves of PUM (c) and the prepared composites: shell–PUM (a + c) and gel–PUM (b + c).

3.4. TG/DTG in Oxidative Atmosphere

The TG/DTG curves recorded during heating of the investigated silicone implants (shell and gel) in an oxidative atmosphere are presented in Figure 9. The TG/DTG curves of PUM and the prepared composites are shown in Figure 10. The corresponding data are summarized in Table 3. In the presence of air, the thermal stability of both silicone implant layers is comparable, with $T_{5\%}$ values above 343–344 °C. As observed, radiotherapy does not affect the $T_{5\%}$ values. The oxidative decomposition behavior of both implant layers is more complex than under inert conditions. The decomposition occurs in at least three to four stages, identified as T_{max1}, T_{max2}, T_{max3}, and T_{max4} (Figure 9). These differences are directly related to the influence of oxygen on the activation energy of bond cleavage within the material structure. The presence of oxygen generally lowers the activation energy of degradation reactions, which is reflected in a shift in T_{max} values to lower temperatures compared to those observed under inert atmospheres [39–50]. At the same time, oxygen

 https://doi.org/10.3390/ma19091798

may initially promote the formation of more stable oxygen-containing intermediate species (oxygen–polymer interactions), which can temporarily increase thermal resistance at early degradation stages. However, with increasing temperature, oxidative processes dominate, leading to bond scission at lower T_{max} values. According to the literature, silicone materials heated above approximately 300 °C in the presence of oxygen undergo thermo-oxidative degradation, resulting in loss of physical properties and chemical breakdown. The organic side groups of the silicone chains are oxidized, the siloxane backbone is cleaved, and volatile and solid residues are formed [51,52]. At temperatures above approximately 450 °C, silicone materials may undergo auto-ignition, ultimately forming thermally stable residues such as silica (SiO_2) or silicon monoxide (SiO). The present results confirm that radiotherapy does not significantly affect the thermo-oxidative decomposition behavior of the investigated silicone layers. Both shell and gel samples, before and after radiotherapy, exhibit similar thermal stability and decomposition stages occurring at comparable T_{max} values.

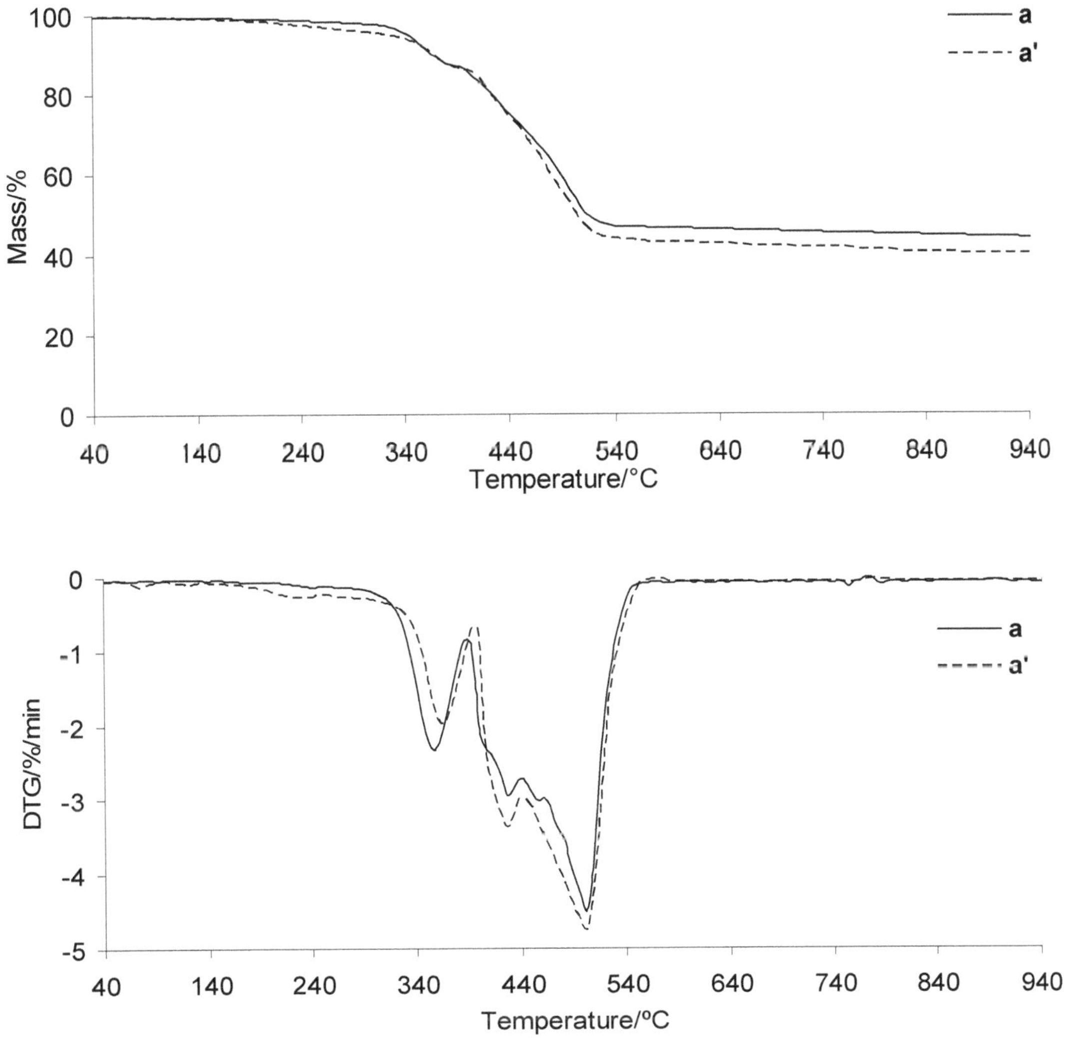

Figure 9. *Cont.*

https://doi.org/10.3390/ma19091798

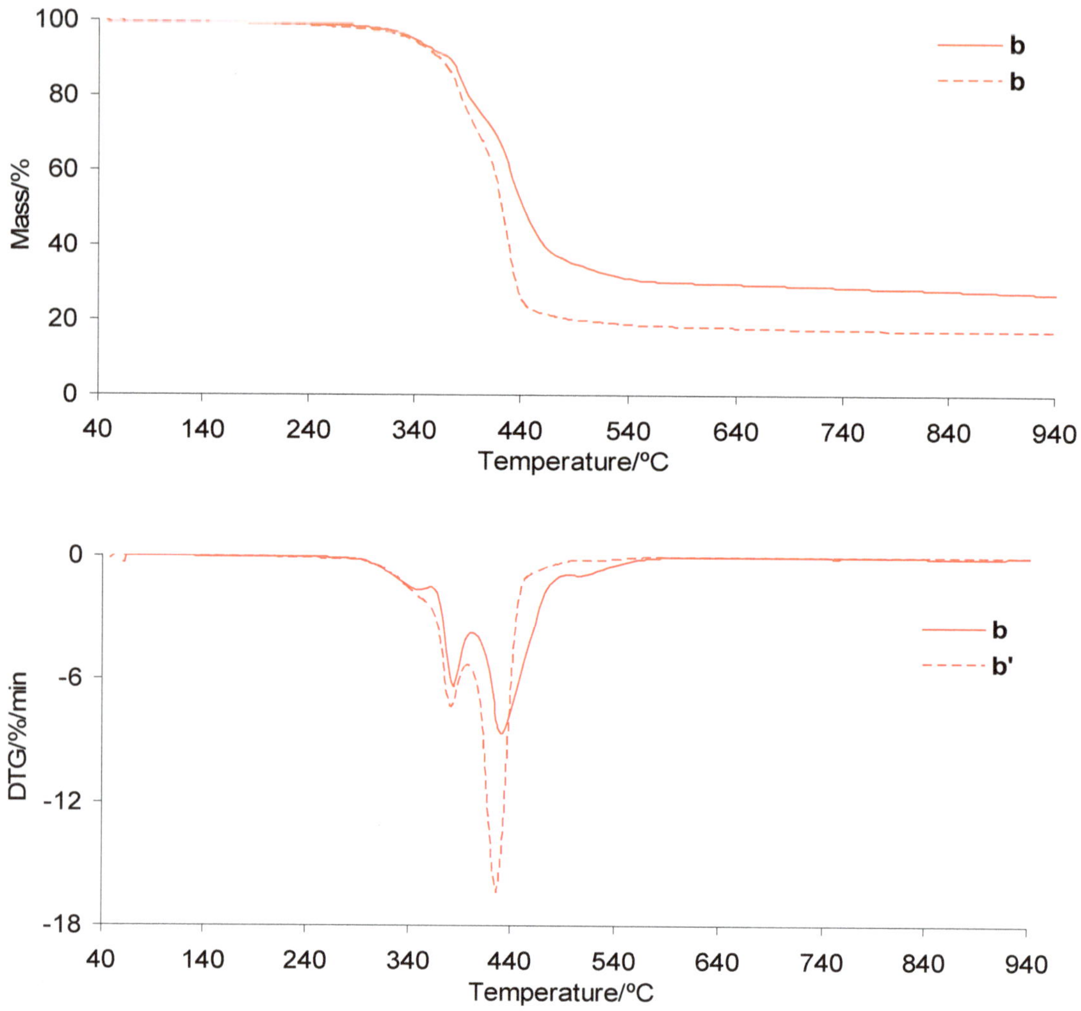

Figure 9. TG/DTG curves of silicone implants: shell before (a) and after radiotherapy (a′), and gel before (b) and after radiotherapy (b′) in oxidative atmosphere.

Table 3. TG/DTG data in oxidative atmosphere.

Sample	$T_{5\%}$/°C	T_{max1}/T_{max2}/T_{max3}/T_{max4}/T_{max5}/°C	Δm_1/%	m_r/%
a	344.2	356.8/429.8/499.7	56.1	43.9
a′	344.5	368.4/428.3/500.1	59.9	40.1
b	343.2	351.2/383.2/430.8/507.3	72.8	27.2
b′	343.0	352.8/383.7/428.5/519.2	82.9	18.1
a + c	237.2	200.2/327.3/394.1/456.5/539.2	77.5	22.5
b + c	235.8	220.4/340.1/380.2/460.3/540.5	89.8	10.2
c	190.0	215.1/310.4/386.6/565.7	100.0	0

m_r—residual mass at 1000 °C.

https://doi.org/10.3390/ma19091798

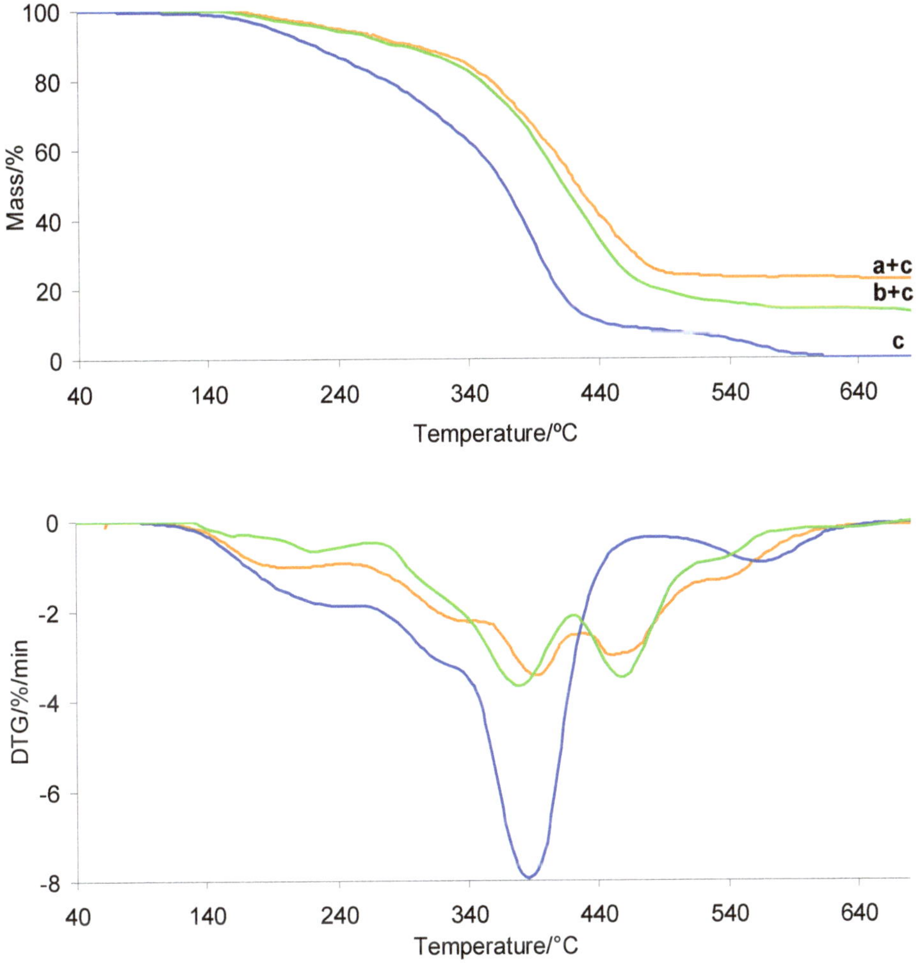

Figure 10. TG/DTG curves of PUM (c) and the prepared composites: shell–PUM (a + c) and gel–PUM (b + c).

In contrast, the thermal stability of the prepared composites is significantly lower than that of the neat silicone implant wastes under oxidative conditions, by approximately 110 °C (Table 3, Figure 10). This decrease is directly associated with the presence of the PUM matrix, which exhibits moderate thermal stability (approximately 190 °C). The incorporation of PUM accelerates the thermo-oxidative degradation of the composites. On the other hand, the composites exhibit higher thermal stability than neat PUM by approximately 45 °C, indicating that the presence of silicone waste enhances the thermal resistance of the polymer matrix. PDMS acts as a thermal stabilizer, delaying the decomposition of the PUM matrix by forming a protective silica layer on its surface, which hinders the diffusion of oxygen into the polymer matrix. The simultaneous presence of PUM and silicone phases results in a more complex thermo-oxidative degradation process, occurring in at least five poorly resolved stages (T_{max}). This suggests comparable energies of bond cleavage within the composite structure. Moreover, unlike PUM, which fully decomposes under oxidative conditions at around 600 °C, the composites do not completely decompose upon heating to 1000 °C. This is attributed to the formation of thermally stable residues resulting from

https://doi.org/10.3390/ma19091798

the oxidative degradation of the silicone phases. However, the residual mass (m_r) of the composites is lower than that of the neat shell or gel silicone materials. This indicates that PUM promotes further oxidation of silicone-derived residues at elevated temperatures. This effect is more pronounced for the shell–PUM composites, where the residual mass at 1000 °C is reduced by approximately 50%.

3.5. Gaseous FTIR in Oxidative Atmosphere

The FTIR spectra of gaseous products released during the oxidative decomposition of shell and gel silicone implants (before and after radiotherapy) (Figure 11) and the prepared composites and PUM (Figure 12) were collected to identify the emitted volatile compounds and assess their potential toxicity. The evaluation of gaseous products formed during heating in the presence of oxygen, particularly for the novel shell–PUM and gel–PUM composites, is important for assessing their behavior during combustion-based disposal after service life. As clearly observed, radiotherapy does not influence the type of volatile products emitted during the thermo-oxidative decomposition of either silicone implant layer. Both shell and gel fractions generate similar gaseous products. The FTIR spectra indicate that degradation involves oxidation of side groups, cross-linking and rearrangement reactions, and cleavage of siloxane bonds, leading mainly to the emission of low-molecular-weight cyclic siloxanes, silica-related species, CO, CO_2, and water vapor (Figure 11), which is consistent with literature data [43–55].

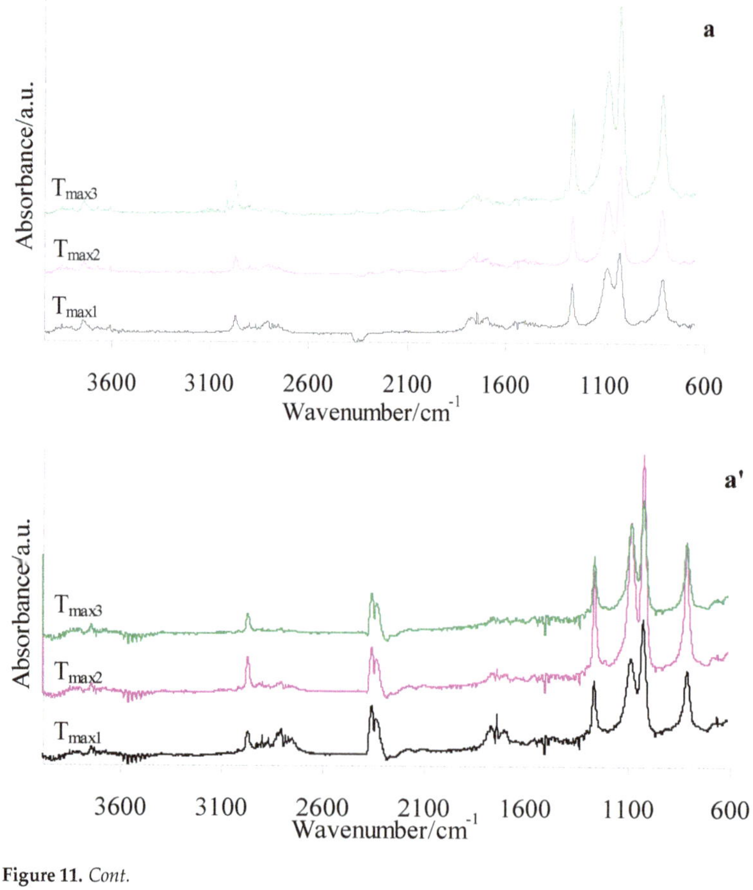

Figure 11. *Cont.*

https://doi.org/10.3390/ma19091798

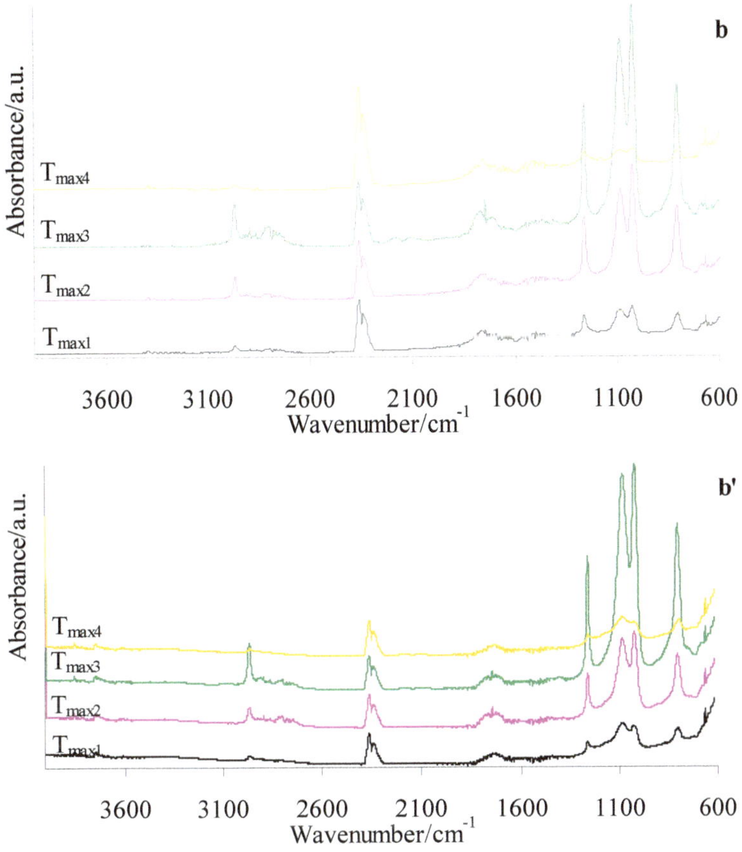

Figure 11. FTIR spectra of gaseous products released during heating of silicone implants in an oxidative atmosphere: shell before (a) and after radiotherapy (a′); gel before (b) and after radiotherapy (b′).

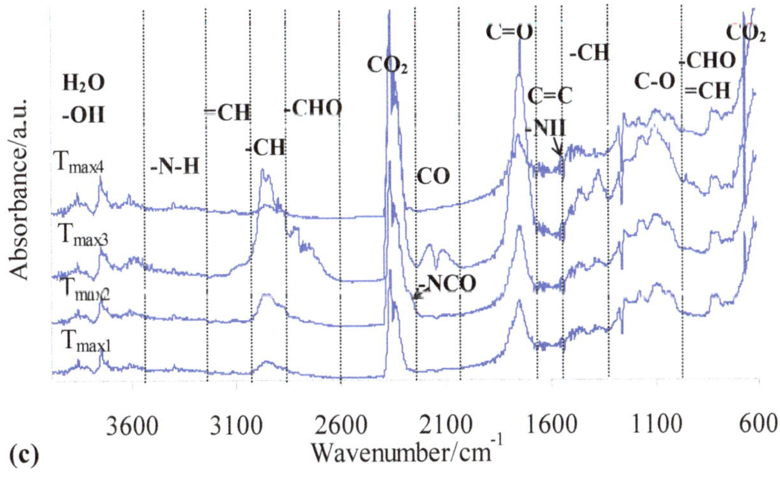

Figure 12. *Cont.*

https://doi.org/10.3390/ma19091798

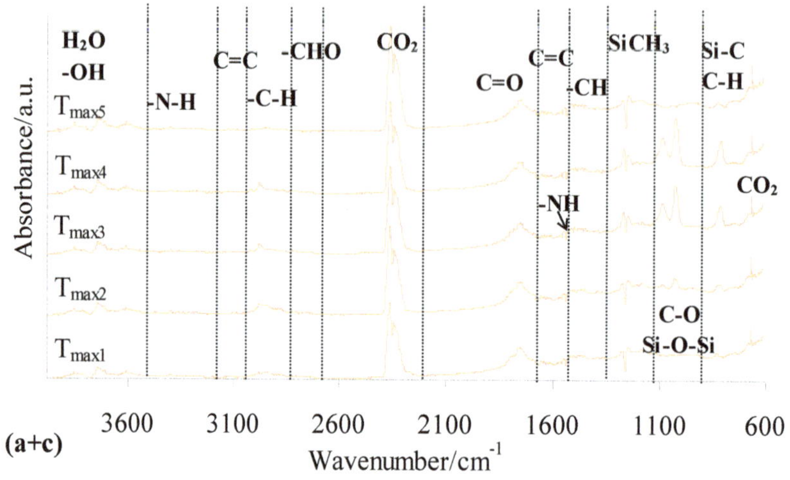

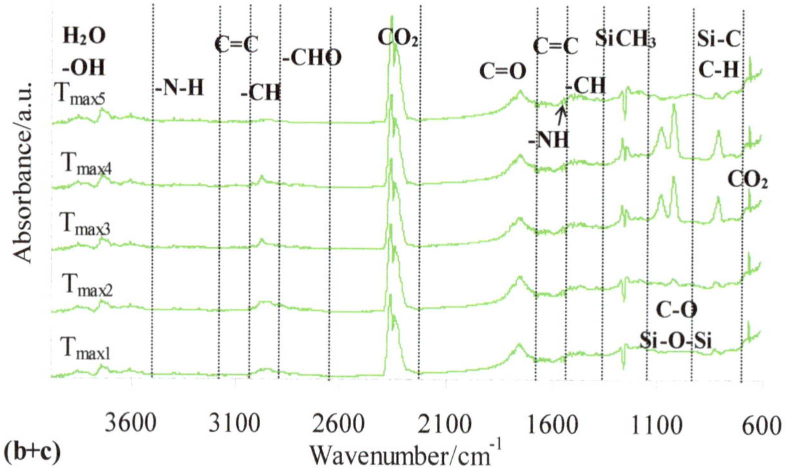

Figure 12. FTIR spectra of gaseous products released during heating of PUM (c) and the prepared composites: shell–PUM (a + c) and gel–PUM (b + c).

In contrast, the FTIR spectra of PUM during oxidative degradation show similar absorption bands over the entire decomposition range, with maximum emission intensity observed at T_{max3}. This indicates a simultaneous breakdown of bonds within the PUM structure, likely associated with its cross-linked network. Based on Figure 12 (c), the following characteristic vibrations are identified during oxidative decomposition of PUM: stretching vibrations of –OH and H_2O (above 3600 cm^{-1}), N–H stretching (3320–3390 cm^{-1}), =CH stretching (3072–3091 cm^{-1}), C–H stretching (2863–2964 cm^{-1}), C–H stretching in aldehyde groups (2724–2760 cm^{-1}), CO_2 emission bands (2359–2310 cm^{-1} and 669 cm^{-1}), isocyanate bands (2256 cm^{-1}), CO emission bands (2092 cm^{-1} and 2163 cm^{-1}), C=O stretching (1745 cm^{-1}), C=C stretching (1627–1630 cm^{-1}), N–H bending (1529 cm^{-1}), C–H deformation (1388–1440 cm^{-1}), C–O stretching (1020–1253 cm^{-1}), and out-of-plane deformation vibrations of =CH and –CH groups in –CHO fragments (727–946 cm^{-1}). The presence of these bands confirms the formation of urethane, aldehyde, alcohol, amide,

 https://doi.org/10.3390/ma19091798

organic acid, and isocyanate fragments, as well as inorganic gases (CO_2, CO, and H_2O) during oxidative decomposition of the PUM matrix used in the composites.

The FTIR profiles of both composite types (shell–PUM and gel–PUM) (Figure 12 (a + c and b + c)) are highly similar at corresponding T_{max} values, indicating the release of volatile products with comparable composition. At T_{max1} (200–220 °C), the emitted gases are primarily associated with the decomposition of the PUM matrix. In this region, only characteristic bands related to PUM degradation are observed, while no signals attributable to silicone-derived volatiles are detected. This suggests that, at this stage, mainly low-molecular-weight species and side groups of the PUM undergo oxidative decomposition. At higher temperatures (T_{max2}), volatile products originating from both PUM and silicone implant waste are detected. The maximum emission intensity occurs at T_{max3} (380–394 °C) and T_{max4} (456–460 °C). At these stages, the spectra show bands corresponding to water vapor (above 3600 cm^{-1}), low-intensity urethane-derived species (N–H at 3320–3390 cm^{-1} and 1529 cm^{-1}), aldehyde fragments (C=O at 1745 cm^{-1} and C–H in aldehydes at 2724–2760 cm^{-1}), and acid-related broad bands (CO and O–H around 3720 cm^{-1}). Weak signals of unsaturated species (=CH at 3072–3091 cm^{-1} and C=C at 1627–1630 cm^{-1}) are also observed, along with emissions of cyclic siloxanes and silica-related species (Si–CH_3 at 1267 cm^{-1}, Si–O–Si at 1022–1076 cm^{-1}, and Si–CH_3 at 815 cm^{-1}), as well as CO_2 [19]. Notably, no significant formation of CO or isocyanates is detected in this temperature range (Figure 12).

In summary, the oxidative decomposition of the prepared composites is a complex process involving multiple simultaneous reactions, including bond cleavage, secondary reactions between intermediate products and oxygen, and the formation of CO_2 and H_2O. In addition, high-temperature sintering processes of the residual solid phase occur, leading to the formation of thermally stable residues. Compared to the residues formed from neat silicone implants, the residual mass of the composites is reduced, indicating enhanced degradation efficiency in the composite systems.

3.6. Dynamic Mechanical Analysis (DMA) and Brinell Hardness

The results obtained from DMA and Brinell hardness measurements are presented in Table 4. Additionally, the storage modulus and tan δ curves as a function of temperature are shown in Figure 13. The DMA study indicates that both the PUM matrix and the prepared shell–PUM and gel–PUM composites are flexible materials. However, the incorporation of silicone breast implant waste into the polymer composites (shell–PUM and gel–PUM) further reduces their stiffness. The incorporation of silicone waste into the composites leads to a decrease in the storage modulus at room temperature ($E_{20°C}$) compared to that of the pure PUM matrix. The PUM matrix exhibits an $E_{20°C}$ value of 280.2 MPa, whereas the composites show values of 206.7 MPa (shell–PUM) and 139.2 MPa (gel–PUM). This suggests that silicone breast implant waste acts as an internal lubricant, particularly the gel layer, enhancing polymer chain mobility and thereby increasing the flexibility of the resulting composites.

Table 4. DMA data and Brinell hardness for PUM, shell–PUM and gel–PUM composites.

Sample	$E'_{20°C}$/MPa	tgδ/°C	HK/MPa
a + c	206.7	−107.5/−39.2	10.8 ± 0.5
b + c	139.2	−107.3/−39.0	9.8 ± 0.3
c	280.2	−44.2	23.3 ± 1.2

https://doi.org/10.3390/ma19091798

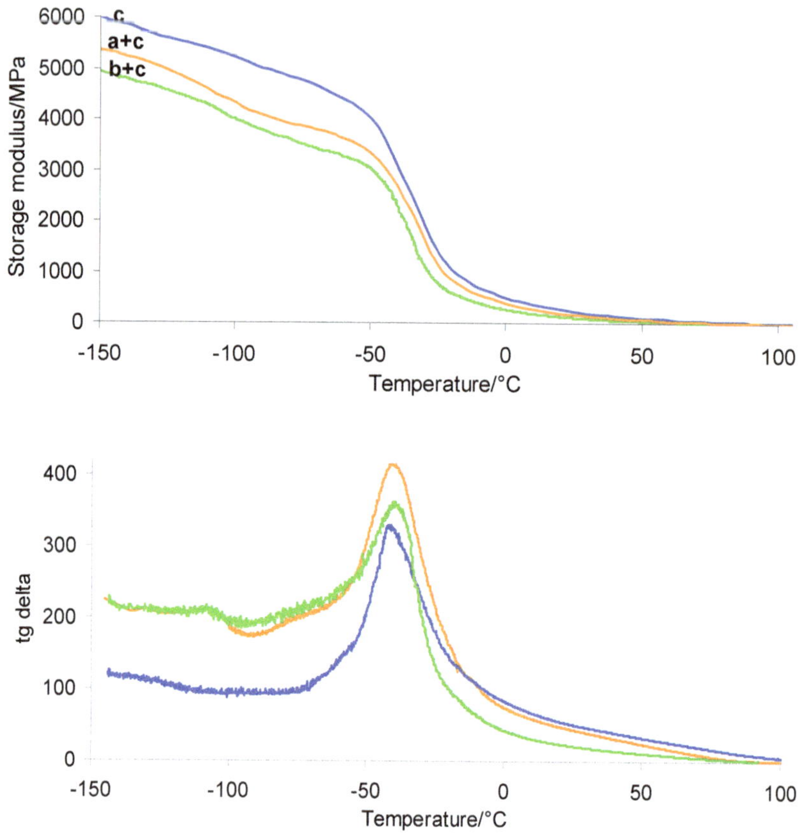

Figure 13. Storage modulus and tg delta of PUM (c) and the prepared composites: shell–PUM (a + c) and gel–PUM (b + c).

DMA also confirms phase separation, as two distinct relaxation processes are observed in the tan δ curves, corresponding to the glass transitions of the silicone phase (at lower temperatures) and the polymer matrix (at higher temperatures). However, the tan δ peak maximum of the PUM phase in the composites is slightly shifted toward higher temperatures (ca. −39 °C) compared to that of the pure PUM matrix (−44.2 °C). This indicates the presence of interactions between the polymer matrix and the silicone phase, i.e., modification of polymer chain mobility at the phase boundary.

The Brinell hardness of pure PUM is 23.3 ± 1.2 MPa. In contrast, the composites exhibit significantly lower hardness values: 10.8 ± 0.5 MPa for the shell–PUM composite and 9.8 ± 0.3 MPa for the gel–PUM composite. Brinell hardness analysis reveals a substantial decrease in indentation resistance following the incorporation of dispersed shell and gel silicone phases into the PUM matrix. This reduction of over 50% confirms the plasticizing effect of the silicone waste on the polymer network. Silicone acts as a flexible filler, reducing the overall stiffness of the polymer matrix, which in turn lowers both the surface hardness and the glass transition temperature of the shell–PUM and gel–PUM composites. Flexible silicone chains interact with the matrix by penetrating between polymer chains, leading to a loosening of the PUM structure. Consequently, the presence of shell or gel phases enhances the elasticity of the composite material.

https://doi.org/10.3390/ma19091798

3.7. Microscopic and Profilometric Analysis

Figure 14 shows that the three-dimensional surface topography analysis reveals noticeable differences in the morphology and roughness of the tested samples. The unmodified PUM sample exhibited the lowest average surface roughness (Ra = 177 nm), whereas the shell–PUM and gel–PUM composites exhibited increased roughness values of 271 and 309 nm, respectively. These results suggest that incorporating shell and gel modifiers results in a more heterogeneous and rougher surface structure.

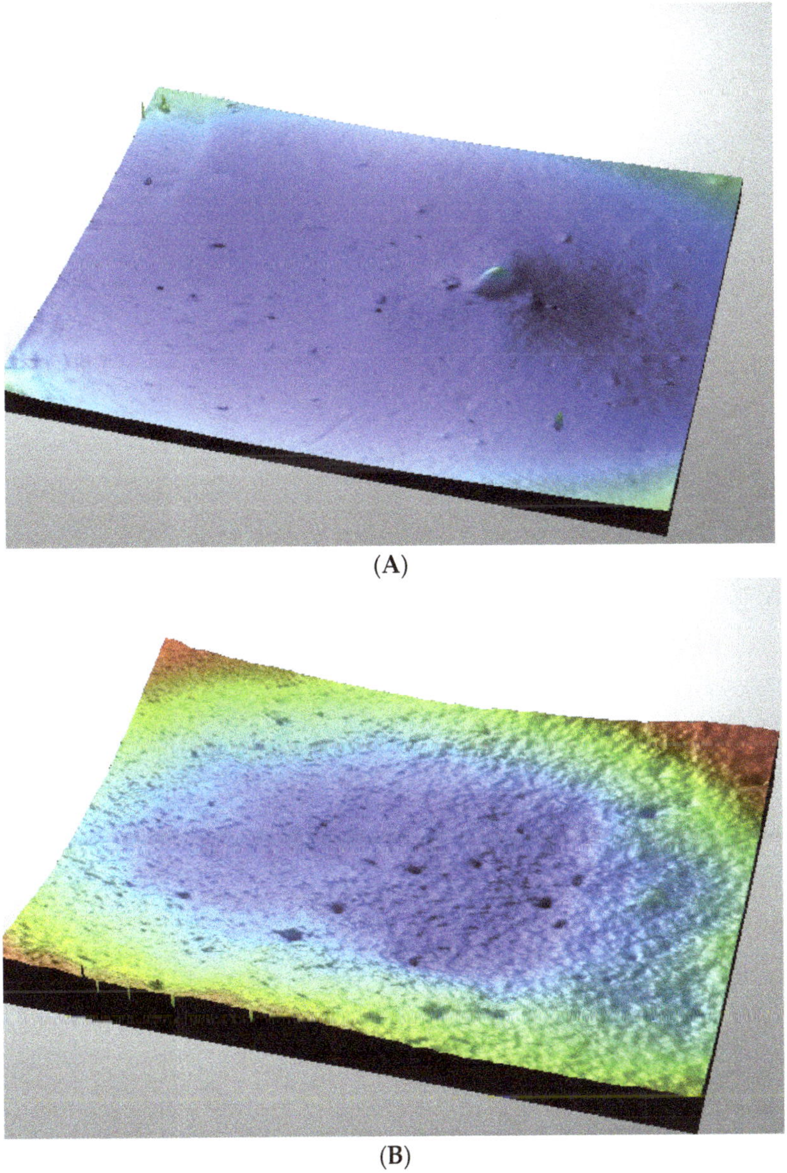

Figure 14. *Cont.*

https://doi.org/10.3390/ma19091798

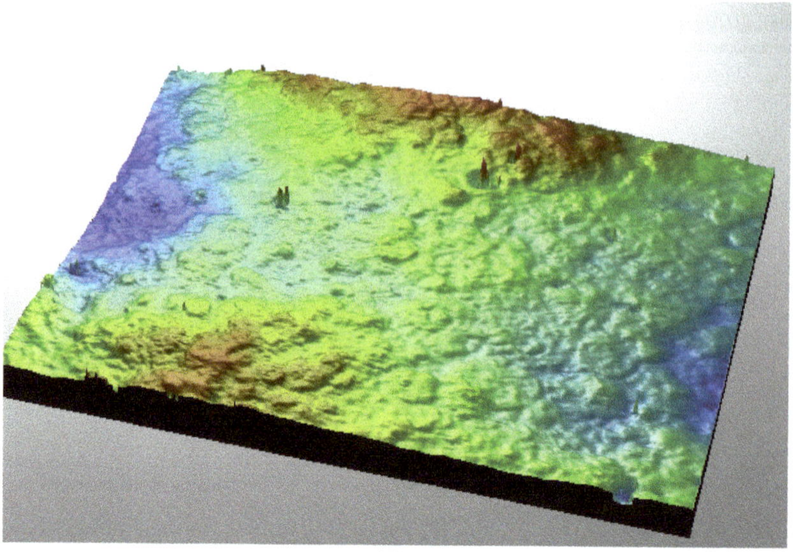

(C)

Figure 14. Three-dimensional surface topography of the sample obtained using a Contour GT-K1 optical profilometer (Veeco). (**A**) PUM (c); the prepared composites: (**B**) shell–PUM (a + c) and (**C**) gel–PUM (b + c).

4. Conclusions

In this work, the effect of radiotherapy on the structure and properties of silicone implants composed of two layers (shell and gel) was investigated. Moreover, the reuse of silicone implants as a component of layered composites was proposed. The conducted analyses confirmed that radiotherapy does not affect the structure, glass transition temperature, melting and crystallization behavior, degree of crystallinity, thermal stability, or thermo-oxidative degradation pathway of the investigated silicone implants. The shell and gel layers of silicone implant waste were successfully incorporated as components of layered composites with PUM. The resulting composite materials exhibited two glass transition regions, both below room temperature: one at approximately −117 to −119 °C and the other at approximately −61 °C. The composites showed a more amorphous structure compared to the silicone implant waste. The ability to crystallize decreased significantly, from approximately 49–50% (shell) to 22.1% (shell–PUM composite) and from approximately 87.1–87.5% (gel) to 33.4% (gel–PUM composite). This effect is attributed to steric hindrance imposed by the PUM matrix, interfacial interactions, and the reduced proportion of silicone waste in the overall composition. DMA confirmed the increased elasticity of the obtained composites compared to the pure PUM matrix. The determined Brinell hardness of the silicone waste-based composites was lower than that of neat PUM, which is attributed to the plasticizing effect of the shell and gel phases within the composites. The introduction of shell and gel additives increases the surface roughness of PUM composites, suggesting a less uniform surface structure than the unmodified material. The thermal stability of the composites was approximately 40 °C higher than that of the PUM matrix. The thermal stability of the shell–PUM and gel–PUM composites was approximately 222.5–226.8 °C under inert conditions and 235–237 °C under oxidative conditions. The thermo-oxidative degradation behavior of the prepared silicone waste-based composites was more complex compared to that of both silicone implant waste and PUM. As confirmed, oxidative degradation involves multiple simultaneous processes leading to the emission of a mixture of inorganic and

https://doi.org/10.3390/ma19091798

organic volatiles, including urethane, aldehyde, alcohol, amide, organic acid, unsaturated fragments, CO_2, and H_2O. These results not only demonstrate the effect of radiotherapy on the structure and properties of silicone implants but, more importantly, present a potential approach for the valorization of silicone implant waste and the development of composite materials with tailored properties for practical applications.

Author Contributions: Conceptualization, B.P., B.T., A.K. and M.W.; methodology, M.W. and B.T.; validation, M.P., B.T. and M.W.; formal analysis, B.P.; investigation, B.T., M.W. and M.P.; writing—original draft preparation, A.K. and M.W.; writing—review and editing, M.W. and B.P.; visualization, B.P. and B.T.; supervision, B.P. All authors have read and agreed to the published version of the manuscript.

Funding: This research received no external funding.

Institutional Review Board Statement: Not applicable.

Informed Consent Statement: Not applicable.

Data Availability Statement: The original contributions presented in this study are included in the article. Further inquiries can be directed to the corresponding author.

Conflicts of Interest: The authors declare no conflicts of interest.

References

1. Sbitani, H.; Gomez-Sanches, C.; Piper, M. Prepectoral breast reconstruction in the setting of postmastectomy radiation therapy: An assessment of clinical outcomes and benefits. *Plast. Reconstr. Surg.* **2019**, *143*, 10–20. [CrossRef]
2. Gradishar, W.J.; Moran, M.S.; Abraham, J.; Abramson, V.; Aft, R.; Agnese, D.; Allison, K.H.; Anderson, B.; Bailey, J.; Burstein, H.J.; et al. NCCN Guidelines Insight: Breast Cancer, version 5:2025: Featured updates to the NCCN guidelines. *J. Natl. Compr. Cancer Netw.* **2025**, *23*, 426–436. [CrossRef]
3. Jimnez, R.; Abdou, Y.; Anderson, P. Post-mastectomy radiation therapy: An ASTRO-ASCO-SSO clinical practice guideline. *J. Clin. Oncol.* **2025**, *43*, 3292–3311. [CrossRef]
4. Chakraborty, S.; Chatterjee, S. HYPORT adjuvant acute toxicity and patient dosimetry quality assurance results—Interim analysis. *Radiother. Oncol.* **2022**, *174*, 59–68. [CrossRef] [PubMed]
5. Poppe, M.M.; Le-Rademacher, J.; Haffty, B.G. A randomized trial of hypofractionated post-mastectomy radiation therapy (PMRT) in women with breast reconstruction (RT CHARM, Alliance A221505). *Int. J. Radiat. Oncol. Biol. Phys.* **2024**, *120*, S11. [CrossRef]
6. Sinnot, C.; Persing, S.; Pronovost, N. Impact of post-mastectomy radiation therapy in prepectoral versus subpectoral implant-based breast reconstruction. *Ann. Surg. Oncol.* **2018**, *25*, 2899–2908. [CrossRef]
7. Wilkins, E.; Hamill, J.; Kim, H. Complications in postmastectomy breast reconstruction. One-year outcomes of the mastectomy reconstruction outcomes consortium (MROC) study. *Ann. Surg.* **2018**, *267*, 164–170. [CrossRef]
8. Smarrito, S. Ecological impact of plastics used in breast augmentation surgery. In *Annales de Chirurgie Plastique Esthétique*; Elsevier: Amsterdam, The Netherlands, 2020; Volume 65, pp. 326–331.
9. Ahmed, Z.; Zargaran, A.; Marudamuthu, D.; Sousi, S.; Zargaran, D.; Hamilton, S.; Mosahebi, A. Sustainability in breast surgery: An eco-audit of the breast implant pathway. *J. Plast. Reconstr. Aesthetic Surg.* **2025**, *105*, 219–229. [CrossRef] [PubMed]
10. Kalmar, C.L.; Evans, A.; Kassis, S.H.; Thayer, W.P.; Drolet, B.C.; Perdikis, G. Environmental sustainability of breast implant supply chains. *Plast. Reconstr. Surg. Glob. Open.* **2022**, *10*, 130. [CrossRef]
11. Iakovou, D.; Sousi, S.; Glynou, S.P.; Ahmed, Z.; Zargaran, A.; Zargaran, D.; Mosahebi, A. A systematic review of sustainability practices in plastic surgery. *J. Plast. Reconstr. Aesthetic Surg.* **2025**, *102*, 104–113. [CrossRef]
12. Tarasiuk, B.; Podkościelny, W. Urethane acrylates as main components of lacquers for protective coating of some materials. *Mol. Cryst. Liq. Cryst. Sci. Technol. Sect. A. Mol. Cryst. Liq. Cryst.* **2000**, *354*, 49–53. [CrossRef]
13. Hamouni, S.; Arous, O.; Abdessemed, D.; Nezzal, G.; Van der Bruggen, B. Alcohol and alkane organic extraction using pervaporation process. *Macromol. Symp.* **2019**, *386*, 1800247. [CrossRef]
14. Le, T.N.; Lee, C.K. Surface functionalization of poly(N-vinylpyrrolidone) onto poly(dimethylsiloxane) for anti-biofilm application. *Appl. Biochem. Biotechnol.* **2020**, *191*, 29–44. [CrossRef]
15. Verma, S.; Das, S.; Mohanty, S.; Nayak, S.K. A facile preparation of epoxy-polydimethylsiloxane (EP-PDMS) polymer coatings for marine applications. *J. Mater. Res.* **2019**, *34*, 2881–2894. [CrossRef]

 https://doi.org/10.3390/ma19091798

16. Wu, S.; Zhang, X.; Sun, Y.; Yang, H.; Lin, B.; Han, Y.; Chen, P. Study on the influence of crosslinking density and free polysiloxan chain length on oxygen permeability and hydrophilicity of multicomponent silicone hydrogels. *Colloid Polym. Sci.* **2021**, *299*, 1327–1335. [CrossRef]
17. Dong, Z.; Zhu, H.; Hang, Y.; Liu, G.; Jin, W. Polydimethylsiloxane (PDMS) composite membrane fabricated on the inner surface of a ceramic hollow fiber: From single-channel to multi-channel. *Engineering* **2020**, *6*, 89–99. [CrossRef]
18. Adrees, M.; Iqbal, S.S.; Ahmad, A.; Jamshaid, F.; Haider Haskim Khan, M.H.; Khan, R.; Zahid Butt, M.T.; Bahadar, A. Characterization of novel polydimethylsiloxane (PDMS) and copolymer polyvinyl chloride-co-vinyl acetate (PVCA) enhanced polymer blend membranes for CO_2 separation. *Polym. Test.* **2019**, *80*, 106163. [CrossRef]
19. Available online: http://webbook.nist.gov (accessed on 6 March 2026).
20. Biron, M. *8-Thermal Properties, Material Selection for Thermoplastic Parts, Practical and Advanced Information*; William Andrew: Norwich, NY, USA, 2016; pp. 339–375.
21. Zalewski, K.; Chyłek, Z.; Trzciński, W.A. A review of polysiloxanes in terms of their application in explosives. *Polymers* **2021**, *13*, 1080. [CrossRef]
22. Utrera-Barrios, S.; Yu, L.; Skov, A.L. Revisiting the thermal transitions of polydimethylsiloxane (PDMS) elastomers: Addressing common misconceptions with comprehensive data. *Macromol. Mater. Eng.* **2025**, *310*, 2500075. [CrossRef]
23. Morawska-Chochół, A.; Szumera, M.; Młyniec, A.; Pielichowska, K. The effect of aging process conditions on the thermal properties of poly(dimethylsiloxane)-based silicone rubber. *Materials* **2024**, *16*, 5608. [CrossRef] [PubMed]
24. Foroushani, F.T.; Dzobo, K.; Khumalo, N.P.; Mora, V.Z.; Mezerville, R.; Bayat, A. Advances In surface modification of the silicone breast implant and impact on its biocompatibility and biointegration. *Biomater. Res.* **2022**, *14*, 80. [CrossRef]
25. Shimura, T.; Jin, Y.; Kubyama, K.; Araki, T.; Kamimura, N.; Masai, E.; Nakamura, M.; Michinobu, T. Improved thermal properties of polydimethylsiloxane by copolymerization and thiol–ene crosslinking of 2-pyrone-4,6-dicarboxylic acid moiety. *Polym. Chem.* **2025**, *16*, 45–51. [CrossRef]
26. Yang, H.; Nguyen, Q.T.; Ping, Z.; Long, Y.; Hirata, Y. Characterization of solvent—Filler-poly(dimethylsiloxane) membranes by Differential Scanning Calorimetry. *Mater. Res. Innov.* **2001**, *5*, 81–86. [CrossRef]
27. Horodecka, S.; Strachota, A.; Mossety-Leszczak, B.; Strachota, B.; Šlouf, M.; Zhigunov, A.; Vyroubalová, M.; Kaňková, D.; Netopilík, M.; Walterová, Z. Low-temperature meltable elastomers based on linear polydimethylsiloxane chains alpha, omega-terminated with mesogenic groups as physical crosslinkers: A passive smart material with potential as viscoelastic coupling. Part I: Synthesis and phase behavior. *Polymers* **2020**, *12*, 2476. [CrossRef]
28. Gupta, N.S.; Lee, K.W.; Labouriau, A. Tuning thermal and mechanical properties of polydimethylsiloxane with carbon fibers. *Polymers* **2021**, *13*, 1141. [CrossRef]
29. Arrighi, V.; Gagliardi, S.; Zhang, C.; Ganazzoli, F.; Higgins, J.S.; Ocone, R.; Telling, M.T.F. A unified picture of the local dynamics of poly(dimethylsiloxane) across the melting point. *Macromolecules* **2003**, *36*, 8738–8748. [CrossRef]
30. Wang, C.; Wu, Y.; Zheng, J. Polydimethylsiloxane-based phase change composites with excellent low-temperature flexibility for battery thermal management. *Mater. Today Commun.* **2025**, *44*, 111958. [CrossRef]
31. Franca, D.C.C.; de Castro, A.L.; Soubhia, A.M.P.; Coelho Avila de Aguir, S.M.H.; Goiato, M.C. Evaluation of the biocompatibility of silicone gel implants-histomorphometric study. *Acta Inform. Medica* **2013**, *21*, 93–97. [CrossRef]
32. Edwards, M.C.; Gabriel, A.; Hammer, J.; Jewell, H.L.; Jewell, M.L. Gel cohesivity and breast augmentation: Applications to clinical practice. *Aesthetic Surg. J. Open. Forum* **2022**, *7*, ojac088. [CrossRef] [PubMed]
33. Kuo, A. Polydimethylsiloxane. In *Polymer Data Handbook*; Mark, J.E., Ed.; Oxford University Press: Oxford, UK, 1999.
34. Teixeira, L.S.; Ribeiro, J. Polydimethylsiloxane mechanical properties: A systematic review. *Mater. Sci.* **2021**, *8*, 952–973. [CrossRef]
35. Umeda, M.; Wakabayashi, T.; Kamiyama, T.; Suzuki, H. Thermodynamic investigation on melting and recrystallization of poly(dimethylsiloxane) rubbers under strain. *Polymer* **2022**, *254*, 125105. [CrossRef]
36. Vasilakos, S.P.; Tarantili, P.A. Evaluation of thermal stability and thermal transitions of hydroxyl-terminated polysiloxane/montmorillonite nanocomposites. *Materials* **2025**, *18*, 1226. [CrossRef] [PubMed]
37. Deshpande, G.; Rezac, M.E. Kinetic aspects of the thermal degradation of poly(dimethyl siloxane) and poly(dimethyl diphenyl siloxane). *Polym. Degrad. Stab.* **2002**, *76*, 17–24. [CrossRef]
38. Camino, G.; Lomakin, S.M.; Lazzari, M. Polydimethylsiloxane thermal degradation Part 1. Kinetic aspects. *Polymer* **2001**, *42*, 2395–2402. [CrossRef]
39. Camino, G.; Lomakin, S.M.; Lageard, M. Thermal polydimethylsiloxane degradation. Part 2. The degradation mechanisms. *Polymer* **2002**, *43*, 2011–2015. [CrossRef]
40. Lewicki, J.P.; Liggat, J.J.; Patel, M. The thermal degradation behavior of polydimethylsiloxane/montmorillonite nanocomposites. *Polym. Degrad. Stab.* **2009**, *94*, 1548–1557. [CrossRef]
41. Lewis, R.N. Methylphenylpolysiloxanes. *J. Am. Chem. Soc.* **1948**, *70*, 1115–1117. [CrossRef]
42. Thomas, T.H.; Kendrick, T.C. Thermal analysis of polydimethylsiloxanes. I. Thermal degradation in controlled atmospheres. *J. Polym. Sci. Part A-2 Polym. Phys.* **1969**, *7*, 537–549. [CrossRef]

43. Ogliani, E.; Yu, L.; Mazurek, P.; Skov, A.L. Thermal degradation mechanisms of silicone elastomer. In Proceedings of the 9th European Silicon Days, Saarbrucken, Germany, 9–12 September 2018.
44. Wang, J.; Li, G.; Zhang, Z.; Huang, Q.; Niu, B.; Zhang, Y.; Long, D. Detailed insights of polydimethylsiloxane (PDMS) degradation mechanism via ReaxFF MD and experiments. *Chem. Eng. J.* **2024**, *488*, 150728. [CrossRef]
45. Grassie, N.; Francey, K.F.; Macfarlane, L.G. The thermal degradation of polysiloxanes-Part 4: Poly(dimethyl/diphenyl siloxane). *Polym. Degrad. Stab.* **1980**, *2*, 67–83. [CrossRef]
46. Grassie, N.; Macfarlane, I.G. The thermal degradation of polysiloxanes-I. Poly(dimethylsiloxane). *Eur. Polym. J.* **1978**, *14*, 875–884. [CrossRef]
47. Jananovic, J.D.; Govedarica, M.N.; Dvornic, P.R.; Popovic, I.G. The thermogravimetric analysis of some polysiloxanes. *Polymer* **1998**, *61*, 87–93.
48. Rucigaj, A.; Krajnc, M.; Sebenik, U. Kinetic study of thermal degradation of polydimethylsiloxane: The effect of molecular weight on thermal stability in inert atmosphere. *Polym. Sci.* **2017**, *3*, 1–7.
49. Chenoweth, K.; Cheung, S.; Duin, A.C.T., Goddard, W.A.; Kober, E.M. Simulations on the thermal decomposition of a poly(dimethylsiloxane) polymer using the ReaxFF reactive force field. *J. Am. Chem. Soc.* **2005**, *127*, 7192–7202. [CrossRef]
50. Grassie, N.; Mcfarlane, I.G.; Francey, K.F. The thermal degradation of polysiloxanes-II. Poly(methylphenylsiloxane). *Eur. Polym. J.* **1979**, *15*, 415–422. [CrossRef]
51. Grassie, N.; Francey, K.F. The thermal degradation of polysiloxanes-Part 3: Poly(dimethyl/methyl phenyl siloxane). *Polym. Degrad. Stab.* **1980**, *2*, 53–66. [CrossRef]
52. Tasic, A.; Pergal, M.V.; Antic, M.; Antic, V.V. Synthesis, structure and thermogravimetric analysis of α, ω-techelic polydimethylsiloxanes of low molecular weight. *J. Ser. Chem. Soc.* **2017**, *82*, 1395–1416. [CrossRef]
53. Hamdani, S.; Longuet, C.; Perrin, D.; Lopez-Cuesta, J.M.; Ganachaud, F. Flame retardancy of silicone-based materials. *Polym. Degrad. Stab.* **2009**, *94*, 465–495. [CrossRef]
54. Hanu, L.G.; Simon, G.P.; Cheng, Y.B. Thermal stability and flammability of silicone polymer composites. *Polym. Degrad. Stab.* **2006**, *91*, 1373–1379. [CrossRef]
55. Scala, L.; Hickam, W. Thermal and oxidative degradation of silicones. *Ind. Eng. Chem.* **1958**, *50*, 1583–1584. [CrossRef]

https://doi.org/10.3390/ma19091798

materials

MDPI

Review

Research Progress on the Application of Soda Residue in Cementitious Materials

Ying Gong [1], Kaiyue Zhao [1,*], Gang Liu [2], Ying Ba [1], Yaoyao Wu [1], Zijian Liu [1] and Yong Yang [1]

[1] College of Architectural Engineering, North China University of Science and Technology, Tangshan 063210, China; yinggong0502@163.com (Y.G.)
[2] College of Materials Science and Engineering, North China University of Science and Technology, Tangshan 063210, China
* Correspondence: zhaokaiyuel@ncst.edu.cn

Abstract

Soda residue (SR) is an industrial waste produced by the ammonia-soda process. The unique structural characteristics and cementitious activity of soda residue enable it to be used as a mineral admixture for cementitious materials, which is an important way for its resource utilization and also the development direction of green building materials. This paper reviews its potential as a supplementary cementitious material in cementitious materials from physicochemical properties, microstructural influence, and macro performance impact perspectives. Key findings indicate that soda residue enhances cementitious materials' compactness and early strength through physical filling and chemical activation, yet it concurrently impairs workability and poses durability risks due to chloride content and salt crystallization. An optimized application requires dosage control, chemical modification, and combined use with mineral admixtures. Future research should focus on developing composite binder systems, innovating solid waste-based material preparation, and advancing desalination technologies to enable large-scale, environmentally sound utilization.

Keywords: soda residue; cementitious materials; microscopic properties; macroscopic properties; shrinkage

1. Introduction

As a commonly used building material, the production and use of concrete have a significant impact on the environment [1]. Traditional production relies on cement, which generates a large amount of carbon dioxide, conflicting with the goals of low-carbon and sustainable development [2–5]. In the context of pursuing carbon peak and carbon neutrality goals, reducing carbon emissions in the building materials sector has become a critical part of the broader transition toward a green economy, requiring fundamental changes at the material level [6,7]. Therefore, researching new types of concrete is a development trend. Industrial solid wastes such as fly ash (FA), slag, steel slag (SS), and soda residue can serve as supplementary cementitious materials to replace a portion of cement, enhancing concrete performance while promoting resource recycling, aligning with sustainable development requirements [8,9]. By utilizing these industrial by-products, the carbon footprint per unit of concrete can be significantly reduced, contributing to emission reductions across the entire life cycle of buildings. Soda residue is a strongly alkaline by-product of the ammonia-soda process (Figure 1) [10], and its production volume is closely related to the output of soda ash. China emits 7.8–10 million tons of soda residue annually, but its utilization rate in China is only 3–4% [11]. Untreated soda residue not only occupies land resources, leading to land degradation, but also causes soil salinization, posing a threat to plant growth [12]. The large-scale open storage of soda residue has already triggered environmental issues [13]. Furthermore, soda residue contains a large amount of soluble salts, particularly chlorides [14,15] and magnesium salts [16], which can pollute groundwater and threaten human health and the ecosystem [17]. Therefore, the recycling and utilization of soda residue brooks no delay. Incorporating such industrial by-products into low-carbon cementitious material systems offers a promising strategy to address environmental challenges while expanding the technological pathways available for carbon emission reduction.

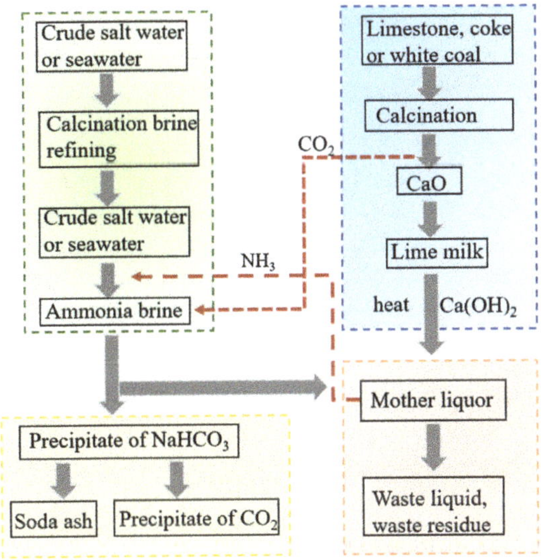

Figure 1. A schematic diagram of soda ash production via the ammonia-soda method [10].

The porous structure and micro-expansion characteristics of soda residue give it significant advantages in concrete applications. The porous structure of soda residue endows it with superior water absorption and release properties, enabling it to establish a balanced humidity state within the concrete, effectively mitigating cracks caused by autogenous shrinkage and drying shrinkage of the concrete [18]. The micro-expansion characteristics of soda residue can effectively compensate for the shrinkage stress of concrete, inhibit crack development, and thereby enhance the durability of the structure [19]. In addition, its alkali-activated effect can synergize with additives to form a high-performance cementitious system, exhibiting significant advantages in replacing traditional cement-based materials [20]. This paper systematically summarizes the physicochemical characteristics of soda residue and classifies SR according to its pretreatment state. On this basis, it systematically reviews the influence mechanisms of SR on the microstructural characteristics (hydration, pore structure, and microscopic morphology) and macroscopic performance (workability, mechanical properties, durability, and shrinkage) of cement-based materials.

The novelty and original contributions of this review are threefold: (1) Classification by pretreatment state. The effects of SR are systematically classified according to its pretreatment condition (untreated (RSR), water-washed (WSR), wet-ground (WMSR), and calcined (CSR)), providing a reconciling framework for the seemingly contradictory optimal dosages and experimental results reported in the paper. The roles and mechanistic

 https://doi.org/10.3390/ma19112228

contributions of each SR type across different cementitious matrices are critically differentiated. (2) Critical synthesis of optimal dosages. A structured synthesis table is established that integrates matrix type, SR pretreatment state, replacement level, evaluated properties, and reported optimal ranges, revealing the underlying condition-dependence of reported optimal dosages and explaining why these values cannot be directly compared without explicitly matching their boundary conditions. (3) The Instability of Friedel's salt (Fs), identifying underappreciated durability risks. This paper critically examines durability concerns that have been previously underexplored, including the long-term stability of Fs under coupled environmental exposures, the fundamental distinction between free and bound chloride, and efflorescence risk, which collectively define the boundary conditions for the safe structural application of SR in cementitious materials. Through this systematic and critical synthesis, this review aims to provide a coherent theoretical foundation for the high-value, safe, and sustainable utilization of SR in the construction materials sector.

In this review, unless otherwise specified, the discussion encompasses SR behavior across paste, mortar, concrete, and alkali-activated cementitious systems.

2. Literature Search and Screening Methodology

A systematic literature search was conducted to identify studies on the application of soda residue in cementitious systems. The following databases were searched: Web of Science, Scopus, and CNKI. The search employed the following keyword combinations: ("soda residue" OR "ammonia-soda residue" OR "alkali waste" OR "distiller waste") AND ("concrete" OR "cement" OR "mortar" OR "paste" OR "cementitious"). The search covered publications from January 2000 to April 2026.

Inclusion criteria were: (1) peer-reviewed journal articles published in English or Chinese; (2) studies directly investigating SR as a component in cementitious materials (including paste, mortar, concrete, and alkali-activated systems); and (3) studies reporting at least one of the following: physicochemical properties, hydration mechanisms, microstructural characteristics, mechanical properties, durability, or shrinkage behavior.

Exclusion criteria were: (1) conference abstracts, dissertations, and non-peer-reviewed reports; (2) studies using SR solely as a soil stabilizer or fill material without discussion of hydration mechanisms; and (3) studies in which SR was not a primary variable (e.g., SR mentioned only as one of many wastes without dedicated analysis).

3. Physicochemical Properties of Soda Residue

3.1. Physical Properties

Soda residue exhibits remarkable lightweight characteristics (Table 1) [16,21–29] with a porous honeycomb-like microstructure (Figure 2) [29,30], endowing it with high porosity (>60%), high specific surface area (SSA), and small particle size distribution characteristics [16,31,32]. During the concrete mixing process, fine particles of soda residue can be evenly distributed in the cement mix, improving material properties through the following synergistic effects: (1) physical filling, where fine particles effectively fill micro pores, enhancing the compactness of the system [33]; (2) interfacial strengthening, promoting the densification of the aggregate–paste transition zone, thereby improving the mechanical strength of concrete [34]; and (3) internal curing, where porous structure stores and release water, facilitating later hydration [35]. It is worth noting that the high specific surface area of soda residue mainly originated from its unique microstructural characteristics. Cao and Zhang [36] confirmed that soda residue mainly consists of loosely porous aggregates composed of well-crystallized $CaSO_4$ and $CaCO_3$, with ~50 nm nano-pores significantly enhancing the specific surface area. Wang et al. [37] found that this porous structure comprises nanoscale pores and particle voids within micrometer-sized agglomerates, and the

https://doi.org/10.3390/ma19112228

nanoscale pores have a decisive impact on adsorption and reaction activity. Wu et al. [38] further pointed out that the proportion of particles with a diameter of ≤25 μm in the soda residue exceeds 95%. This characteristic distribution of fine particles, coupled with the porous structure, endows the soda residue with excellent surface activity.

Table 1. Basic physical properties of soda residue [16,21–29].

ρ (g/cm^3)	SSA (m^2/kg)	Moisture Content (%)	Plastic Limit (%)	Liquid Limit (%)	Particle Size (μm)
2.25~2.35 [16,22,25–27]	200~600 [16,21,23,25]	40~60 [16,21,23,25,29]	38~54 [24,28,29]	60~90 [24,28,29]	≤25 μm [21,25,26,29]

Note: ρ = particle density; SSA = specific surface area; all percentages reported in Table 1 (e.g., moisture content, plastic limit, liquid limit) represent mass percentages (% *w*/*w*).

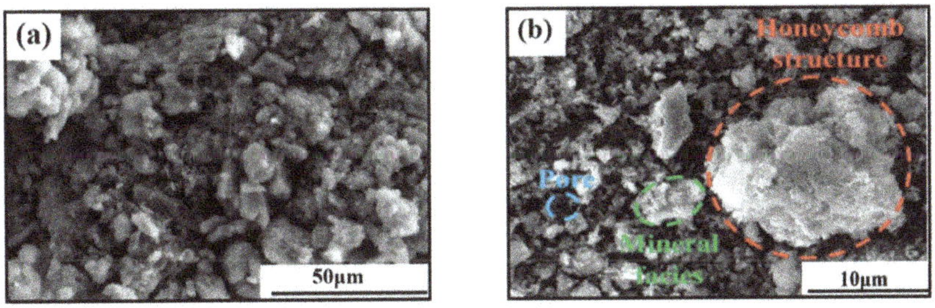

Figure 2. SEM images of soda residue: (**a**) Porous structure [30]; (**b**) honeycomb-like agglomerate [29].

Meanwhile, Yang et al. [39] established a "three-level model" of the microstructure of soda residue through SEM and TEM analysis: (1) The primary structure consists of 2~5 μm radiating spherical agglomerates formed by the self-stacking of nanoscale calcium carbonate crystals (Figure 3a); (2) The secondary structure is formed by further overlapping of primary agglomerates into 5~10 μm aggregates, with micron-scale through-pores inside (Figure 3b); and (3) The macrostructure is a porous skeleton formed by the accumulation of aggregates and dihydrate gypsum particles (Figure 3c). Cao et al. [40] further confirmed through microscopic characterization that this skeleton is mainly composed of well-crystallized dihydrate gypsum ($CaSO_4.2H_2O$) and $CaCO_3$ porous agglomerates, exhibiting significant powder characteristics. It is worth noting that this multi-level assembly structure not only helps to reduce the porosity of concrete but also provides structural support for the soda residue–cement composite system, thereby enhancing the macroscopic properties of the material [41]. In addition, Zhao et al. [42] and Yin et al. [43] observed that the surface of soda residue particles exhibits rough and porous characteristics, and the pore connectivity on the surface and inside of the soda residue particle aggregates is relatively strong. This structural characteristic affects the water absorption performance and mechanical properties of concrete. Meanwhile, Bai et al. [44] found that the honeycomb structure of the soda residue is gradually filled with cementitious products after the hydration reaction, and the structural compactness is enhanced.

3.2. Chemical Properties

The chemical composition characteristics of soda residue have become a key factor in analyzing its reaction mechanism in cement-based materials and optimizing its application in building materials. Yang et al. [45] characterized the element content and chemical composition of soda residue and the results are shown in Figure 4a, where the content of elements such as Ca, Cl, and Mg is significant, and the main components are calcium salts ($CaCO_3$, $CaSO_4$, $CaCl_2$ and $Ca(OH)_2$, etc.), with a small amount of oxides such as

https://doi.org/10.3390/ma19112228

SiO_2, MgO, Al_2O_3, Fe_2O_3 and Na_2O (Table 2). However, XRD analysis indicated that soda residue itself does not contain $Ca(OH)_2$ or $CaCl_2$; these components (primarily $CaCO_3$) were detected in the filtrate and precipitate after stirring with water (Figure 4b). Furthermore, the study found that increasing the water-to-solid ratio enhances the solubility of Cl^-, and the solution remains highly alkaline, with a pH value exceeding 11.4 even at high water-to-solid ratios. Their significantly high alkaline characteristics (pH > 10) make them particularly suitable as alkali-activated materials [22].

Figure 3. A "three-level model" of the microstructure of soda residue [39].

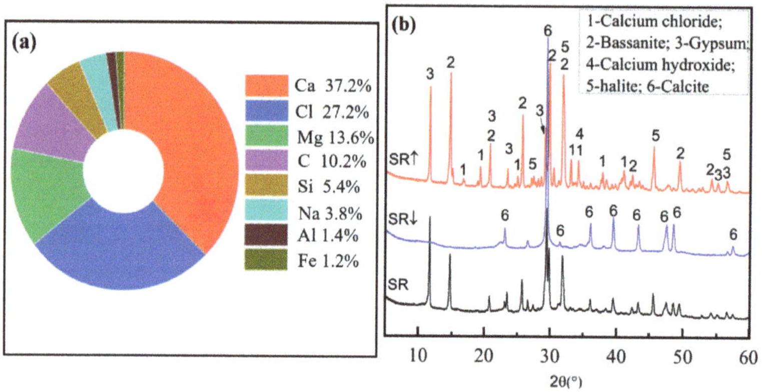

Figure 4. Composition of soda residue (SR): (**a**) Element content of soda residue (all percentages are expressed as mass percent (% *w*/*w*)) [45]; (**b**) XRD patterns of supernatant (SR↑), precipitate (SR↓), and SR [22].

According to the GB175–2023 [46] standard, the equivalent alkali content of cement should be controlled below 0.6%. However, as shown in Table 2, the alkali content of most soda residue exceeds this limit, which may induce alkali–aggregate reaction. In addition, elements such as Mg^{2+} and Cl^- in soda residue have a significant impact on concrete properties. When the Mg^{2+} content is too high, its corrosivity and crystalline expansion effects can lead to significant deterioration in the strength of cement-based materials [47]. The Cl^- content (often exceeding 9%) far exceeds the limit of 0.06% specified in the standard GB/T50476-2008 [48]. Microscopic analysis shows that the Cl^- are mainly present in the interstices between particles and in the micron-sized pores of $CaCO_3$ crystals [39]. High Cl^- content may trigger steel corrosion and volume expansion, ultimately leading to concrete cracking [49,50].

The main alkaline components in soda residue are $CaCO_3$ and $Ca(OH)_2$, whose hydrolysis reaction can create an alkaline environment with a pH of 10~12 [38]. This

 https://doi.org/10.3390/ma19112228

alkaline medium can effectively stimulate the pozzolanic activity of materials such as fly ash and promote the formation of hydration products, thereby enhancing the mechanical properties of composite cementitious materials [51]. Although $CaCl_2$ and $CaSO_4$ increase the complexity of soda residue treatment, studies have shown that $CaCl_2$ can significantly promote early-stage hydration of cement, enhancing the early strength of cement-based materials [52]. Meanwhile, $CaSO_4$ can react with alkaline components to form ettringite (AFt), whose moisture retention properties contribute to the self-curing of concrete [53,54]. It is worth noting that although a $CaCl_2$ solution is suitable for antifreeze engineering due to its low freezing point, corresponding anti-corrosion measures must be taken during its application [55].

Table 2. Chemical composition of soda residue %.

Literature	CaO	SiO_2	Al_2O_3	MgO	Fe_2O_3	SO_3	K_2O	Na_2O	Cl^-
An et al. [56]	39.20	5.92	1.92	6.73	0.44	5.33	0.31	4.65	35.40
Guo et al. [21]	43.20	9.87	3.25	9.77	0.91	5.57	0.29	3.93	23.00
Qi et al. [57]	40.30	9.02	2.09	10.00	0.59	6.41	0.31	3.58	27.40
Cheng et al. [58]	52.25	4.06	1.76	2.33	1.17	16.97	0.15	2.46	18.39
Pang et al. [59]	43.51	6.48	1.77	6.91	0.69	7.63	0.47	0.1	32.25
Xu et al. [60]	52.88	10.19	3.25	8.35	1.23	8.87	0.38	1.84	11.95
Li et al. [61]	42.70	9.67	3.45	9.56	0.94	0.11	030	3.84	24.00
Chen et al. [62]	44.70	10.10	2.55	8.09	0.89	5.97	0.43	4.40	20.30
Song et al. [63]	35.12	4.17	2.06	2.97	0.29	2.8	0.33	13.04	39.11

Note: SO_3 refers to the total sulphate content in cement; all percentage values in Table 2 represent mass percent (% w/w).

3.3. Fabrication Method

To overcome the application bottlenecks of raw soda residue, namely its excessive chloride content, low reactivity, and high compositional variability, researchers have developed various pretreatment strategies. By selectively modulating key properties such as chloride concentration, specific surface area, and mineralogical composition through physical, chemical, or thermal pathways, these treatments fundamentally alter the functional role and behavioral characteristics of soda residue in cementitious systems.

3.3.1. Raw Soda Residue (RSR)

Raw soda residue (RSR) refers to the waste material collected directly from the discharge outlet of the ammonia-soda process without any subsequent industrial treatment. Its characteristics directly inherit the final state of the alkali production process. The primary advantages of RSR are its low processing cost and operational simplicity; however, its inherent drawbacks, namely, high chloride content, substantial compositional variability, and limited reactivity, severely restrict its scope of application.

From a property perspective, the principal concern with RSR lies in its high soluble salt content, particularly Cl^-. When RSR is incorporated into concrete as a cementitious component, a large quantity of Cl^- dissolves into the pore solution, far exceeding the critical chloride threshold required to maintain steel passivation. This poses a fundamental risk of reinforcement corrosion, which represents the most critical barrier to its large-scale utilization [49]. Furthermore, RSR particles exhibit irregular morphology and a wide particle size distribution. Although these particles can provide a limited micro-filler effect, they tend to increase the water demand of the mixture, thereby adversely affecting workability. In addition, the primary mineral phases in RSR (such as $Ca(OH)_2$ and $CaCO_3$) possess negligible pozzolanic activity. When RSR is used directly as a cement replacement, its contribution to strength development is minimal and may even cause strength loss due to a dilution effect [60].

3.3.2. Washed Soda Residue (WSR)

Water washing is currently the most widely adopted and effective method for removing soluble salts, particularly chlorides from soda residue. This process involves mixing the residue with water, mechanical agitation, and subsequent solid–liquid separation, whereby a substantial fraction of soluble salts such as NaCl and, to a lesser extent, $CaCl_2$ are dissolved and removed, thereby yielding washed soda residue (WSR) with significantly reduced chloride content.

The core function of water washing lies in the purification of the chemical composition. Studies have shown that thorough washing can reduce the chloride content of soda residue by more than 90%, rendering it suitable for use as a mineral admixture in ordinary Portland cement concrete. When the chloride content of WSR falls below a critical threshold (0.30%), its incorporation at recommended replacement levels (10%) does not induce discernible reinforcement corrosion. Under these conditions, WSR can partially substitute cement or fly ash to produce concrete that meets both strength and workability requirements [49]. Moreover, the washing process also removes a portion of water-soluble sodium salts, which moderately lowers the pH and reduces the alkalinity of WSR, which is beneficial in mitigating its potential to promote alkali–aggregate reaction (AAR).

3.3.3. Wet-Milled Soda Residue (WMSR)

Wet-milling modification is a physical refinement method in which soda residue is co-ground with water, typically in a ball mill or similar equipment. This process employs mechanical forces to break down agglomerates and reduce particle size, thereby producing wet-milled soda residue (WMSR) with substantially enlarged specific surface area and enhanced reactivity.

Studies have demonstrated that incorporating WMSR as a nucleation seeding additive in Portland cement yields pronounced benefits. The nano or micron-sized WMSR particles provide abundant heterogeneous nucleation sites for the precipitation of cement hydration products, greatly accelerating the precipitation and growth of C–S–H gel. This in turn substantially shortens setting time and significantly boosts early-age compressive strength (a 1.69–fold increase in 12–h strength) [57,64]. Moreover, these fine particles exert a micro-filler effect that refines the pore structure and densifies the hardened cement paste. A notable advantage of wet-milling is that the process is relatively mature and requires no chemical additives during grinding, thereby avoiding secondary pollution. Its principal drawback, however, is high energy consumption. Grinding soda residue to the nanoscale demands considerable electrical energy, which, to a certain extent, limits the economic viability of its large-scale application.

Moderate grinding helps to enhance the activity of soda residue and maintain the integrity of its agglomerate structure [36]. Zhang et al. [65] ground the soda residue and incorporated it into concrete, finding that a 10% soda residue content could significantly reduce the internal micropores of the concrete, enhancing its compactness and wear resistance. Water washing can effectively remove soluble chlorides (such as $CaCl_2$, NaCl), but it may result in the loss of active ingredients. Recently, Yang et al. [64] achieved selective leaching of ions from soda residue through a wet grinding process. Their XRD and TG analysis results indicated that this process promoted the dissolution of SO_4^{2-} and Cl^-, leading to the weakening of characteristic peaks of $CaCO_3$ and $Ca(OH)_2$, while the production of $NaSO_4$ increased. XPS results further confirmed that wet grinding treatment

https://doi.org/10.3390/ma19112228

reduced the binding energies of S, Cl, and Ca, verifying the dissolution of crystal phases such as $CaSO_4 \cdot 2H_2O$, $Ca(OH)_2$ and $CaCO_3$ in soda residue.

3.3.4. Calcined Soda Residue (CSR)

The effectiveness of calcination modification depends to a large extent on the calcination temperature. If the temperature is too low, decomposition remains incomplete, and the activation effect is limited. Conversely, excessively high temperatures may induce sintering of the activated phases (CaO), leading to a reduction in specific surface area and a consequent decline in reactivity. Identifying the optimal calcination temperature is therefore of critical importance. As the calcination temperature increases, soda residue undergoes a significant structural evolution process, with its nano-scale porous structure gradually being destroyed and its specific surface area decreasing. Meanwhile, a gradual increase in pore size is evident. At the same time, the agglomerates' densification degree increases, their particle size distribution concentrates in the range of 0.01 mm to 0.074 mm, and the pore morphology transitions from interconnected pores to large explosive pores [40]. Thermal analysis reveals that moderate calcination (600~800 °C) can effectively retain the active component $Ca(OH)_2$, promoting its hydration reaction with pozzolanic materials (e.g., FA, Ground Granulated Blast Furnace Slag(GGBS)) to generate high-strength hydrated calcium silicate [38]. When the temperature exceeds 1200 °C, the active component becomes inactive and is accompanied by an increase in free calcium oxide (f–CaO) content, requiring a combination with a water quenching process to restore its activity [66,67]. It is worth noting that during calcination at 600 °C, the surface of soda residue particles undergoes melting, forming a porous solid structure with a structural framework function, significantly enhancing the mechanical stability of the material [40].

3.3.5. Chemically Modified Soda Residue

Chemical modification refers to a class of methods in which one or more chemical reagents are added to soda residue, or the residue is blended with other industrial wastes, with the aim of altering its chemical composition and reaction characteristics. The objectives of these approaches are diverse and include further reduction in chloride content, activation of latent reactivity, or the generation of synergistic effects with co-existing components.

A more prevalent form of chemical modification is synergistic activation, in which soda residue is combined with other industrial wastes to form a novel cementitious system. Wang et al. [68] blended soda residue with carbide slag (CS) and demonstrated that these two materials supply a substantially higher concentration of OH^-, generating a "synergistic alkali–calcium activation" effect that powerfully mobilizes the reactivity of slag and enables the production of high-performance, clinker-free cementitious materials. Within such a system, Cl^- and SO_4^{2-} originating from the soda residue can react with Ca^{2+} and $(Al(OH)_6)^{3-}$ to form Fs and AFt. The mutual filling and interlocking of these hydration products create a dense, high-strength microstructure. The key advantage of this synergistic modification strategy lies in its principle of "treating waste with waste"; it realizes the resource recovery of multiple solid wastes and often generates a synergistic effect where the combined performance exceeds the sum of individual contributions. Nevertheless, chemical modification also introduces new challenges, including the cost of chemical reagents, the increased complexity of mix design for multi-component systems, and the potential introduction of additional impurities.

To more clearly illustrate the regulatory effects of different modification methods, Table 3 provides a comprehensive comparison of the five primary processes.

In summary, the optimization of the microstructure of soda residue is mainly achieved through the following mechanisms: (1) High-temperature treatment promotes the melt-

https://doi.org/10.3390/ma19112228

ing and consolidation between particles, forming a denser agglomerated structure [69]; (2) Preparing nano-micron soda residue seed crystals by wet grinding method significantly improves the pore structure, reduces the porosity and critical pore size, and increases the content of gel pores [64]; and (3) Supplementing alkaline activators promotes the participation of active components in the soda residue in the polymerization reaction, and improves the composition and structure of the gel phase [63,70,71].

Table 3. Comparative analysis of methods for soda residue modification.

Type	Regulatory Mechanism	Key Property Changes	Typical Application Scenarios
Raw	Bone	High Cl^-, low reactivity; Complex composition; Low cost, simple process.	Clinker-free binder activator; S/S solidifying agent.
Washed	Dissolution–separation	Cl^- content significantly reduced; pH decreased; Effectively removes harmful ions.	Mineral admixture for cement concrete.
Wet-Milled	Mechanical comminution	Reduced particle size; Increased SSA; Enhanced reactivity; Uniform dispersion.	Cement early–strength agent; seeding, high-performance concrete.
Calcined	Thermal decomposition –phase transition	Formation of reactive $CaO/Ca(OH)_2$; loosened structure; Greatly enhanced pozzolanic activity.	High–activity cementitious material component.
Chemically Modified	Chemical reaction, synergistic effect	Composition reconstruction; Multiple activation; Strongly designable performance; synergistic enhancement.	Alkali–calcium synergistically activated binders, specialty functional materials.

4. Soda Residue Effect on Concrete Micro Characteristics

4.1. Cement Hydration

4.1.1. Soda Residue–Cement Binary System

The hydration mechanism of the binary system of soda residue–cement can be divided into five stages (Figure 5a) [23]: (1) The rapid exothermic reaction: The cementitious material reacts rapidly with water and releases heat, which is primarily caused by the dissolution of raw material minerals in the liquid phase. The addition of SR significantly increases the exothermic heat of this initial stage, indicating that SR accelerates the dissociation of raw material particles, with higher SR amounts leading to faster dissociation rates. (2) The dormant period: As the reaction progresses, the main hydration products begin to form, and Calcium Silicate Hydrate(C–S–H)/Calcium Aluminate Hydrate(C–A–H) initially fill the pores. (3) The acceleration stage. Raw materials release more heat. The presence of SR causes the fast exothermic peak to become narrow and advances the maximum exothermic point. The gel phase became denser, and its strength jumped. (4) The deceleration stage: The reaction rate gradually slows down, and the hydration products interweave to form a dense microstructure. (5) The steady stage: The reaction tends towards equilibrium, and the rapid precipitation of Fs forms a dense barrier layer on the surface of unreacted particles, creating a barrier effect. It impedes the diffusion of reactants and restricts the further formation of Calcium Aluminosilicate Hydrate(C–A–S–H) gel and restricts the release of Cl^-, thereby exerting an ion–fixation effect. The structure is optimized, and the porosity decreases. Although excessive dosage (20%) increases the early hydration heat,

 https://doi.org/10.3390/ma19112228

the high cement substitution rate limits the total amount of hydration products in the later stage (Figure 5b) [23].

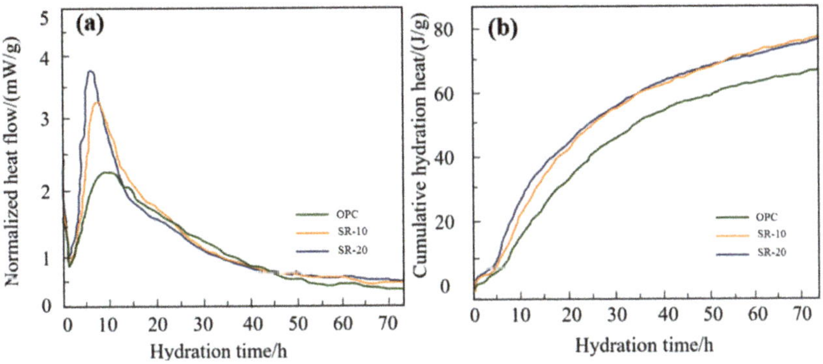

Figure 5. Hydration flow (**a**) and cumulative hydration heat curve (**b**) of the cement–soda residue binary cementitious material system at 72 h (OPC refers to ordinary Portland cement) [23].

In the SR–cement binary system, the regulatory mechanisms governing hydration kinetics differ fundamentally depending on the type of soda residue employed.

Ionic acceleration effect is primarily associated with RSR. RSR, the core mechanism driving hydration acceleration, lies in its ionic effect. The abundant soluble chloride salts ($CaCl_2$, NaCl) and $Ca(OH)_2$ present in RSR dissolve rapidly upon contact with water, releasing substantial quantities of Cl^-, Na^+, Ca^{2+}, and OH^- into the liquid phase. This process accelerates hydration at three distinct levels: (1) It enhances the ionic strength and electrical conductivity of the pore solution, thereby accelerating the surface dissolution kinetics of the cement clinker minerals (Tricalcium Silicate (C_3S)) and Dicalcium Silicate (C_2S)) [60]; (2) $CaCl_2$, a well-established cement accelerator, reacts with Tricalcium Aluminate (C_3A) to form Fs ($3CaO{\cdot}Al_2O_3{\cdot}CaCl_2{\cdot}10H_2O$), consuming the initial hydration layer on the C_3A surface and exposing fresh reaction interfaces [72–74]; and (3) elevated Ca^{2+} supersaturation promoting C–S–H nucleation. The high concentration of dissolved Ca^{2+} raises the degree of supersaturation with respect to C–S–H, effectively lowering the nucleation barrier and facilitating heterogeneous precipitation of C–S–H gel on available particle surfaces [75].

Physical nucleation effect, WMSR: For WMSR produced through wet grinding treatment, the dominant mechanism promoting hydration shifts from an ionic effect to a physical nucleation effect. The wet-milling process refines the SR particles to the nano-/micron-scale (D_{50} < 1 μm) and drastically increases their specific surface area [64]. When uniformly dispersed in the paste, these ultrafine particles serve as heterogeneous nucleation substrates for the precipitation of C–S–H gel, substantially lowering the nucleation barrier for hydration products. The study by Yang et al. [64] provides direct evidence for this mechanism: the incorporation of only 6% WMSR increased the 12-h compressive strength of Portland cement by a factor of 1.69, reduced the total porosity from 34.95% to 29.19%, and sharply decreased the critical pore diameter from 155.67 nm to 21.29 nm. That such pronounced effects can be achieved at a remarkably low dosage stands in clear contrast to RSR, which requires a dosage in excess of 10% to produce comparably significant effects, a distinction that clearly reveals the fundamentally different mechanisms of action between the two types.

Reactive calcium effect, CSR: When CSR is produced by calcination at 600~800 °C, its principal mineral, $CaCO_3$, undergoes partial decomposition into reactive CaO, while $Ca(OH)_2$ also dehydrates to form CaO [38]. This reactive CaO hydrates rapidly upon con-

https://doi.org/10.3390/ma19112228

tact with water to generate $Ca(OH)_2$, thereby contributing not only directly to the formation of hydration products but also enhancing the alkalinity of the system and accelerating the hydrolysis of C_3S and C_2S. This represents a combined mechanism of chemical and physical activation. It should be noted, however, that excessively high calcination temperatures (beyond approximately 1100~1200 °C) cause CaO crystal sintering and densification, commonly referred to as "dead-burning" (over-burning), which drastically reduces its specific surface area and hydration reactivity [66,67].

4.1.2. Soda Residue–Solid Waste Multi-Component System

SR, as the core component of the multi-component solid waste collaborative technology, utilizes its chemical properties rich in alkali metal oxides and silicate minerals (providing active components and strong alkali environment), porous rough surface expansion reaction interface, and suitable particle size guarantee uniformity to synergistically activate the cementitious system with solid wastes such as ground-granulated blast–furnace slag [56,73], fly ash [20,76], and carbide residue [72,77]. Table 4 shows that soda residue can stimulate the potential activity of mineral admixtures in binary systems, promote the generation of dense microstructures from multi-component hydration products, provide a structural basis for high-strength alkali-activated cementitious materials, and significantly improve the mechanical strength of concrete [61,63]. Table 5 further reveals that the multi-component system constructed by soda residue synergistic slag, iron tailings, and carbide residue forms a multi-dimensional reaction network through ion exchange and network polymerization, and other multi-component physical and chemical coupling, constructing a multi-level pore structure, ultimately achieving a significant improvement in the performance of cementitious materials and the efficiency of solid waste resource utilization.

Table 4. The synergistic effect of the soda residue–single mineral admixture binary system.

Binary System	Synergistic Reaction with Soda Residue
SR–Fly ash	Pozzolanic effect: The inherent alkalinity of the soda residue effectively activates the reactive SiO_2 within fly ash, facilitating the formation of cementitious hydration products [29]. Ion exchange reaction: In alkaline conditions, Ca^{2+} from soda residue participates in ion exchange with alkali metal ions (Na^+, K^+) on fly ash particle surfaces. This exchange facilitates depolymerization and dissolution of the fly ash vitreous matrix, releasing reactive Si^{4+} and Al^{3+} that form precursors for polymerization reactions [78]. Gel polymerization reaction: The released reactive ions undergo polycondensation to form oligomeric gels, which further polymerize into three–dimensional network gel phases (C–A–S–H, Sodium Aluminosilicate Hydrate(N–A–S–H). As key strength contributing phases, these gels densify the microstructure by pore filling, thereby enhancing overall compactness [79].
SR–GGBS	Alkali-activated reaction: Alkali activators (Na_2O, K_2O) derived from soda residue react with reactive SiO_2 and Al_2O_3 in slag, generating hydraulic cementitious phases [30]. Cl^- solidification: Slag can combine with Cl^- from soda residues to form Fs, effectively immobilizing chloride. This process not only addresses the issues caused by high chloride concentration in soda residue but also improves the performance of the material [63]. Hydration promotion: Soda residue activates slag hydration, facilitating the formation of C–S–H, C–A–S–H, and Fs. Supplementary $Ca(OH)_2$ and $CaCl_2$ from soda residue elevate Ca^{2+} concentration in the system, accelerating hydration kinetics and promoting C–S–H gel nucleation and growth [62].

https://doi.org/10.3390/ma19112228

Table 4. *Cont.*

Binary System	Synergistic Reaction with Soda Residue
SR–SS	Complexation reaction: During later hydration stages, Calcium (Alumino–) Silicate Hydrate(C–(A)–S–H) gels integrate with soda residue, undergoing microstructural densification that refines pore networks. This cohesive transformation enhances compactness and structural integrity, ultimately elevating mechanical performance and durability [80]. AFt Formation: Calcium species from soda residue enable early–stage AFt formation. Deposited on steel slag particles, AFt concurrently modulates hydration kinetics through surface inhibition while functioning as a structural framework to densify the matrix. This dual mechanism refines pore architecture and significantly reduces drying shrinkage [81]. Alkali-activated reaction: The inherent alkalinity of soda residue provides OH^- ions that effectively activate latent reactive minerals in steel slag (C_3S, C_2S). This activation accelerates mineral dissolution and subsequent hydration, generating strength-enhancing C–S–H gel to optimize cementitious performance [82].

Table 5. The synergistic effect of soda residue–multi-mineral admixture system.

Multi-Component System	Synergistic Reaction with Soda Residue
SR-Slag-Gypsum	The hydration products, including C–S–H, AFt, and Fs, form a dense three-dimensional network structure. Concurrently, gel phases cohesively integrate unreacted particles within this network while infilling interstitial spaces, inducing microstructural densification that elevates concrete compressive strength [56].
SR-GGBS-iron tailings	Primary cementitious phases C–S–H, C–A–S–H, and Fs exhibit progressive accumulation with extended curing, ensuring strength development from early to mature stages in the binder system [57].
SR-Carbide slag-red mud–Fly ash	The alkaline environment derived from carbide slag, soda residue, and red mud drives congruent dissolution of reactive phases in fly ash and red mud. This process facilitates the formation of a cohesive three-dimensional network comprising N–A–S–H, C–(A)–S–H, and Fs, significantly enhancing early strength development. Concurrently, Cl^- from soda residue is chemically immobilized by red mud components, mitigating permeability risks through chloride penetration [83].

Although Tables 4 and 5 systematically document the synergistic mechanisms of SR with single and multiple mineral admixtures, a critical horizontal comparison reveals distinct design philosophies and inherent trade-offs among the three multi-component systems: (1) SR–slag–gypsum: achieves the highest densification through synergistic C–S–H–AFt–Fs network formation, but its high sulfate content raises concerns regarding delayed AFt formation under wet–dry cycling; (2) SR–GGBS–iron tailings: benefits from the additional physical filling of tailings particles with sustained strength development, yet its long-term chloride binding stability remains unverified beyond short-term laboratory tests; (3) SR–carbide slag–red mud–fly ash: uniquely combines Fs formation with red mud adsorption for dual chloride immobilization and excellent early strength, but its quaternary complexity poses significant challenges for mix design standardization and field quality control.

These three systems respectively embody "densification-first", "filler-reinforced" and "chemical immobilization-first" design strategies. Their divergent strengths and limitations underscore that system selection must balance multiple criteria (early strength, chloride binding capacity, long-term volumetric stability, and practical workability), rather than relying on compressive strength alone.

In summary, soda residue has dual functions of physical filling and chemical activation in alkali-activated systems. The chloride and hydroxide contained in it deeply affect the microstructure evolution path by regulating the concentration of liquid-phase ions and

https://doi.org/10.3390/ma19112228

participating in hydration reaction (Figure 6): soluble alkaline substances (NaCl, $CaCl_2$, $Ca(OH)_2$) dissociate and release Na^+ and Ca^{2+} upon contact with water to form a highly alkaline environment, directly activating the activity of slag and other additives [84]; at the same time, it promotes pozzolanic reaction of fly ash and slag to generate characteristic crystal phases such as C–S–H, C–A–H, C–A–S–H amorphous gels and Fs, significantly enhancing chemical cementation [57,85]; finally, the gel and crystals interweave to form a spatial network skeleton, and the amorphous phase transforms into a stable crystalline phase, driving the microstructure to mature and stabilize [84].

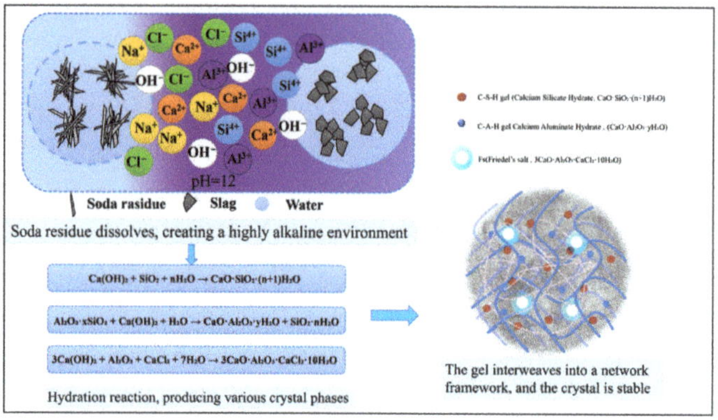

Figure 6. Microstructural evolution of soda residue–slag system.

Numerous studies have demonstrated that raw soda residue (RSR) can be directly employed as an activator for the production of alkali-activated slag cement (AASC) systems: (1) Provision of an alkaline environment: Upon dissolution, RSR releases OH^- at concentrations sufficient to disrupt the glassy network structure of slag, inducing its depolymerization [26]. (2) Provision of a calcium source: The $Ca(OH)_2$ and $CaCO_3$ constituents present in RSR supply Ca^{2+} to the system. As Ca^{2+} is an essential cation for the formation of C–S–H and C–A–S–H gels, this dual "alkali–calcium" action yields an activation effect superior to that achieved with sodium-based alkaline activators alone [63]. (3) Participation in the formation of hydration products: The Cl^- and $SO_4{}^{2-}$ ions originating from RSR can combine with aluminate phases in the hydration products to form Fs and AFt [63]. These hydration products not only contribute to strength development but also effectively immobilize Cl^-, reducing its free concentration in the pore solution and thereby mitigating, to a certain extent, the deleterious effects of chloride ions.

In AASC systems, the synergistic performance between soda residue and other components is a complex process. Although the application of RSR embodies the principle of "treating waste with waste", it also introduces new challenges. The first concern is workability: (1) The strong alkalinity and fine particle characteristics of RSR typically lead to high viscosity and poor fluidity of AASC pastes, posing difficulties for construction operations [25]; (2) volumetric stability: the formation of AFt and Fs is accompanied by volume expansion, while a moderate degree of expansion can fill pores, excessive expansion can induce structural cracking [86]; (3) long-term durability issues, including the risk of alkali–silica reaction (ASR) and carbonation stability, all of which require in-depth investigation [87].

The application of all-solid-waste cementitious systems not only eliminates the carbon emissions associated with cement production but also enables the resource recovery of multiple industrial waste streams, representing an important direction for the future development of building materials. However, the mix design of such systems is exceedingly

 https://doi.org/10.3390/ma19112228

complex, imposing stringent requirements on the chemical composition and reactivity of the raw materials. Furthermore, their long-term performance, particularly with respect to carbonation resistance and sulfate attack resistance, awaits validation through further engineering practice.

4.2. Pore Structure

Soda residue optimizes the pore structure of concrete through physical filling and chemical activity, significantly reducing total porosity, refining pore size distribution, and improving pore morphology and connectivity, thereby enhancing material mechanical properties and durability [88–91]. This effect is regulated by the dosage of soda residue, the type of cementitious system, and the synergistic effect of multiple materials [88–91].

Within the low dosage range, the pore structure refinement afforded by soda residue is pronounced; however, the underlying optimization mechanism varies with the SR type. Initially, Xu et al. [23,60] demonstrated through MIP analysis that when the RSR dosage is maintained below 6%, the cumulative pore volume of the hardened paste reaches its minimum, and the proportion of innocuous pores (<20 nm) increases significantly (Figure 7a). This is caused by the synergistic effect of physical filling (micro particles filling the gap to reduce the initial porosity [92]) and chemical coagulation (stimulating the generation of hydration products such as C–S–H gel, AFt and Fs, filling pores and reducing connectivity [93]). Building on these findings regarding porosity reduction, the research of Yang et al. [64] shows that adding 6% wet grinding soda residue reduces the total porosity from 34.95% to 29.19%, and the critical pore size sharply reduces from 155.67 nm to 21.29 nm, the pore size pattern of the cement mortar with SR seed was significantly shifted to the left, with most of the pore size within 50 nm, This effect originates from the nucleation seeding action of WMSR, which promotes the formation of a more uniform and denser C–S–H gel, together with the filling of submicron pores by the nano-sized particles. Furthermore, multi-system validation shows that soda residue can refine pore size distribution in increasingly complex matrices. Lin et al. [22] obtained the optimal pore structure (Figure 7b) in the alkali-activated slag system (soda residue: GGBS = 16:84), the population of large pores was minimized, the pore structure was optimized, and the compressive strength reached its highest value. In this system, the pore refinement arises from two contributions: the GGBS hydration activated by RSR generates C–S–H and AFt that fill the pores, while the appropriate proportion of GGBS supplies sufficient aluminate phases to bind Cl^- as Fs, thereby avoiding the negative effects of excess chloride. Xu et al. [94] optimized the pore size distribution and early strength improvement in the soda residue–steel–slag system, revealing the intrinsic relationship between pore structure and durability. Soda residue reduced harmful pore volume fraction, resulting in a decrease in Cl^- diffusion coefficient and improved impermeability.

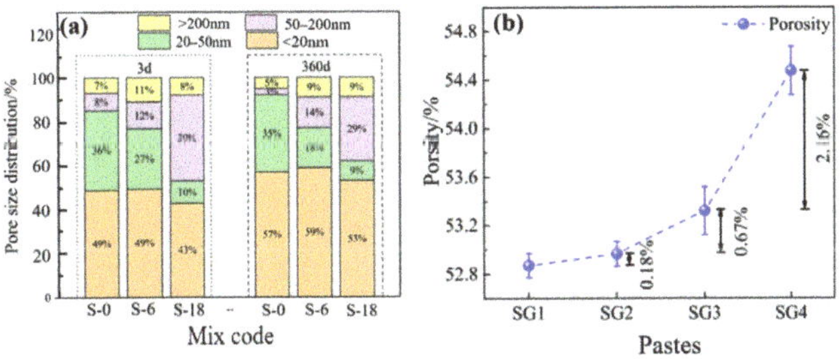

Figure 7. Porosity and pore size distribution of soda residue multi–component system: (**a**) cement mortar with soda residue (no pretreatment) [23]; (**b**) SR–GGBS–cement system at 28 days [22].

 https://doi.org/10.3390/ma19112228

However, there is a critical threshold for the optimization of pore structure by soda residue, and excessive addition will lead to negative effects. Lin et al. [22] found that when the RSR dosage was increased from 8% to 32%, both the total porosity and water absorption rose continuously, and the MIP study by Xu et al. [23] confirmed that the proportion of harmful pores significantly increases when the RSR content exceeds 10%. The mechanism is attributed to the increase in liquid phase ion strength caused by high soluble salts, which leads to flocculation, excessive inert components ($CaCO_3/SiO_2$) diluting the cementitious material to hinder densification [95–97], and Cl^-/SO_4^{2-} inducing excessive AFt generation (crystal pressure deteriorates pore structure) [98]. Although the densification effect of WMSR at low dosages is markedly superior to that of RSR, its dosage threshold is correspondingly lower. This is because the ultrafine WMSR particles require more water for wetting; at high dosages, non-uniform dispersion may lead to local agglomeration, which paradoxically undermines the filler effect [64]. Meanwhile in the RSR–GGBS system, excessive RSR dilutes the GGBS and reduces the formation of C–S–H gel. Simultaneously, the excess Cl^- cannot be fully bound by the available aluminate phases, leading to an elevated free chloride concentration. The resulting imbalance between the Fs and C–S–H gel fractions reduces the stiffness of the matrix [94]. The SEM results of Zhang et al. [26] showed a significant increase and uneven distribution of AFt crystals in the high alkalinity system, with an increase in unreacted particles and large pores (Figure 8). In addition, the dissolution recrystallization process of Fs in high salt environments can generate volume expansion stress, damage pore structure, and reduce strength [94], so it is necessary to avoid such environmental applications.

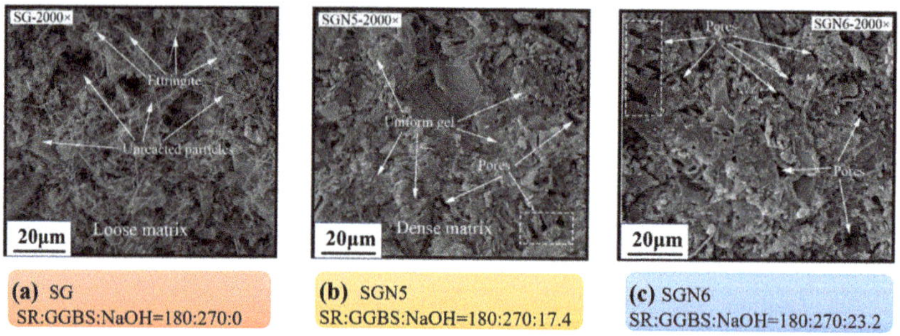

Figure 8. SEM images of 28 days multi-component system [26].

1.3. Microscopic Morphology

The active components such as Ca^{2+}, SiO_2, and Al_2O_3 in soda residue are reconstructed into the microstructure of cement-based materials through secondary reactions: (1) Product phase transition, promoting the formation of new phases such as calcium chloroaluminate ($3CaO{\cdot}Al_2O_3{\cdot}CaCl_2{\cdot}10H_2O$ namely Fs) and calcium chloride silicate sulfate ($Ca_{10}(SiO_4)_3(SO_4)_3Cl_2$) [37], while reducing the content of $Ca(OH)_2$ and AFt (Figure 9a) [23,63]. (2) Morphological regulation: SEM showed that Friedel salt (flower like) (Figure 9b) $Ca_{10}(SiO_4)_3Cl_2$ (layered) and other characteristic phases interlaced with C–S–H gel to form a dense complex (Figure 9c) [37,63]; at the kinetic level, $CaCl_2$ in soda residue accelerates the dissolution and hydration rate of C_3S and C_2S [52,64,99], and its nanoscale size and high specific surface area provide rich reaction interfaces [100]. In terms of microstructural evolution, at the microscopic level, Friedel salts and $Ca(OH)_2$ synergistically fill C–S–H pores, constructing low porosity dense structures, ultimately improving mechanical strength and durability [23,101].

https://doi.org/10.3390/ma19112228

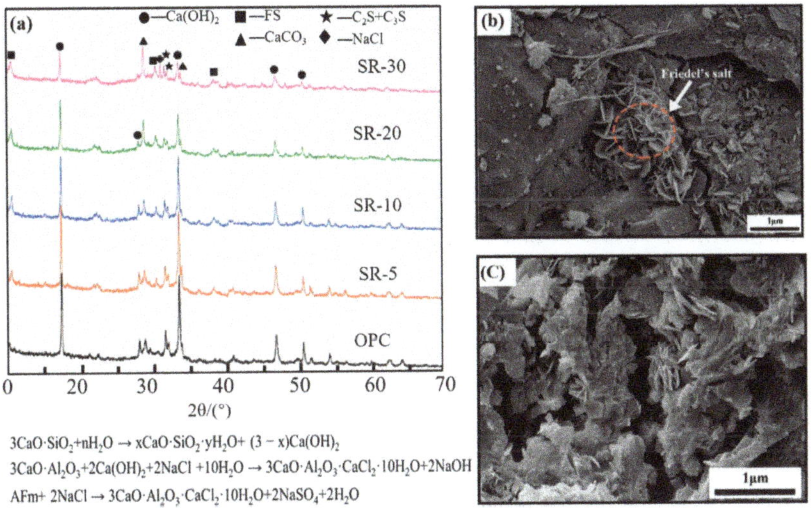

Figure 9. Soda residue–cement binary system: (**a**) XRD patterns of hydration products of produced soda residue–cement at 28 days (OPC refers to ordinary Portland cement) [23,63]; (**b**) SEM image of Fs [23]; (**c**) SEM–EDS analysis of hydration products from cement pastes at curing ages of 28 days [63].

Soda residue significantly changes the morphology of hydration products by regulating the chemical environment and reaction process of the cementitious system. In Portland cement, C–S–H is amorphous flocculent or fibrous, while $Ca(OH)_2$ is hexagonal plate-like [102,103]. After adding soda residue, the C–S–H structure of the Portland cement system transforms into a honeycomb-like pore-filling structure [23], while in the high alkali-activated slag system, C–S–H evolves into a dense, thin sheet or foil-like stacked morphology [86]. Beyond these morphological changes in individual phases, the synergistic interaction among multiple hydration products has been increasingly recognized. Xu et al. [60] observed that needle-like AFt and flake Fs were interpenetrated in the C–(A)–S–H gel matrix in the soda residue–slag–steel slag–desulfurization gypsum quaternary system, wherein AFt crystal played a role of micro reinforcement, and Fs fixed some harmful Cl^-. The XRD pattern confirmed that the characteristic peaks of C–S–H, AFt and Fs matched the standard spectrum, revealing the synergistic enhancement mechanism of heterogeneous products. Jiang et al. [104] observed that needle-like AFt and flake Fs were interpenetrated in the C–(A)–S–H gel matrix in the SR–GGBS–phosphogypsum(PG) ternary system, wherein AFt crystals acted as micro-reinforcement, and Fs immobilized harmful Cl^-. The XRD pattern confirmed that the characteristic peaks of C–(A)–S–H, AFt and Fs matched the standard spectra, revealing the synergistic enhancement mechanism of heterogeneous products. Additionally, the dense microstructure with reduced porosity further accounted for the superior mechanical performance, where the highest proportion of mesopores (<50 nm, 67.62%) and the lowest total pore volume (0.121 cm^3/g) were achieved at the optimal mix proportion (S10G81P9), correlating well with the 28-day compressive strength of 46.15 MPa.

Soda residue can significantly improve the performance of the concrete interface transition zone (ITZ). Soda residue micro particles aggregate on the aggregate surface to provide nucleation sites, promote heterogeneous nucleation of C–S–H gel, reduce the size and orientation of $Ca(OH)_2$, and improve the compactness of ITZ [93]. In addition, the soda residue activator improves the alkalinity of the system, accelerates the hydration reaction

 https://doi.org/10.3390/ma19112228

in the ITZ region, and forms a better structure [82,105]. Xu et al. [23] found that when the soda residue was greater than 10%, excessive Cl^- generated Fs and consumed $Ca(OH)_2$, resulting in the reduction in C–S–H gel and matrix stiffness. Ren et al. [106] claimed that when the soda residue was greater than 80%, the C–S–H gel was fragmented, the porosity increased, the micro crack propagation path was shortened, the fracture toughness was reduced, and the interfacial bonding strength was significantly deteriorated.

5. Soda Residue Effect on Concrete Macro Performance

5.1. Working Performance

Because of its unique physical and chemical properties, soda residue has a complex impact on the working performance of concrete. Yang et al. [49] confirmed that the slump of soda residue concrete was significantly reduced under a low water–binder ratio. Zhao et al. [43] found that the fluidity and density of soda residue–fly ash geopolymer mortar decreased with the increase in dosage, while the porosity increased synchronously, which was attributed to the physical water absorption characteristics of soda residue. Liu et al. [107] pointed out that its strong water absorption can enhance water retention, effectively inhibit bleeding during transportation and construction, and improve uniformity. Li et al. [61] and Ren et al. [106] further revealed that when the particle fineness of soda residue is optimized, and the grading is good, it can fill the micro gap and synergistically improve the cohesion and water retention capacity. The effect of soda residue on the workability of concrete is regulated by multiple factors: its alkaline component and $CaCl_2$ accelerate the hydration of cement, resulting in early setting, and reduce the fluidity and plasticity [52]; fine particle size increases water demand [44], irregular morphology increases particle friction resistance and inhibits fluidity [22]; and strong water absorption adsorbs free water to reduce lubrication and significantly reduce fluidity [26].

5.2. Mechanical Properties

Proper addition of soda residue can improve the mechanical properties of cementitious materials. However, when the amount of soda residue exceeds a certain proportion range, the strength of the test piece will decrease (Figure 10a) [30,37,108–113]. In cement-based systems, Yang et al. [49] pointed out that the strength can be improved by replacing 10% cement with soda residue, but a water-reducing agent should be added in the case of a low water-to-binder ratio system. Furthermore, in multi-component solid-waste systems (e.g., soda residue–fly ash–slag), similar optimal dosage patterns have been observed. Li et al. [114] showed that the compressive and flexural strength reached the peak when the ratio of soda residue to fly ash was 40:60. Xu et al. [111] found that the strength of the soda residue–fly ash mineral powder mixed system increased first and then decreased when the soda residue content was 20–30%. Building on these findings, Zhang et al. [72] further refined the optimal dosage range in a SR–CS activated GGBS–FA system. They demonstrated that the 28-day compressive strength reached a maximum of 30.4 MPa when the soda residue content was 22%, and the carbide slag content was 8%; increasing the soda residue to 26% or decreasing it to 18% both led to strength reduction (Figure 10b). Moreover, the addition of 4% desulfurization gypsum further enhanced the 28-day strength to 34.2 MPa, confirming a synergistic sulfate activation effect (Figure 10c). Beyond the influence of the soda residue proportion, its physical fineness and activator concentration also significantly affect mechanical performance. Liu et al. [115] further revealed that, among the three factors examined (soda residue-to-fly ash ratio, liquid–solid ratio, and sodium silicate concentration), the sodium silicate concentration exerted the most significant influence on compressive strength, followed by the liquid–solid ratio, while the soda residue-to-fly ash ratio had the least effect. The optimal compressive strength (17.778 MPa) was obtained

 https://doi.org/10.3390/ma19112228

at a sodium silicate concentration of 3.0 mol/L, a liquid–solid ratio of 1.1, and a soda residue-to-fly ash ratio of 2:3. They also found that reducing the sieving size of soda residue from <2 mm to <0.25 mm enhanced the 28-day compressive strength by approximately five times, and a reduction to <0.5 mm led to an increase of about 51%. In addition, Guo et al. [78,116] showed that the splitting tensile strength and flexural strength of the solid waste-based concrete made of SR–CS–FA cementitious material combined with iron tailings aggregate are equivalent to those of ordinary concrete, which is due to the formation of dense ITZ structure between the cementitious system and iron tailings to reduce interface defects, but the elastic modulus of the system is low and the brittleness characteristics after peak value are significantly increased. Soda residue has dual effects on concrete. When properly mixed, its hydrate forms a compact structure through cemented aggregates and closed pores, which significantly improves the short-term strength. However, the high salt content of excessive soda residue is easy to induce long-term efflorescence and degrade the durability. The optimization strategy includes the use of soda residue–carbide slag wet basis, combined with precise temperature control curing at 60~75 °C for 12 h, which can accelerate the hydration reaction, compact the microstructure, and improve the mechanical properties, but it is necessary to avoid overheating or overtime curing [21,108].

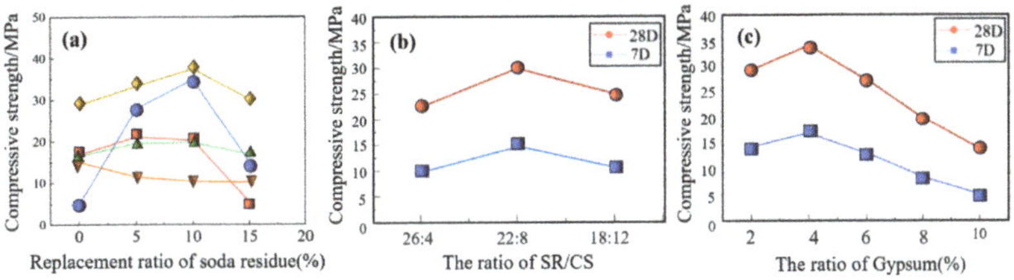

Figure 10. The effect of soda residue on the mechanical properties of multi-cementing systems: (**a**) Compressive strength by soda residue ratio [30,37,108–110]; (**b**) compressive strength by soda residue/carbide slag ratio [72]; (**c**) compressive strength by gypsum ratio [72].

Table 6 summarizes matrix type, soda residue status, substitution level, evaluation characteristics, and the optimal ranges reported, aiming to visually illustrate the variability and patterns of optimal incorporation levels under different study conditions. Synthesizing the data in the table above, we can deeply analyze the key factors influencing the optimal dosage of soda residue from the following two dimensions.

Matrix type is the primary factor determining the optimal dosage range of soda residue, as it fundamentally defines the role that soda residue plays within the system: (1) In Portland cement-based materials, the role of soda residue is "supplementary" or "functional". As a blending material, its optimal dosage is constrained by the hydration product environment of the cement. The highly alkaline pore solution formed by cement hydration is the core for maintaining the cementitious nature of the system. Excessive soda residue dilutes this environment and introduces large amounts of non-reactive or low reactivity Cl^- and SO_4^{2-}, which may disrupt the microstructure of the hardened cement paste and lead to strength reduction. Therefore, its optimal dosage is typically low, such as 6% [64], 10% [49]. (2) In alkali-activated materials (AAMs), the role of soda residue transforms into a "core" or "dominant" one. It can be the primary alkaline activator (e.g., 70% soda residue activating slag [86]) or a cementitious main body alongside activated components such as slag (e.g., 20% soda residue with 80% slag [63]). In such a system, the alkalinity and ionic environment of the entire cementitious system are built around the characteristics of the soda residue, thus substantially raising the upper limit of its dosage. (3) In autoclaved aerated concrete (AAC), the role of soda residue is that of a "raw material". It provides a CaO source and participates in the autoclave hydrothermal synthesis reaction

 https://doi.org/10.3390/ma19112228

to generate high-strength crystalline phases, such as tobermorite. The optimal dosage in this case must be precisely matched with the SiO_2 source (e.g., fly ash) to satisfy the stoichiometric ratio and ensure a reasonable reaction process. A dosage of 20–30% is a commonly reported optimal range [117,118], reflecting its efficiency in providing calcium.

Replacement level itself is not an independent influencing factor; rather, it is a quantitative outcome that directly reflects the "concentration" of soda residue in the system. As discussed above, the optimal replacement level varies dramatically depending on the matrix type: (1) <5%: In high-performance cement-based materials requiring high dimensional stability and long-term durability, the dosage of soda residue is typically restricted to extremely low levels to avoid its adverse effects [37,60]. (2) A range of 5–15%: This is the most common optimal range reported for ordinary Portland cement-based materials (e.g., concrete, mortar, cement–stabilized soil). Within this range, the "micro–aggregate effect" and "pozzolanic effect" of soda residue are fully exerted, effectively filling pores and improving the interfacial transition zone, thereby reducing material costs and enhancing certain early-age properties without significantly compromising long-term performance [63]. (3) A range of 15–40%: This dosage range is more commonly encountered in autoclaved aerated concrete, where soda residue serves as the primary calcareous raw material [117], or in clinker-free cements where it is an important component of the composite cementitious system [119]. Here, the contribution of soda residue extends beyond physical filling; more critically, its chemical components participate directly in the principal hydration or hydrothermal reactions [43]. (4) Greater than 40%: In alkali-activated systems, when soda residue plays the role of an activator, its dosage can far exceed conventional limits. A dosage as high as 70% signifies that soda residue constitutes the main body of the system, with its function shifting from "replacement" to "dominance" [56,86].

Table 6. Optimal mixing amounts of variously modified soda residues in different cement-based materials.

Matrix Type	Soda Residue State	Replacement Level	Evaluated Properties	Reported Optimal Range
Autoclaved Sand Aerated Concrete (AAC) [120]	SR (Not specified)	SR replacing lime: 10–30%	Volumetric expansion, dry density, compressive strength	SR replacing lime: 20%
Autoclaved Aerated Concrete (AAC) [118]	Not specified	SR dosage: 0–30%	Compressive strength	Approx. 20–30%
Autoclaved Aerated Concrete (AAC) [117]	Not specified	SR: 20% (of total material)	Dry density, compressive strength	20% (SR + construction waste + lime + cement + gypsum system)
Alkali-Activated Slag Geopolymer [86]	Not specified	SR: 70% (of total mass of precursor)	Compressive strength	70% SR
Clinker-free composite cementitious material [63]	Not specified	SR: Slag = 1:4 (mass ratio)	Compressive strength, microstructure	SR: Slag = 1:4 (SR $\approx$ 20% of total binder)
Composite Portland cement [23]	Untreated (RSR)	SR: 5%, 10%, 20% (replacing cement)	Pore structure, setting time, compressive strength	<6% (pore structure); 5–10% acceptable; >10% reduces 28 d strength
Alkali-activated slag cement (AASC) [22]	Untreated (RSR)	SR: 8–34% (replacing GGBS)	Compressive strength	SR: GGBS $\approx$ 16:84 (SR $\approx$ 16%)
Clinker-free concrete (SR-GGBS-SS-FGD) [60]	Untreated (RSR)	SR: 6% (of total binder)	Compressive strength (up to 360 d)	6% SR (360 d strength 66.31 MPa)
Portland cement concrete [49]	Washed (WSR)	WSR: 10% (replacing cement or fly ash)	Compressive strength, Cl^- content	10% WSR when Cl^- < 0.3%

https://doi.org/10.3390/ma19112228

Table 6. *Cont.*

Matrix Type	Soda Residue State	Replacement Level	Evaluated Properties	Reported Optimal Range
Portland cement [64]	Wet-milled (WMSR)	WMSR: 6% (as nucleation seeding additive)	12 h compressive strength, hydration heat	6% WMSR (12 h strength increased by 1.69 times)
Alkali-activated slag geopolymer [121]	Chemically modified (SR + CS)	SR + CS as composite activator	28 days compressive strength	≥27 MPa at 28 days (specific proportion depends on system)
Geopolymer [68]	Chemically modified (SR + PG)	SR + PG as composite activator	Compressive strength, Microstructure Compressive strength, microstructure	Significant synergistic enhancement (AFt + C–S–H)

5.3. Chloride-Related Risks

5.3.1. Steel Corrosion

The most immediate and severe challenge posed by introducing SR into concrete is the enormous discrepancy between its chloride content and the limits stipulated in relevant national codes and standards. Table 7 summarizes the permissible limits for chloride ions (as a percentage of cement weight) in concrete, as specified in various standards for typical environmental categories. Cl^- in concrete is the primary culprit responsible for inducing reinforcement corrosion. The soda residue is rich in soluble chloride, which significantly increases the concentration of Cl^- in the concrete and forms an initial chlorine reservoir [37]. When the critical value is exceeded, it destroys the passive film of the reinforcement and causes corrosion [122–126]. In particular, it accelerates the migration of Cl^- in the humidity fluctuation environment, resulting in weakening of the reinforcement section and cracking of the protective layer [23,49], degradation of the bond between steel and concrete, and ultimately compromising the safety and durability of the structure.

Table 7. Chloride content limits in concrete are specified by different design codes.

Code/Standard	Exposure Class	Chloride Limit (%)
China Code for Design of Concrete Structures (GB 50010 [127])	Class I (Indoor dry environment)	≤0.30%
	Class IIa (Humid outdoor environment)	≤0.20%
	Class IIIa (Marine salt spray zone)	≤0.10%
USA Building Code Requirements for Structural Concrete (ACI 318 [128])	–	≤0.07% (Prestressed) /0.15% (Non-prestressed)
Europe Design of Concrete Structures (EN 1992-1-1 [129])	XD1/XC4 (Moderate corrosion)	≤0.40%

Note: Chloride limits are expressed as percent by mass of cementitious binder (% w/w). In the ACI 318 row, "≤0.07% (Prestressed)/0.15% (Non-prestressed)" means "≤0.07% for prestressed concrete and ≤0.15% for non-prestressed concrete."

To control this risk, stringent limits on chloride ion content have been established in design codes worldwide. The codes require the chloride ion content to be controlled below 0.3%. In contrast, the chloride content of SR generally exceeds 10%. Even when replacing cement at a modest substitution rate of 10%, the chloride ion content in the concrete would instantaneously exceed the 0.3% limit several times over. Lin et al. [22] demonstrated that when SR constituted 8% of the total cementitious material, the free chloride ion content by mass in the hardened cement paste at 28 days remained as high as 0.13%, far exceeding the stringent limit of 0.06%. Although the wet base SR:CS:GGBS:FA (1:0.5:3:0.5) quaternary cementitious material developed by Zhao et al. [43] meets the P.O42.5 strength standard, it is still not suitable for reinforced concrete due to its high chlorine characteristics. In practical application, the soda residue must be strictly controlled in dosage, dechlorination pretreatment or avoiding the reinforcement structure to ensure that the total chlorine content is lower than the specification limit to ensure durability [22,37].

5.3.2. Free Versus Bound Chloride Thresholds

The Critical Chloride Threshold Level (CCTL) is defined as the minimum concentration of free chloride ions in the pore solution required to induce depassivation and initiate corrosion of reinforcing steel under specific conditions. The determination of CCTL is central to corrosion risk assessment and service life prediction. However, extensive research demonstrates that CCTL is not a fixed constant but a complex variable influenced by numerous factors, including material properties [130], environmental conditions, and testing methodologies.

Cement type and mineral admixtures are one of the most significant factors influencing CCTL. For ordinary Portland cement concrete, CCTL is generally considered to range from 0.2% to 1.0% (by weight of cement). However, the situation becomes considerably more complex when substantial amounts of mineral admixtures (e.g., FA, slag) are incorporated. On the one hand, these admixtures can decrease the pH of the pore solution (particularly with high-volume fly ash), weakening the passive film stability and thus lowering the CCTL [130]. On the other hand, they enhance chloride binding capacity through pozzolanic reactions or the formation of hydration products. Zhang et al. [131] noted that reported CCTL values exhibit an extremely wide range of variation, from below 0.1% to over 1.0%, which is closely related to the type and dosage of mineral admixtures involved. Currently, there are few papers specifically addressing the effects on CCTL. However, based on extensive fundamental research and corrosion mechanism studies on CCTL, it can be reasonably inferred that increased temperature reduces the CCTL value, accelerates the corrosion reaction, while the type of reinforcing steel also influences the critical chloride ion concentration [132].

The determination of CCTL itself also faces considerable challenges. Traditional long-term immersion methods require months or even years, failing to meet the demands of rapid assessment [130]. Consequently, researchers have developed various accelerated testing methods. However, different testing methods, criteria (e.g., abrupt change in corrosion potential, current density exceeding a threshold), solution compositions, and experimental setups can all lead to significant discrepancies in the measured CCTL values. Table 8 summarizes the significant variations in CCTL across different studies, providing an intuitive visualization of its uncertainty.

Table 8. CCTL in different studies.

Concrete Type	CCTL (% by Weight of Cement)	Test Method
Conventional concrete (normal) [130]	0.44%	RCI accelerated test
High-volume fly ash concrete [130]	0.069%	RCI accelerated test
Marine concrete (marine environment) [131]	0.07–0.70%	Review, influenced by multiple factors
Conventional concrete (literature values) [131]	0.20–1.0%	Review, wide fluctuation range
Concrete affected by sulfates [133]	Reduced	Chloride threshold lowered due to sulfate presence

For high-chloride SR concrete, the assessment of CCTL becomes exceptionally difficult. On the one hand, its extremely high initial chloride content means that the reinforcing steel is in a "high-risk" environment from the outset. On the other hand, the complex chemical

https://doi.org/10.3390/ma19112228

composition of SR itself (e.g., high pH, potential presence of sulfates) and the instability of Fs make its true CCTL difficult to predict. Simply applying CCTL values derived for ordinary concrete may severely underestimate its corrosion risk.

5.3.3. Limitations of Chloride Binding and Long-Term Stability Concerns

Although the formation of Fs is widely regarded as a form of chloride "immobilization", this immobilization is not permanent [134]. Under specific physicochemical conditions, the crystal structure of Fs can be disrupted, leading to the release of the once "captive" chloride ions, forming what is termed an "internal source of chlorides". This dynamic process of "immobilization–release" is key to the long-term performance degradation of high-chloride SR concrete and constitutes its high corrosion risk.

First, chloride ion release induced by sulfate attack, SO_4^{2-}, possesses an ionic radius and charge similar to those of Cl^-, and their chemical properties are also relatively close, making them one of the most "effective" ions capable of destabilizing Fs. Studies have demonstrated that when sulfate ions are present in the environment, they undergo an anion exchange reaction with Fs, displacing the chloride ions (Cl^-) from their structure [135]. The reaction process can be represented as follows:

$$3CaO{\cdot}Al_2O_3{\cdot}CaCl_2{\cdot}10H_2O\ (Fs) + SO_4^{2-} \rightarrow 3CaO{\cdot}Al_2O_3{\cdot}CaSO_4{\cdot}10H_2O\ (AFt) + 2Cl^-$$

The reaction product is AFt, whose structure can accommodate the larger sulfate ion more readily than that of Fs. The essence of this process is that, thermodynamically, sulfate ions exhibit a stronger affinity for the layered AFm phase than chloride ions. Shen et al. [135] elucidated this mechanism at the atomic level: the incorporation of sulfate ions, through its charge redistribution effect, weakens the electrostatic interaction between chloride ions and the $(Ca_2Al(OH)_6)^+$ layers, and simultaneously disrupts the hydrogen bond network between chloride ions and water molecule layers, thereby substantially lowering the energy barrier for chloride ions to detach from the crystal structure and facilitating their release into the pore solution.

This constitutes a potential internal risk source. Even if concrete successfully immobilizes the majority of chloride ions within Fs at an early stage, once it encounters external sulfate attack (e.g., from seawater, groundwater, or sulfates in soil), these "bound" chloride ions may be "reactivated" into a free state, causing the chloride ion concentration in the pore solution to resurge, thereby posing a persistent threat to the safety of reinforcing steel. When concrete has a high soda residue content, which may introduce sulfate impurities, the threat to the safety of reinforcing steel is enhanced.

Secondly, changes in alkalinity (pH) affect the stability of Fs [136]. The stable existence of Fs requires a highly alkaline environment. The strong alkalinity of hardened cement paste (with pH typically ranging from 13 to 14) is critical for maintaining the passive film on reinforcing steel and the stability of various hydration products, including Fs [137,138]. However, during the long-term service life of concrete, its internal alkalinity may decrease for several reasons. The first is carbonation. Atmospheric carbon dioxide (CO_2) reacts with calcium hydroxide ($Ca(OH)_2$) in the pore solution to form calcium carbonate ($CaCO_3$) and water, leading to a sharp decline in the pH of the pore solution [139]. When the carbonation front reaches the surface of the reinforcing steel, the pH may plummet from above 13 to below 9, which will directly result in the breakdown of the steel's passive film [140,141]. Meanwhile, acid rain erosion or the metabolic activity of certain microorganisms also consumes the alkalinity within concrete [142]. Therefore, carbonation and neutralization of the concrete cover represent another important cause leading to the loss of the Fs stability and the subsequent re-release of chloride ions.

https://doi.org/10.3390/ma19112228

Wet–dry cycles and temperature variations also affect the stability of Fs and the speciation of chloride ions. In environments with alternating wetting and drying, concrete undergoes repeated cycles of water absorption and moisture loss. During the drying process, salts in the pore solution become concentrated and may exceed the solubility limit of Fs at that temperature, leading to its precipitation from the supersaturated solution. However, this portion of chloride ions may re-dissolve and become free ions upon subsequent water absorption. Furthermore, the drying process generates capillary pressure within the concrete, potentially inducing microcracks that provide rapid pathways for chloride ion transport [143]. Previously "bound" chloride ions may have a higher probability of re-entering the pore solution, increasing the local chloride concentration.

In summary, the "immobilization" of chloride ions by Fs is relative and conditional. Under the combined or sequential action of various environmental factors (including sulfate attack, carbonation, wet–dry cycles, and elevated temperatures), this immobilization may be disrupted, forming a persistent, internal source of chloride ion supply. This "immobilization–release" double effect renders the durability issues of high-chloride SR concrete more insidious and complex. Traditional static assessment methods are inadequate to capture their dynamic risks, and this must be given serious attention.

5.3.4. Limitations for Structural Applications

The limitations of high-chloride SR concrete in structural applications are mainly reflected in the following aspects: (1) Inability to meet mandatory code requirements for total chloride content: As discussed, the chloride content of SR far exceeds code limits. Regardless of the mix design, it is difficult to reduce the total chloride content (both free and bound) in the final hardened body to within safe limits. This renders it unsuitable for direct use in various types of reinforced concrete structures defined by codes, especially those located in humid, outdoor, or corrosive environments [131]. (2) Potential degradation of mechanical properties: The high content of soluble salts and chlorides may affect the hydration process and microstructure of the cement. Some studies have observed that although a certain amount of SR can accelerate early hydration, excessively high dosages may lead to reduced long-term strength or deteriorated durability indicators [22]. This performance uncertainty introduces additional risk to structural design. (3) Unpredictability of long-term performance: Owing to the "immobilization–release" mechanism, the long-term performance of high-chloride SR concrete is not static. Under environments involving carbonation [139], sulfate attack [135], etc., its interior may become a continuous "chloride generator", causing the risk of reinforcement corrosion to accumulate progressively over time. This unpredictability of long-term evolution contradicts the century-long design service life requirements typical of structures. (4) Strict limitations on applicable scenarios: For the reasons stated above, the application of high-chloride SR concrete must be strictly limited to non-load-bearing, dry, indoor environments free from groundwater erosion. Research has successfully applied it as a lightweight fill, subgrade cushion, or backfill material in road engineering [144]. In these applications, reinforcement is either completely excluded or is situated in an extremely low–risk environment. However, beyond these specific scenarios, its application is accompanied by extremely high structural safety risks.

5.4. Durability

5.4.1. Chloride Penetration Resistance Performance

The chloride penetration resistance of soda residue concrete is governed by a dual regulatory mechanism: the interplay between short-term penetration resistance (derived from pore structure optimization) and the long-term risk of chloride re-release (due to Fs instability). The active components in the soda residue fill the micropores by generating

https://doi.org/10.3390/ma19112228

dense cementitious products (C–(A)–S–H, AFt, Fs), which significantly improve the chloride ion penetration resistance of concrete [61]. This improvement effect is particularly prominent in the SR–GGBS system, which combines with Cl^- to form Fs to solidify free chloride ions, and the binding rate increases with the curing time [22,30]. The comparative study of multiple systems shows that the chloride ion solidification ability of carbide slag is better than that of GGBS and fly ash, and the impermeability can be enhanced by mixing soda residue and carbide slag [108]. Steel slag and soda residue can improve the ITZ compactness and then improve the overall impermeability [145]. The chloride salt in soda residue has a dual regulation mechanism on the chloride ion permeability resistance of concrete. When a certain amount is added, the calcium component reacts with Cl^- to form Fs to fix free chloride ions [58,106], and Fs fills the pores to enhance compactness and improve the permeability resistance [108]. However, excessive incorporation will lead to sharp reduction in C–S–H gel content and sharp increase in Fs, which will lead to increase in porosity and structural looseness, and significantly reduce the impermeability [106,107].

5.4.2. Sulfate Corrosion Resistance Performance

Soda residue has a complex impact on the sulfate resistance of concrete. Its alkaline substances promote the generation of hydration products and fill the pores to improve compactness, and enhance the corrosion resistance [25,79]. At the same time, Cl^- competes with SO_4^{2-} to inhibit sulfate attack [62]. However, soda residue changes the formation ratio of AFt and Fs. Fs consumes $Ca(OH)_2$, which weakens the buffering capacity of the sulfuric acid environment [30,37]. Excessive AFt causes structural damage and increased porosity due to volume expansion, which significantly reduces the corrosion resistance [29,62].

The impact of soda residue on the sulfate resistance of concrete needs dialectical evaluation. The carbonate composition in the soda residue–cement system is easy to react with sulfate to damage the structure and reduce the corrosion resistance [23], so it is not suitable for high sulfate resistance engineering scenarios. In view of this, soda residue is usually used in combination with other mineral admixtures to improve sulfate resistance. Xu et al. [146] confirmed that the corrosion resistance of the soda residue–GGBS synergistic system is better than that of ordinary Portland cement at an appropriate dosage. Yang et al. [64] performed a 5% Na_2SO_4 erosion test showing that the strength and quality of mortar with 20% soda residue content had zero loss. Zhao et al. [147] also found that C–S–H/N–A–S–H generated by the soda residue fly ash system has the ability to repair acid erosion. However, it should be noted that excessive soda residue will weaken sulfate resistance, and the Na^+/K^+ contained in it may induce alkali–aggregate reaction [148–150].

5.4.3. Efflorescence

Efflorescence refers to the phenomenon where salts dissolved in water migrate to the surface of building materials along with moisture, and upon moisture evaporation, the salts precipitate as crystals, forming a layer of white or colored powdery deposit. For SR–based materials, due to the intrinsic high content of soluble salts (NaCl, Na_2CO_3, Na_2SO_4, etc.), efflorescence is an apparent quality issue that cannot be ignored in its application.

Although research directly targeting efflorescence in SR-based materials is limited, referencing studies on related waste residues (red mud [151], MSWI fly ash [152]), efflorescence can be suppressed through the following approaches: (1) Reducing the source of soluble salts: Pretreating SR before material preparation, such as water washing, can effectively remove most of the water-soluble Cl^- and Na^+ [49]. (2) Reducing porosity and connectivity: By optimizing mix proportions, lowering the water-to-binder ratio, incorporating ultra-fine mineral admixtures such as silica fume, or employing measures like high-pressure steam curing, the material's porosity can be significantly reduced, blocking

https://doi.org/10.3390/ma19112228

capillary water transport pathways. (3) Incorporating mineral admixtures: Incorporating pozzolanic materials (FA, GGBS) can react with alkaline substances such as $Ca(OH)_2$ in secondary reactions, consuming some soluble alkalis and generating C–S–H gel of lower alkalinity, thereby reducing the concentration of alkali metal ions in the liquid phase. (4) Surface treatment: After the material has hardened, applying a hydrophobic agent or waterproof coating can form a protective film on the surface, preventing moisture ingress from the outside and the precipitation of internal salts. Wu et al. [153] found that incorporating phosphogypsum could effectively transform the efflorescence product from Na_2CO_3 to $Na_2SO_4 \cdot 10H_2O$, which has lower solubility, thereby mitigating the degree of efflorescence. This concept is equally applicable as a reference for SR-based materials.

5.4.4. Long-Term Performance

Material durability is a dynamic, continuously changing process. Long-term performance refers to the ability of a material to resist various environmental factors and load effects during its service life, maintaining its functionality and safety. For SR-based materials, due to their relatively short application history in the field of civil engineering, the assessment and prediction of long-term performance are key challenges and focal points of current research. Current understanding is primarily based on short and medium-term accelerated tests and theoretical inferences.

When SR is used as a cementitious component, during the initial curing period, due to the strong alkaline activation effect of SR, material strength increases rapidly [83]. In alkali-activated GGBS–FA systems, early strength mainly originates from the rapid hydration of GGBS [108]. However, the development of long-term strength depends on the completeness of the reaction and the stability of the microstructure. Nanayakkara et al. [154] found that AAS (alkali-activated slag) concrete still exhibits some strength gain under long-term curing (1 year), but the rate of increase gradually slows down. Wardhono et al. [155] observed that due to ongoing carbonation or microcrack propagation, the strength of certain AAS concretes may exhibit a slight decrease at later ages (360 days). Although there is a lack of research experiments specifically investigating the long-term performance of SR composite cementitious materials, reasonable inferences can be drawn based on relevant theories and studies of similar materials: (1) Early strength advantage: The alkaline components in the soda residue may accelerate the early hydration reaction, thereby enhancing the 28-day strength [23]. (2) Long-term performance risk: Prolonged accumulation of soluble salts may induce crystallization pressure, compromising the stability of hydration products.

5.5. Shrinkage Properties

5.5.1. Self-Shrinkage

The autogenous shrinkage of concrete with a low water–binder ratio results from capillary tension caused by cement hydration [156]. The key to crack control is to inhibit self-shrinkage, which can be achieved through internal curing technology or a shrinkage-reducing agent [157–160]. The internal curing material can relieve the self-drying effect by continuously releasing water, maintain the pore water saturation state, and reduce the capillary shrinkage stress, so as to significantly inhibit the self-shrinkage [161,162].

Soda residue can significantly inhibit the autogenous shrinkage of concrete through the triple synergistic mechanism: (1) Internal curing effect: Its micro pore structure continuously releases water to maintain internal humidity and reduce capillary stress [18]. (2) Micro-expansion compensation: $CaSO_4$ reacts with C_3A to produce expansive AFt neutralizing contraction [18,19]. (3) Activity regulation: Alkali residue replacing cement reduces early hydration activity and shrinkage source [19]. Pu [35] confirmed that 10–15% content can improve internal humidity and effectively reduce self-shrinkage. Yue [163]

 https://doi.org/10.3390/ma19112228

further optimized the process and found that 5.25% dechlorinated soda residue can effectively reduce autogenous shrinkage of concrete under the premise of ensuring the strength of 28 days, and the effect of additional underwater synergy is more significant. However, excessive soda residue will damage the mechanical properties, so the dosage needs to be strictly controlled, and the dosage of admixture for its treatment process needs to be further studied [35,163].

5.5.2. Chemical Shrinkage

Chemical shrinkage is an inherent property of the cement hydration process, which results from the reduction in solid volume of hydration products [164–166]. Soda residue regulates the chemical shrinkage of cementitious materials through dual mechanisms: (1) Alkali excitation accelerates hydration, the depolymerization and repolymerization of GGBS and steel slag. In the strong alkali environment provided by soda residue, this process accelerates the formation of C–S–H and C–A–S–H, and promotes the centralized completion of chemical shrinkage in a short time [37,60]; microstructure densification reconstructs later shrinkage dynamics [60]. (2) Ion-induced expansion compensation occurs when SO_4^{2-}/Cl^- react with C_3A to form AFt or Fs. Xu et al. [94] confirmed that Fs can compensate for the expansion gap of AFt reduction. Lin et al. [84] verified that AFt and C–S–H cooperatively filled the pores and brought volume expansion effect in the soda residue–carbide slag system. Liu et al. [82] pointed out that the soda residue activator can realize contraction/expansion directional switching.

Based on the dual role of soda residue as alkali activator and expansion source, under the appropriate dosage and compound conditions, it can effectively control chemical shrinkage and significantly improve the volume stability of concrete by accelerating hydration densification and expansion phase volume compensation. It is necessary to further quantify the expansion–contraction equilibrium point of the composite system to optimize the long-term performance.

5.5.3. Drying Shrinkage

The drying shrinkage of concrete is closely related to the pore structure of cement paste [167]. The higher the total porosity or the larger the proportion of macropores (>50 nm), the more significant the drying shrinkage of cement-based materials [168,169]. Drying shrinkage can be effectively inhibited by refining pore size and reducing pore connectivity [170,171]. Ning et al. [172] found that in the concrete strengthened by multiple solid wastes, with the increase in the amount of waste residue (including soda residue), the hydration product C–S–H gel/AFt continued to fill the pores, driving the microstructure to evolve from a three-dimensional network to a dense body, thus enhancing its ability to resist external interference. However, Guo et al. [78,116] found that although solid waste concrete has high strength, its elastic modulus is lower than that of ordinary concrete of the same grade. This low modulus characteristic gives the material excellent "stress absorption" ability, and can produce greater deformation without causing excessive stress under constraint conditions, significantly reducing the risk of cracking caused by drying shrinkage stress. This is consistent with the creep theory, that is, under the same shrinkage strain, the higher the elastic modulus, the greater the shrinkage stress, and the low modulus of solid waste concrete can alleviate this effect [173].

Soda residue can regulate the drying shrinkage of concrete through two paths: (1) Optimize the pore structure: Its hydration products can significantly improve the compactness of cement paste and effectively reduce the drying shrinkage, which is particularly obvious in the alkali-activated system. (2) Expansion compensation: The induced expansion products, such as AFt, provide volume compensation for drying shrinkage, but excessive

https://doi.org/10.3390/ma19112228

mixing may increase the shrinkage risk due to the introduction of bubbles or the increase in evaporation water.

6. Research Gaps and Priority Directions

6.1. Long-Term Stability of Friedel's Salt

Current research has identified key factors influencing the stability of Fs, such as pH [136], temperature, and the presence of sulfate ions [135]. However, systematic understanding of its long-term stability is still lacking: (1) Multi-factor coupling mechanisms remain unclear: Existing studies are predominantly single-factor laboratory tests, lacking investigations into the coupling effects of realistic conditions, such as carbonation–sulfate attack–wet–dry cycling–freeze–thaw cycles. When the carbonation front overlaps with the sulfate ingress zone, the acceleration or inhibition of the decomposition reaction of Fs currently lacks quantitative data. (2) Absence of phase-transition thermodynamic and kinetic databases: Fundamental thermodynamic data, such as the solubility equilibrium constants and Gibbs free energy of phase transitions of Fs at various temperatures (20 °C~80 °C) [174–176], pH values (9~13) [177], and Cl^-/SO_4^{2-} molar ratios [178], are extremely scarce, making it impossible to establish reliable thermodynamic prediction models. (3) Insufficient linkage between microstructural evolution and macroscopic property degradation: How the Cl^- released from Fs decomposition redistributes within the pore microstructure and how this process quantitatively correlates with the breakdown of the steel passive film remains a research gap [135,179].

6.2. Behavior of Free Chloride Versus Bound Chloride

The "immobilization–release" of chloride ions is central to evaluating the durability of SR concrete. However, quantitative research on its dynamic behavior is critically lacking: (1) Lack of systematic studies on chloride binding isotherms [180,181]: Chloride binding isotherm data for different SR contents, various composite cementitious systems (SR–GGBS [63], SR–FA [78], SR–SS [78]), and under various temperatures (5~60 °C) and ionic strengths are virtually non-existent. This makes it impossible to quantify the relative contributions of physical adsorption by C–S–H gel and chemical binding by AFm phases [182,183]. (2) Unclear critical thresholds for desorption behavior: The critical environmental conditions (critical pH, critical SO_4^{2-} concentration) that trigger the desorption of bound chloride under scenarios such as carbonation-induced pH reduction [184], ion exchange due to sulfate ingress [135,185], or salt concentration during wet–dry cycles have not been systematically determined. (3) Absence of a coupled diffusion–binding–desorption kinetic model: Traditional Fick's second law assumes that chloride binding is an instantaneous equilibrium linear process, which cannot describe the nonlinear binding and reversible decomposition behavior of Fs in SR systems [186]. There is an urgent need to develop mass-transfer kinetic models that couple chemical reactions.

6.3. Performance in Reinforced Concrete (RC)

The vast majority of current research remains at the level of paste or mortar, with very few studies addressing the long-term service behavior of real RC elements: (1) Unclear degradation mechanisms of RC elements in soda residue soil (SRS) environments: The chloride concentration in coastal SRS can be 2~3 times that of seawater [187]. When RC structure foundations or backfills consist of SRS, key issues such as the two/three-dimensional transport laws of chloride under unsaturated conditions and the spatiotemporal evolution of steel depassivation have not been systematically investigated. (2) Undetermined Critical Chloride Threshold Level (CCTL) for SR–containing concrete: Due to the high alkalinity and complex ionic composition introduced by SR, the pore solution chemistry differs

 https://doi.org/10.3390/ma19112228

significantly from that of conventional OPC concrete. Reliable experimental data on the CCTL for SR-containing systems are still lacking. (3) Absence of structural–level durability design methods: The establishment of service life prediction models for SR-containing RC elements, based on parameters such as chloride permeability coefficient, CCTL, and concrete cover thickness, represents a bridging gap connecting material research to engineering application.

6.4. Desalination Strategies and Pretreatment Processes

Desalination is the fundamental approach to eliminating the chloride risk of SR [188], but a technical solution suitable for large-scale implementation is yet to be developed: (1) Lack of low-energy, low-water-consumption dechlorination technologies: Traditional water washing is water intensive and generates high salinity wastewater, while electrokinetic methods are energy-intensive and slow. How to integrate technologies such as industrial waste heat utilization, CO_2-assisted water washing, and selective electrodialysis to develop lower-cost, more environmentally friendly dechlorination processes remains a significant technical gap [189]. (2) Unclear mechanisms for preserving and regulating SR activity after desalination: Water washing or chemical desalination processes simultaneously leach out valuable active components such as Ca^{2+} and OH^-, leading to a decrease in the cementitious activity or alkaline activation capacity of the treated SR. How to achieve "selective dechlorination" while retaining useful components is a technical challenge. (3) Lack of a holistic scheme for the resource utilization of desalination wastewater: NaCl and $CaCl_2$ in the wash water are potential industrial salt resources. Current research focuses more on wastewater treatment than resource recovery, lacking an integrated "desalination–salt separation–water recycling–zero discharge" full-process technology.

By conducting principal analysis, performance comparison, cost accounting, and environmental assessment of mainstream technologies such as water washing and electrokinetic methods, combined with frontier explorations like multi-technology coupling and water resource recycling, it is expected that efficient pretreatment technologies possessing both economic viability and environmental friendliness can be developed, thereby fundamentally resolving the concerns posed by chloride ions.

6.5. Environmental Assessment of Pre-Treatment Processes

The evaluation dimensions of existing studies mostly focus on mechanical properties and short-term durability, lacking systematic assessments: (1) Gap in Life Cycle Assessment (LCA) models: For different pretreatment pathways (water washing [49], electrokinetic, CO_2-assisted water washing [188]) and different end-use scenarios (alkali-activated cementitious materials [57], road base fillers [190], foamed concrete [144]), there is a lack of unified LCA models with clearly defined boundaries, making it impossible to quantify comprehensive environmental performance indicators such as Global Warming Potential (GWP) and Acidification Potential (AP); (2) Lack of data for Techno-Economic Analysis (TEA): A huge gap exists between cost estimates at the laboratory scale and actual operational costs at the pilot/industrial scale. There is a lack of complete cost accounting data based on real pilot/demonstration projects, especially regarding the hidden costs of wastewater treatment, equipment depreciation, labor costs, and transportation costs. (3) Absence of comparative assessments with conventional disposal pathways: There is a lack of systematic comparison, from a full life-cycle perspective, between the "pretreatment + building material application "pathway and the "direct stockpiling/landfilling + virgin raw material production" pathway, which is crucial for policy-making and industrial investment decisions.

 https://doi.org/10.3390/ma19112228

6.6. Techno-Economic Feasibility at Scale

SR is typically used in combination with slag, fly ash, steel slag, etc., but the long-term durability behavior of multi-component systems is far more complex than that of single systems: (1) Unclear synergistic/antagonistic effects of multiple solid waste components: In multi-component systems such as SR–GGBS–FA, the quantitative relationships governing the interactions among the activity release rates of each component, the evolution of hydration products, and the pore structure development remain unclear [21,108]. (2) Lack of performance data under extreme service conditions: Long-term performance data under extreme conditions, such as freeze–thaw cycles, coupled carbonation–chloride attack, and microbiologically induced corrosion, are extremely scarce. (3) Absence of standardized accelerated testing methods: Current durability tests mostly employ single-factor accelerated methods, lacking standardized accelerated testing protocols that can effectively simulate multi-factor coupling effects, resulting in poor data comparability between different studies.

Furthermore, the long-term volumetric stability of SR-blended systems under coupled environmental exposures, including drying shrinkage, autogenous shrinkage, and the balance between expansive hydration products (AFt, Fs) and long-term dimensional stability, remains poorly characterized. Dedicated long-term deformation monitoring under varying humidity and temperature conditions is needed. Similarly, the durability of SR-blended materials under wet–dry cycling and freeze–thaw attack has rarely been investigated, and the absence of standardized multi-factor durability testing protocols represents a significant barrier to service life prediction.

7. Conclusions and Prospects

This study summarizes the application of soda residue in cementitious materials from the aspects of its physical and chemical properties, microstructure, and impact on cementitious materials. The main conclusions are as follows:

Physicochemical characteristics and pretreatment-dependent properties: Soda residue is an industrial by-product with a lightweight porous honeycomb structure (high porosity of ~60%, high specific surface area of 200–600 m^2/kg), mainly composed of calcium salts ($CaCO_3$, $CaSO_4$, etc.), with high alkalinity (pH > 10) and Cl^-/Mg^{2+} enrichment characteristics. Its active components and pore structure can be directionally controlled through calcination, grinding, and chemical treatment. Each pretreatment method produces a distinct SR type that exhibits fundamentally different behavior in cementitious systems: untreated SR (RSR) functions primarily as an alkaline activator in alkali-activated systems; water-washed SR (WSR), with reduced chloride content, is more suitable for Portland cement systems; wet-ground SR (WMSR) provides nucleation seeding and micro-filling effects; and calcined SR (CSR) contributes enhanced pozzolanic reactivity.

In Portland cement systems at controlled dosages (typically <10% for untreated SR and <6% for wet-ground SR) under adequate curing conditions, SR refines the pore structure and densifies the matrix through physical filling, nucleation seeding, and chemical activation, promoting the transition from a loosely packed to a densified microstructure and strengthening particle interface bonding. In alkali-activated slag systems, SR serves additionally as an alkaline activator.

Coupled effect on workability and mechanical properties: When used at appropriate dosages (typically <10% in Portland cement systems, and SR:GGBS $\leq$ 16:84 in alkali-activated slag systems) and under adequate wet curing conditions, soda residue improves the compactness and early strength of concrete through the synergistic effect of physics and chemistry, In contrast, in systems employing untreated SR at higher replacement levels, its high water absorption and chloride salt crystallization characteristics lead to the deteriora-

 https://doi.org/10.3390/ma19112228

tion of rheological properties, forming a coupling effect of mechanical enhancement and workability reduction, which is controlled by multiple factors such as dosage, gradation, pretreatment method and curing system.

Soda residue has a bidirectional effect on the durability of concrete. Positively, at controlled dosages in appropriately designed composite systems and under conditions that favor stable Fs formation, SR improves impermeability and crack resistance through chemical chlorine fixation, pore structure optimization, and internal curing functions. In the negative direction, excessive or improper mixing can cause chloride salt crystallization expansion, intensified sulfate corrosion, and steel corrosion risk. Durability optimization needs to be achieved through dosage control, mineral admixture compounding, and dechlorination treatment, combined with the use of high-efficiency water-reducing agents and wet curing to enhance durability. For any reinforced concrete application, the total chloride content must be verified to comply with the limits specified in relevant structural concrete design codes (e.g., $\leq$0.06–0.30% by mass of cement, depending on exposure class).

Adding an appropriate amount of soda residue to cementitious materials can improve their mechanical and durability properties, reduce its self shrinkage, but compared to industrial waste such as slag, fly ash, and steel slag, the large-scale application of soda residue faces bottlenecks such as performance fluctuations, high pretreatment costs, pollutant leaching risks, chloride salt and alkali–aggregate reactions, and potential expansion. Future research needs to focus on breakthroughs in three major directions: the development of soda residue composite high-strength cementitious materials, technological innovation in the preparation of solid waste-based cementitious materials, and collaborative modification technology for efficient dechlorination and desalination processes, ultimately achieving the industrialization goal of balancing performance, cost, and environmental coordination.

Author Contributions: Writing—original draft preparation, Y.G.; writing—review and editing, K.Z., Y.B., Y.W., Z.L., G.L. and Y.Y.; visualization, Y.G. and K.Z.; supervision, K.Z. and G.L.; project administration, K.Z. and Y.Y.; funding acquisition, K.Z. and Z.L.; formal analysis, Y.B., Y.W. and G.L.; resources, Z.L. and Y.Y. All authors have read and agreed to the published version of the manuscript.

Funding: The financial support for ongoing projects by the National Natural Science Foundation of China (52508268), Natural Science Foundation of Hebei Province (E2024209134) and Science and Technology Project of Hebei Education Department (BJ2025165) are greatly acknowledged.

Data Availability Statement: No new data were created or analyzed in this study. Data sharing is not applicable to this article.

Conflicts of Interest: The authors declare no conflicts of interest.

References

1. Junaid, M.F.; Rehman, Z.U.; Kuruc, M.; Medved, I.; Bacinskas, D.; Curpek, J.; Cekon, M.; Ijaz, N. Lightweight concrete from a perspective of sustainable reuse of waste byproducts. *Constr. Build. Mater.* **2022**, *319*, 126061. [CrossRef]
2. Luo, J.; Huang, R.; Zhang, Y.; Wang, J.J. Carbon sequestration by preparing recycled cement, recycled aggregates and recycled concrete from construction and demolition (C&D) wastes. *Materials* **2024**, *17*, 5020. [CrossRef]
3. Kriskova, L.; Ducman, V.; Loncnar, M.; Tesovnik, A.; Zibret, G.; Skentzou, D. Alkali-Activated mineral residues in construction: Case studies on bauxite residue and steel slag pavement tiles. *Materials* **2025**, *18*, 257. [CrossRef]
4. Mosquera, E.M.; Cifuentes, S.; Obando, J.C.; Monteiro, S.N.; Colorado, H.A. Concrete obtained with the viterbo O'reilly method for aggregate gradation: A potential model for sustainable design and reducing development costs. *Materials* **2025**, *18*, 3558. [CrossRef]
5. Tang, J.; Yu, Y.; Bu, Y.Q.; Ma, B.; Zhou, H.; Zhou, R.; Wang, J.Q.; Zhang, H.H. The synergistic effect of calcained coal-series kaolinite and limestone on the hydration of portland cement. *Materials* **2024**, *17*, 4512. [CrossRef] [PubMed]
6. Zou, C.N.; Xue, H.Q.; Xiong, B.; Zhang, G.S.; Pan, S.Q.; Jia, C.Y.; Wang, Y.; Ma, F.; Sun, Q.; Guan, C.X.; et al. Connotation, Innovation and Vision of "carbon Neutrality". *Nat. Gas Ind. B* **2021**, *8*, 523–537. [CrossRef]

7. Niu, X.X.; Ma, Z.; Ma, W.J.; Yang, J.J.; Ma, T. The Spatial Spillover Effects and Equity of Carbon Emissions of Digital Economy in China. *J. Clean. Prod.* **2024**, *434*, 139885. [CrossRef]
8. Dos Anjos, M.A.S.; Camoes, A.; Malheiro, R.; Pederneiras, C.M.; Peixoto, L.K.S. Experimental study of carbonation and chloride resistance of self-compacting concretes with a high content of fly ash and metakaolin, with and without hydrated lime. *Materials* **2025**, *18*, 422. [CrossRef] [PubMed]
9. Li, K.X.; Li, X.L.; Yao, J.; Xue, H.W.; Du, C.Y. Innovative synthesis of low-carbon cemented backfill materials through synergistic activation of solid wastes: An integrated assessment of economic and environmental impacts. *Case Stud. Constr. Mater.* **2024**, *20*, e03203. [CrossRef]
10. Zong, Y.L.; Gong, J.T.; Zhang, J.L.; Su, Y.B.; Hu, C.L.; Li, T.G.; Wu, Y.L.; Jiang, M. Research status of soda residue in the field of environmental pollution control. *RSC Adv.* **2023**, *13*, 28975–28983. [CrossRef]
11. Zhao, X.H.; Zhang, G.Q.; Wang, H.Y.; Ynag, H.Q.; Zhao, X.E.; Zhao, R.L. Integrating Fe_2O_3-rich soda residue into alkali-activated materials:Evaluation of mechanical and structural propertiese. *Cem. Concr. Compos.* **2025**, *162*, 106151. [CrossRef]
12. Wei, W.; Ding, Y.J.; Wang, H.M.; Cao, X.X. Effect of soda residue on Skeleton formation and strength development in soil stabilization. *Eng. Geol.* **2026**, *362*, 108518. [CrossRef]
13. Zhao, X.H.; Liu, C.Y.; Zuo, L.M. Synthesis and Characterization of fly ash geopolymer paste for goaf backfill: Reuse of soda residue. *J. Clean. Prod.* **2020**, *260*, 121045. [CrossRef]
14. Hodgins, G.; Robinson, C.E.; Power, C.; Roy, J.W. Road salt chloride exposure in urban streambeds and links to groundwater-Surface Water Interactions and Salt Source. *Sci. Total Environ.* **2024**, *957*, 177488. [CrossRef] [PubMed]
15. Szklarek, S.; Gorecka, A.; Wojtal-Frankiewicz, A. The effects of road salt on freshwater ecosystems and solutions for mitigating chloride pollution—A Review. *Sci. Total Environ.* **2022**, *805*, 150289. [CrossRef] [PubMed]
16. Li, C.B.; Liang, Y.H.; Jiang, L.F.; Zhang, C.M.; Wang, Q. Characteristics of ammonia-soda residue and its reuse in magnesium oxychloride cement pastes. *Constr. Build. Mater.* **2021**, *300*, 123981. [CrossRef]
17. Yu, D.; Deng, J.Y.; Jiang, Q.; Liu, H.S.; Yu, C.L.; Ma, H.; Pu, S.Y. Evaluation of groundwater quality with multi-source pollution based on source identification and health risks. *Sci. Total Environ.* **2024**, *949*, 175064. [CrossRef]
18. Wang, Z.M.; Lyu, X.J.; Yao, G.; Wu, P.; Wang, J.X.; Wei, J. Preparation of Ca-Si-Al-Mg porous ceramics by Co-operation of Ca&Mg-contained soda residue and altered rock gold tailings. *J. Clean. Prod.* **2020**, *262*, 121345. [CrossRef]
19. Yang, Y.B.; Yue, X.D.; Chen, B.X.; Yang, W.; Guo, W.T.; Wang, H.C. Autogenous shrinkage and cracking of ultra-high-performance concrete with soda residue as an internal curing agent. *Mater. Struct.* **2024**, *57*, 216. [CrossRef]
20. Guo, W.C.; Zhao, Q.X.; Sun, Y.J.; Xue, C.H.; Bai, Y.Y.; Shi, Y.X. Effects of various curing methods on the compressive strength and microstructure of blast furnace slag-fly ash-based cementitious material activated by alkaline solid wastes. *Constr. Build. Mater.* **2022**, *357*, 129397. [CrossRef]
21. Guo, W.C.; Zhang, Z.Y.; Zhao, Q.X.; Song, R.J.; Liu, J.Z. Mechanical properties and microstructure of binding material using slag-fly ash synergistically activated by wet-basis soda residue-carbide slag. *Constr. Build. Mater.* **2020**, *269*, 121301. [CrossRef]
22. Lin, Y.H.; Xu, D.Q.; Zhao, X.H. Properties and hydration mechanism of soda residue-activated ground granulated blast furnace slag cementitious materials. *Materials* **2021**, *14*, 2883. [CrossRef]
23. Xu, D.; Fu, P.; Ni, W.; Wang, Q.H.; Li, K.Q. Characterization and hydration mechanism of ammonia soda residue and portland cement composite cementitious material. *Materials* **2021**, *14*, 4794. [CrossRef]
24. Yang, T.F.; Huang, B.; Zhan, C.N.; Jiang, C.; Zhang, L.P.; Zhao, X.Q.; Zhao, M.R. Mechanical properties and mechanisms of soda residue and fly ash stabilized soil. *Sci. Rep.* **2025**, *15*, 1103. [CrossRef] [PubMed]
25. Zhang, Z.; Xie, C.; Sang, Z.; Li, D.J. Optimizing the mechanical performance and microstructure of alkali-activated soda residue-slag composite cementing materials by various curing methods. *Sustainability* **2022**, *14*, 13661. [CrossRef]
26. Zhang, Z.Y.; Xie, C.; Sang, Z.H.; Li, D.J. Mechanical properties and microstructure of alkali-activated soda residue-blast furnace slag composite binder. *Sustainability* **2022**, *14*, 11751. [CrossRef]
27. Zhao, X.H.; Liu, C.Y.; Zuo, L.M.; Zhu, Q.; Wang, M.; Liu, Y.C. Preparation and characterization of press-formed fly ash cement Incorporating soda residue. *Mater. Lett.* **2020**, *259*, 126852. [CrossRef]
28. Zhao, X.Q.; Yang, T.F.; Liang, T.; Zong, Z.L.; Li, J.W. Mechanical properties and dry–wet stability of soda residue soil. *Buildings* **2023**, *13*, 2407. [CrossRef]
29. Zhao, X.Q.; Yang, T.F.; Xiao, J.Z.; Xie, Q.H.; Zong, Z.L.; Yu, Z.L. Mechanical properties and field test of lime-soda residue stabilized soil for subgrade. *Constr. Build. Mater.* **2024**, *419*, 135489. [CrossRef]
30. Lin, Y.; Xu, D.; Zhao, X. Effect of soda residue addition and its chemical composition on physical properties and hydration products of soda residue-activated slag cementitious materials. *Materials* **2020**, *13*, 1789. [CrossRef]
31. Wang, H.Y.; Zhao, X.H.; Gao, H.; Yuan, T.B.; Zhang, X.J. The effects of salt-loss soda residue and oxalate acid on property and structure of fly ash-based geopolymer. *Constr. Build. Mater.* **2023**, *366*, 130214. [CrossRef]
32. Gan, D.Q.; Sun, H.K.; Xue, Z.L.; Liu, Z.Y.; Yang, X. Damage evolution and strength prediction model of soda residue modified cemented tailings backfill under uniaxial compression. *Constr. Build. Mater.* **2023**, *408*, 133709. [CrossRef]

33. Feng, Y.B.; Qi, W.Y.; Zhao, Q.X.; Huang, Y.L.; Ren, Q.S.; Qi, W.Y.; Kong, F.K. Synthesis and characterization of cemented paste backfill: Reuse of multiple solid wastes. *J. Clean. Prod.* **2023**, *383*, 135376. [CrossRef]
34. Guo, W.C.; Xu, Z.H.; Zhang, Z.Y.; Zhang, Z.Y.; Zhang, C.W.; Bai, Y.Y.; Zhao, Q.X. Mechanical performance and microstructure improvement of soda residue–carbide slag–ground granulated blast furnace slag binder by optimizing Its preparation process and curing method. *Constr. Build. Mater.* **2021**, *302*, 124403. [CrossRef]
35. Pu, Y.Q. Research of Dechlorination Soda Residue Used as Internal Curing Materials on Concrete Performance Influence and Mechanism. Master's Thesis, South China University of Technology, Guangzhou, China, 2017.
36. Cao, J.; Zhang, Y. Study on the fundamental physical properties of industrial soda residue. *Energy Environ.* **2013**, 13–14+20. [CrossRef]
37. Wang, Q.; Li, J.J.; Yao, G.; Zhu, X.T.; Hu, S.G.; Qiu, J.; Chen, P.; Lv, X.J. Characterization of the Mechanical Properties and Microcosmic Mechanism of Portland Cement Prepared with Soda Residue. *Constr. Build. Mater.* **2020**, *241*, 117994. [CrossRef]
38. Wu, P.; Zhang, T.; Geng, Y.; Huang, S.B.; Wang, W.X.; Lv, X.J. Research progress on the physical and chemical properties and applications of soda residue. *China Powder Sci. Technol.* **2022**, *28*, 35–42.
39. Yang, Y.B.; Pu, Y.Q.; Yan, W.J.; Guo, W.Y.; Wang, H.C. Microstructure and chloride ion dissolution characteristics of soda residue. *J. South China Univ. Technol. (Nat. Sci. Ed.)* **2017**, *45*, 82–89.
40. Cao, J.; Jiang, H.; Zhang, Y. Study on basic physical and chemical properties and its engineering characteristic of caustic. *J. Salt Sci. Chem. Ind.* **2016**, *45*, 37–40.
41. Ma, J.X.; Yan, N.; Zhang, M.Y.; Liu, J.W.; Bai, X.Y.; Wang, Y.H. Mechanical characteristics of soda residue soil incorporating different admixture: Reuse of soda residue. *Sustainability* **2020**, *12*, 5852. [CrossRef]
42. Yin, W.; Zhang, K.; Ouyang, S.; Bai, X.M.; Sun, W.J.; Zhao, J.R. Experimental study on gangue backfilling materials improved by soda residue and field measurement of surface subsidence. *Front. Earth Sci.* **2021**, *9*, 747675. [CrossRef]
43. Zhao, X.H.; Wang, H.Y.; Zhou, B.Y.; Gao, H.; Lin, Y.H. Resistance of soda residue–fly ash based geopolymer mortar to acid and sulfate environments. *Materials* **2021**, *14*, 785. [CrossRef]
44. Bai, X.Y.; Ma, J.X.; Liu, J.W.; Zhang, M.Y.; Yan, N.; Wang, Y.H. Field experimental investigation on filling the soda residue soil with liquid soda residue and liquid fly ash. *Int. J. Damage Mech.* **2021**, *30*, 502–517. [CrossRef]
45. Yang, L.Z.; Li, Y.D.; Li, Y.D.; Zhang, L.H. The treatment and utilization of soda slag. *Environ. Prot. Sci.* **2008**, *34*, 70–73.
46. *GB 175-2023*; Common Portland Cement. Standards Press of China: Beijing, China, 2023.
47. Babu, U.R.; Kondraivendhan, B. Corrosion performance of steel reinforcement in concrete admixed with magnesium chloride and sulphate. *Anti-Corros. Methods Mater.* **2020**, *67*, 106–118. [CrossRef]
48. Ministry of Housing and Urban-Rural Development of the People's Republic of China. *Code for Durability Design of Concrete Structures*; China Architecture & Building Press: Beijing, China, 2008.
49. Yang, Y.B.; Pu, Y.Q.; Yan, W.J.; Guo, W.Y.; Wang, H.C. Experimental study on the performance of concrete using soda residue used as mineral admixture. *Concrete* **2017**, 80–83+88. [CrossRef]
50. Guo, Y.; Wang, X.P.; Zhu, Y.F.; Zhang, J.; Gao, Y.B.; Yang, Z.Y.; Du, R.G.; Lin, C.J. Electrochemical and XPS study on effect of Cl^- on corrosion behavior of reinforcing steel in simulated concrete pore solutions. *Int. J. Electrochem. Sci.* **2013**, *8*, 12769–12779. [CrossRef]
51. Wu, Y.J.; Xue, S.G.; Liu, L.P.; Li, F.; Millar, G.J.; Ge, F.; Tian, J. Enhanced dealkalization of bauxite residue through calcium-activated desulfurization gypsum. *Trans. Nonferrous Met. Soc. China* **2024**, *34*, 3412–3424. [CrossRef]
52. Oey, T.; Stoian, J.; Li, J.L.; Balonis, M.; Kumar, A. Comparison of $Ca(NO_3)(2)$ and $CaCl_2$ admixtures on reaction, setting, and strength evolutions in plain and blended cementing formulations. *J. Mater. Civ. Eng.* **2015**, *27*, 04014267. [CrossRef]
53. Xu, L.L.; Wang, P.M.; Zhang, G.F. Formation of ettringite in Portland cement/calcium aluminate cement/calcium sulfate ternary system hydrates at lower temperatures. *Constr. Build. Mater.* **2012**, *31*, 347–352. [CrossRef]
54. Zhang, S.P.; Xu, X.X.; Memon, S.A.; Dong, Z.J.; Li, D.X.; Cui, H.Z. Effect of calcium sulfate type and dosage on properties of calcium aluminate cement-based self-leveling mortar. *Constr. Build. Mater.* **2018**, *167*, 253–262. [CrossRef]
55. Chen, H.L.; Li, Z.H.; Ying, G.B. Improvement of the negative-temperature properties of calcium sulphoaluminate cement by three multifunctional chemical admixtures. *Dev. Built Environ.* **2024**, *20*, 100537. [CrossRef]
56. An, Q.; Pan, H.M.; Zhao, Q.X.; Du, S.; Wang, D.L. Strength development and microstructure of recycled gypsum-soda residue-GGBS based geopolymer. *Constr. Build. Mater.* **2022**, *331*, 127312. [CrossRef]
57. Qi, W.Y.; Ren, Q.S.; Zhao, Q.X.; Feng, Y.B.; Qi, W.Y.; Han, Y.J.; Huang, Y.L. Effectiveness of soda residue-activated GGBS as alternative binder on compressive strength and workability of cemented paste backfills: Reuse of multi-source solid wastes. *Constr. Build. Mater.* **2022**, *348*, 128594. [CrossRef]
58. Cheng, W.B.; Yin, W.; Wang, X.Y.; Xu, Q.; Wang, G.D.; Cao, J.; Zhu, S.X. Multi-factor orthogonal experiments and enhancement mechanisms of unconfined compressive strength of soda residue cement lime soil. *Buildings* **2024**, *14*, 2189. [CrossRef]

59. Pang, H.T.; Qi, W.Y.; Huang, Y.L.; Zhao, Q.X.; Zhang, J.H.; Zhao, Z.D.; Yu, J.C. Investigation of mechanical, microscopic, and leaching properties of coal-based solid waste geopolymer mortar activated by soda residue and phosphogypsum. *Sci. Total Environ.* **2024**, *947*, 174459. [CrossRef]
60. Xu, D.; Ni, W.; Wang, Q.H.; Xu, C.W.; Li, K.Q. Ammonia-soda residue and metallurgical slags from iron and steel industries as cementitious materials for clinker-free concretes. *J. Clean. Prod.* **2021**, *307*, 127262. [CrossRef]
61. Li, H.; Li, M.H.; Guo, W.C.; Zhao, Q.X.; Zhang, H.J. Bond behavior between steel and solid waste concrete activated by soda residue and carbide slag in concrete-filled steel tubes (CFSTs). *Case Stud. Constr. Mater.* **2025**, *22*, e04383. [CrossRef]
62. Chen, H.Z.; Zhang, Z.Y.; Li, J.Y.; Guo, W.C.; Li, Y.; Zhao, Q.X. Impact of curing conditions on the mechanical properties and micro characteristics of alkali-activated material synthesized from industrial waste soda residue. *Case Stud. Constr. Mater.* **2024**, *21*, e03675. [CrossRef]
63. Song, R.J.; Zhao, Q.X.; Zhang, J.R.; Liu, J.Z. Microstructure and composition of hardened paste of soda residue-slag-cement binding material system. *Front. Mater.* **2019**, *6*, 211. [CrossRef]
64. Yang, J.; Lu, S.Y.; He, X.Y.; Su, Y.; Zeng, J.Y.; Tian, C.; Tang, D.K.; Wang, Y.F.; Li, Y.B. Nano-micron soda residue as green nucleation seeding additive for silicate cement. *Constr. Build. Mater.* **2024**, *419*, 135563. [CrossRef]
65. Zhang, Y.; Liu, X.; Cao, J.; Chen, B.H.; Dong, Y.; Zhang, Y.; Zong, R. Experimental research on abrasion resistance of alkaline residue concrete. *China Concr. Cem. Prod.* **2013**, 81–84. [CrossRef]
66. Gu, W.F.; Diao, J.; Tao, H.R.; Deng, J.Y.; Yu, H.F.; Iwma, T.; Elizaveta, C.; Li, H.Y.; Xie, B.; Ueda, S. Effects of Cooling Methods on Phase Evolution, Microstructure, and Stability of Steelmaking Slag. *Metall. Mater. Trans. B* **2025**, *56*, 3970–3979. [CrossRef]
67. Tanaçan, L.; Ersoy, H.Y.; Arpacioglu, Ü. Effect of high temperature and cooling conditions on aerated concrete properties. *Constr. Build. Mater.* **2009**, *23*, 1240–1248. [CrossRef]
68. Wang, D.; Zhang, Z.Y.; Guo, W.C.; Li, J.Y.; Li, X.W.; Zhao, Q.X. A novel all-solid-waste binder prepared by salt-alkali synergistic activation system constructed from phosphogypsum, soda residue and calcium carbide slag. *Cem. Concr. Compos.* **2025**, *155*, 105841. [CrossRef]
69. Zhang, Y.; Cao, J.; Dong, Y.; Chen, B.H.; Zhang, Y.; Zong, R. Experimental investigation of influence of temperature on microstructure of alkaline residue. *J. Lanzhou Univ. Technol.* **2013**, *39*, 126–129.
70. Wang, D.L.; Na, Q.; Liu, Y.K.; Feng, Y.; Zhang, Q.L.; Chen, Q.S. Hydration process and fluoride solidification mechanism of multi-source solid waste-based phosphogypsum cemented paste backfill under CaO modification. *Cem. Concr. Compos.* **2024**, *154*, 105804. [CrossRef]
71. Lin, Y.H.; Xu, D.Q.; Ji, W.G.; Zhao, X.H. Experiment on the Properties of soda residue-activated ground granulated blast furnace slag mortars with different activators. *Materials* **2022**, *15*, 3578. [CrossRef]
72. Zhang, H.Y.; Li, Y.C.; Wei, D.B.; Wu, X.; Wang, Y.P. Mechanistic Insights into Full Solid-Waste Activators for Enhancing the Performance of Blast Furnace Slag-Fly Ash Cementitious Composites. *Materials* **2025**, *18*, 3275. [CrossRef]
73. Li, X.; Qin, X.L.; Wei, Y.C.; Yang, T.; Gao, X.; Zhu, H.J.; Wu, Q.S. Hydration Kinetics and Microstructural Evolution of Soda Residue-Blended Cement Binders: From Synergistic Acceleration to the Barrier Effect. *Constr. Build. Mater.* **2026**, *520*, 146006. [CrossRef]
74. Jupe, A.C.; Wilkinson, A.P.; Luke, K.; Funkhouser, G.P. Slurry consistency and in situ synchrotron X-Ray diffraction during the early hydration of portland cements with calcium chloride. *J. Am. Ceram. Soc.* **2007**, *90*, 2595–2602. [CrossRef]
75. Ouyang, X.W.; Koleva, D.A.; Ye, G.; Van Breugel, K. Insights into the mechanisms of nucleation and growth of C–S–H on fillers. *Mater. Struct.* **2017**, *50*, 213. [CrossRef]
76. Zhao, X.Q.; Zhang, L.P.; Huang, B.; Yang, T.F.; Zhou, Z.M.; Xie, Q.H. Physical and Mechanical Properties and Microscopic Mechanisms of Soda Residue-Fly Ash-Based Geopolymer Filling Paste. *Powder Technol.* **2026**, *469*, 3. [CrossRef]
77. Pang, H.W.; Huang, Y.L.; Qi, W.Y.; Zhao, Q.X.; Wang, Y.Y.; Zhao, D.Z.; Kong, F.K.; Han, J.Y. Preparation of sustainable backfill cementitious materials: Mix proportion optimization, hydration mechanism, and sustainability assessment. *Green Smart Min. Eng.* **2026**, *3*, 41–53. [CrossRef]
78. Guo, W.C.; Bai, Y.Y.; Xu, Z.H.; Zhang, J.R.; Zhao, Q.X.; Wang, D.L. Stress-strain behavior of low-carbon concrete activated by soda residue-calcium carbide slag under uniaxial and triaxial compression. *J. Build. Eng.* **2022**, *55*, 104678. [CrossRef]
79. Zhao, X.H.; Liu, C.Y.; Wang, L.; Zuo, L.M.; Ma, W. Physical and mechanical properties and micro characteristics of fly ash-based geopolymers incorporating soda residue. *Cem. Concr. Compos.* **2019**, *98*, 125–136. [CrossRef]
80. Wu, J.; Jiang, J.P.; Ding, S.; Duan, P. Hydration and microstructure of steel slag as cementitious material and fine aggregate in mortar. *Molecules* **2020**, *25*, 4456. [CrossRef]
81. Wang, M.Q.; Qian, B.B.; Jiang, J.; Liu, H.C.; Cai, Q.; Ma, B.; Hu, Y.Y.; Wang, L.M. The reaction between Ca^{2+} from steel slag and granulated blast-furnace slag system: A unique perspective. *Chem. Pap.* **2020**, *74*, 4401–4410. [CrossRef]
82. Liu, J.P.; Hu, Z.Y.; Liu, P.; Wang, S.Y.; Wang, Y.S.; Zhang, W.X.; Li, X.H.; Pang, S.Y. Properties of alkali-activated furnace slag/steel slag cementitious materials mixed with silica fume and soda residue. *J. Shenyang Univ. Technol.* **2023**, *45*, 594–600.

83. Bai, Y.Y.; Guo, W.C.; Zhao, Q.X.; Shen, H.Y.; Qiu, Y.X. Strength formation mechanism and curing system optimization of low-carbon cementitious materials prepared by synergistic activation of multiple alkaline solid wastes. *Constr. Build. Mater.* **2023**, *402*, 132931. [CrossRef]
84. Lin, Y.H.; Xu, D.Q.; Zhao, X.H. Experimental research on mechanical property and microstructure of blast furnace slag cementitious materials activated by soda residue. *Bull. Chin. Ceram. Soc.* **2019**, *38*, 2876–2881+2889.
85. Wang, M.Q.; Xu, J.; Zhang, X.J.; Tan, L.Z.; Mei, Y. Mechanical performance optimization and microstructural mechanism study of alkali-activated steel slag–slag cementitious materials. *Buildings* **2024**, *14*, 1204. [CrossRef]
86. Wang, Q.; Zhang, T.; Wu, P.; Lv, X.J. Recovered soda residue as alkaline activator of furnace slag: Geopolymer prepared from furnace slag incorporated with soda residue and its reuse in composite cement. *Clean Technol. Environ. Policy* **2022**, *25*, 1525–1537. [CrossRef]
87. Duan, W.W.; Zhuge, Y.; Chow, C.W.K.; Keegan, A.; Liu, Y.; Merta, I. Mitigation of alkali-silica reaction in blast-furnace slag-based alkaline activated material through incorporation of alum water treatment residue. *Constr. Build. Mater.* **2023**, *406*, 133383. [CrossRef]
88. Vishavkarma, A.; Venkatanarayanan, H.K. Assessment of pore structure of foam concrete containing slag for improved durability performance in reinforced concrete applications. *J. Build. Eng.* **2024**, *86*, 108939. [CrossRef]
89. Das, B.B.; Kondraivendhan, B. Implication of pore size distribution parameters on compressive strength, permeability and hydraulic diffusivity of concrete. *Constr. Build. Mater.* **2012**, *28*, 382–386. [CrossRef]
90. Zhang, M.H.; Li, H. Pore structure and chloride permeability of concrete containing nano-particles for pavement. *Constr. Build. Mater.* **2011**, *25*, 608–616. [CrossRef]
91. Li, L.J.; Liu, Q.F.; Tang, L.P.; Hu, Z.; Wen, Y.; Zhang, P. Chloride penetration in freeze–thaw Induced cracking concrete: A numerical study. *Constr. Build. Mater.* **2021**, *302*, 124291. [CrossRef]
92. Ma, Y.; Gong, J.; Ouyang, X.W.; Li, Z.J.; Wang, H.; Fu, J.Y. Roles of recycled concrete powder on the properties of alkali-activated slag: Reaction products and microstructure development. *Compos. B Eng.* **2025**, *301*, 112493. [CrossRef]
93. Liang, G.; Liu, T.; Li, H. Shrinkage mitigation, strength enhancement and microstructure improvement of alkali-activated slag/fly ash binders by ultrafine waste concrete powder. *Compos. B Eng.* **2022**, *231*, 109570. [CrossRef]
94. Xu, C.W.; Ni, W.; Li, K.Q. Effect of ammonia-soda residue on the strength and chloride-resistance performance of steel slag-granulated blast furnace slag-based concrete after immersion in artificial seawater. *Materials* **2021**, *14*, 6048. [CrossRef]
95. Snellings, R. Assessing, Understanding and unlocking supplementary cementitious materials. *RILEM Tech. Lett.* **2016**, *1*, 50–55. [CrossRef]
96. Al-Saffar, F.Y.; Wong, L.S.; Paul, S.C. An elucidative review of the nanomaterial effect on the durability and calcium-silicate-hydrate (C-S-H) gel development of concrete. *Gels* **2023**, *9*, 613. [CrossRef] [PubMed]
97. Altawaiha, H.; Alhomaidat, F.; Eljufout, T. A review of the effect of nano-Silica on the mechanical and durability properties of cementitious composites. *Infrastructures* **2023**, *8*, 132. [CrossRef]
98. Shen, Q.; Li, B.X.; He, W.; Meng, X.; Shen, Y.L. Associated effects of sodium chloride and dihydrate gypsum on the mechanical performance and hydration properties of slag-based geopolymer. *Buildings* **2023**, *13*, 1285. [CrossRef]
99. Peterson, V.K.; Whitten, A.E. Hydration Processes in Tricalcium Silicate: Application of the Boundary Nucleation Model to Quasielastic Neutron Scattering Data. *J. Phys. Chem. C* **2009**, *113*, 2347–2351. [CrossRef]
100. Ding, W.; Liu, X.; Hu, F.; Zhu, H.L.; Luo, Y.X.; Li, S.; Li, H. How the Particle Interaction Forces Determine Soil Water Infiltration: Specific Ion Effects. *J. Hydrol.* **2019**, *568*, 492–500. [CrossRef]
101. Mutisya, S.M.; Almeida, J.M.D.; Miranda, C.R. Probing the dynamics of water over multiple pore scales in cement by atomistic simulations. *Appl. Surf. Sci.* **2021**, *565*, 150426. [CrossRef]
102. Chen, J.J.; Sorelli, L.; Vandamme, M.; Ulm, F.J.; Chanvillard, G. A coupled nanoindentation/SEM-EDS study on low water/cement ratio portland cement paste: Evidence for C-S-H/$Ca(OH)_2$ nanocomposites. *J. Am. Ceram. Soc.* **2010**, *93*, 1484–1493. [CrossRef]
103. Walker, C.S.; Sutou, S.; Oda, C.; Mihara, M.; Honda, A. Calcium silicate hydrate (C-S-H) gel solubility data and a discrete solid phase model at 25 °C based on two binary non-Ideal solid solutions. *Cem. Concr. Res.* **2015**, *79*, 1–30. [CrossRef]
104. Jiang, Q.; Zhan, Q.; Hu, H.; Wang, X.Y. Solid waste-based cementitious material for ground granulated blast furnace slag based on soda residue activation: Preparation, properties and mechanism. *J. Environ. Chem. Eng.* **2025**, *13*, 120302. [CrossRef]
105. Fu, Q.; Bu, M.X.; Zhang, Z.R.; Xu, W.R.; Yuan, Q.; Niu, D.T. Hydration characteristics and microstructure of alkali-activated slag concrete: A review. *Engineering* **2023**, *20*, 162–179. [CrossRef]
106. Ren, Q.S.; Qi, W.Y.; Zhao, Q.X.; Jia, Y.L.; Feng, Y.B.; Han, Y.J. Preparation and characterization of low-carbon cementitious materials based on soda-residue-activated ground granulated blast-furnace slag: A case study on cemented paste backfills. *Metals* **2023**, *13*, 694. [CrossRef]
107. Liu, J.J.; Zha, F.S.; Xu, L.; Kang, B.; Yang, C.B.; Feng, Q.; Zhang, W.; Zhang, J.W. Strength and microstructure characteristics of cement-soda residue solidified/stabilized zinc contaminated soil subjected to freezing-thawing cycles. *Cold Reg. Sci. Technol.* **2020**, *172*, 102992. [CrossRef]

 https://doi.org/10.3390/ma19112228

108. Guo, W.C.; Zhang, Z.Y.; Bai, Y.Y.; Zhao, G.Q.; Sang, Z.H.; Zhao, Q.X. Development and characterization of a new multi-strength level binder system using soda residue-carbide slag as composite activator. *Constr. Build. Mater.* **2021**, *291*, 123367. [CrossRef]
109. Han, Y.J.; Qi, W.Y.; Pang, H.T.; Zhao, Q.X.; Huang, Y.L.; Zhao, D.Z.; Zhu, W.H.; Zhang, J.H. A novel coal gasification coarse slag-based geopolymer: Influences of physico-chemical coupling activation on its properties, microstructure, and hazardous material immobilization. *Constr. Build. Mater.* **2024**, *420*, 135591. [CrossRef]
110. Qi, W.Y.; Duan, G.; Han, Y.J.; Zhao, Q.X.; Huang, Y.L.; Zhu, W.H.; Pang, H.T.; Zhang, J.H. Comparison of mechanical properties and microstructure of GGBS-based cementitious materials activated by different combined alkaline wastes. *Constr. Build. Mater.* **2024**, *422*, 135784. [CrossRef]
111. Xu, S.S.; Yang, Q.; Wei, J.J.; Du, X.S. Study on the Effect of Distiller Waste Mud Content on the Properties of Cementitious Materials. *Soda Ind.* **2024**, 13–16. [CrossRef]
112. Qing, L.; Li, X.C.; Li, C.M.; Wang, J.X.; Lyu, X.J. Synthesis and optimization of green one-part geopolymer from mine tailings and slag: Calcium carbide residue and soda residue as supplementary alkali sources. *Constr. Build. Mater.* **2022**, *353*, 129013. [CrossRef]
113. Liu, X.; Hou, P.; Chen, H. Effects of nanosilica on the hydration and hardening properties of slag cement. *Constr. Build. Mater.* **2021**, *282*, 122705. [CrossRef]
114. Li, B.J.; Lou, D.L.; Zhang, L.; Li, G.Z. Study on the mechanical properties of soda residue/fly ash composite cementitious materials. *Brick-Tile* **2009**, 35–37. [CrossRef]
115. Liu, Y.F.; Liu, C.Y.; Zuo, M.L.; Pang, Y.Z.; Zhao, X.H. Preparation of soda residue-fly ash-based grouting surry and its influencing factors. *Bull. Chin. Ceram. Soc.* **2017**, *36*, 1518–1523.
116. Guo, W.C.; Zhang, Z.Y.; Xu, Z.H.; Zhang, J.R.; Bai, Y.Y.; Zhao, Q.X.; Qiu, Y.X. Mechanical properties and compressive constitutive relation of solid waste-based concrete activated by soda residue-carbide slag. *Constr. Build. Mater.* **2022**, *333*, 127352. [CrossRef]
117. Zhong, L.M.; Yang, Y.S.; Sheng, P.C. Influence of alkali residue content and autoclaved aeration parameters on compressive strength of building concrete. *Contemp. Chem. Ind.* **2021**, *50*, 812–815. [CrossRef]
118. Zhang, H.L.; Xu, K.M.; Chen, Y.L. Preparation of autoclaved aerated concrete with construction waste and alkali residue. *Chin. J. Environ. Eng.* **2019**, *13*, 441–448.
119. Xu, D.; Ni, W.; Wang, Q.H.; Xu, C.W.; Jiang, Y.Q. Preparation of clinker-free concrete by using soda residue composite cementitious material. *J. Harbin Inst. Technol. (New Ser.)* **2020**, *52*, 151–160.
120. Liang, Y.N.; Xu, C.C.; Ji, T.; Shi, Y.Y. Performance study of neutral Sodium salt alkali slag cement autoclaved aerated concrete blocks. *Constr. Wall Innov. Build. Energy-Sav.* **2019**, 25–30. [CrossRef]
121. Song, H.; Qi, W.; Cao, L.Q.; Wu, Q.; Huang, Y.L.; Zhao, Q.X.; Kong, F.D.; Zhang, I.H. Effects of compounding soda residue with high-calcium activators on the performance of coal gasification coarse slag-based cementitious materials. *Energy* **2026**, *342*, 139531. [CrossRef]
122. Tian, Y.; Xu, J.C.; Shao, Y.Z.; Zhang, G.Y.; Yan, D.M.; Liu, Y. Dynamic analysis of chloride-induced passive film breakdown mechanisms in carbon steel within simulated concrete pore solution. *Case Stud. Constr. Mater.* **2025**, *22*, e04409. [CrossRef]
123. Gao, Y.N.; Yu, H.F.; Ma, H.Y.; Feng, T.T.; Xu, M.; Song, S.S.; Li, L.Y. Electrochemical characters of chloride induced reinforcement corrosion in concrete structures: Initial pitting criterion and the law of mass transfer resistance of concrete cover. *Structures* **2025**, *74*, 108635. [CrossRef]
124. Jiang, H.S.; Jin, Z.Q.; Chen, Y.H.; Liu, Y.; Zhang, X.Y. The Influence of passivation temperature on the corrosion Resistance of passivation film for Q235 steel in simulated pore solution of concrete with chloride or/and sulfate ions. *J. Sustain. Cem.-Based Mater.* **2024**, *13*, 1362–1375. [CrossRef]
125. Kenny, A.; Katz, A. Steel-concrete interface influence on chloride threshold for corrosion–empirical reinforcement to theory. *Constr. Build. Mater.* **2020**, *244*, 118376. [CrossRef]
126. Wu, L.J.; Jiang, H.; Ju, X.L.; Liu, M.W.; Di, Y.T.; Zhao, Y.; Xiang, Z.Y. Service life evaluation of marine reinforced concrete structures in coastal soda residue soil subjected to chloride attack. *Constr. Build. Mater.* **2024**, *455*, 139222. [CrossRef]
127. *GB 50010-2010*; Code for Design of Concrete Structures. China Architecture & Building Press: Beijing, China, 2011.
128. *ACI 318-25*; Building Code Requirements for Structural Concrete (ACI 318-25): An ACI Standard: Commentary on Building Code Requirements for Structural Concrete (ACI 318R-25), an ACI Report. American Concrete Institute: Farmington Hills, MI, USA, 2025.
129. *EN 1992-1-1:2023*; Eurocode 2: Design of Concrete Structures—Part 1-1: General Rules and Rules for Buildings, Bridges and Civil Engineering Structures. European Committee for Standardization (CEN): Brussels, Belgium, 2023.
130. Fakhri, H.; Fishman, K.L.; Ranade, R. Rapid determination of critical chloride content in cement-based composites. *Constr. Build. Mater.* **2021**, *268*, 121148. [CrossRef]
131. Zhang, M.H.; Sun, S.H.; Liu, K.; Li, T.; Yang, H.X. Research on the critical chloride content of reinforcement corrosion in marine concrete—A review. *J. Build. Eng.* **2023**, *79*, 107838. [CrossRef]
132. Zhou, H.; Zhang, Q.; Liu, J. Study of chloride threshold level for corrosion of steel in concrete. *Concrete* **2015**, 13–16,20. [CrossRef]

133. Li, C.Z.; Jiang, L.H. Effect of flue gas desulfurization gypsumaddition on critical chloride content for rebar corrosion in fly ash concrete. *Constr. Build. Mater.* **2021**, *286*, 122963. [CrossRef]
134. Long, W.J.; Zhang, X.H.; Feng, G.L.; Xie, J.; Xing, F.; Dong, B.Q.; Zhang, J.R.; Khayat, K.H. Investigation on chloride binding capacity and stability of Friedel's salt in graphene oxide reinforced cement paste. *Cem. Concr. Compos.* **2022**, *132*, 104603. [CrossRef]
135. Shen, F.M.; Liu, G.J.; Zhang, Y.S.; Liu, C. Molecular insights into sulfate-induced chloride release from Friedel's salt. *Cem. Concr. Res.* **2026**, *199*, 108042. [CrossRef]
136. Teymouri, M.; Shakouri, M.; Vaddey, N.P. Ph-Dependent chloride desorption isotherms of portland cement paste. *Constr. Build. Mater.* **2021**, *312*, 125415. [CrossRef]
137. Machner, A.; Hemstad, P.; Weerdt, K.D. Towards the understanding of the Ph dependency of the chloride binding of portland cement pastes. *Nord. Concr. Res.* **2018**, *58*, 143–162. [CrossRef]
138. Avet, F.; Scrivener, K. Influence of Ph on the chloride binding capacity of limestone calcined clay cements (LC3). *Cem. Concr. Res.* **2020**, *131*, 106031. [CrossRef]
139. Pu, Q.; Jiang, L.H.; Xu, J.X.; Chu, H.Q.; Xu, Y.; Zhang, Y. Evolution of Ph and chemical composition of pore solution in carbonated concrete. *Constr. Build. Mater.* **2012**, *28*, 519–524. [CrossRef]
140. Wan, X.M.; Wittmann, F.H.; Zhao, T.J.; Fan, H. Chloride content and pH value in the pore solution of concrete under carbonation. *J. Zhejiang Univ.-Sci. A* **2013**, *14*, 71–78. [CrossRef]
141. Wang, Y.; Nanukuttan, S.; Bai, Y.; Basheer, P.A.M. Influence of combined carbonation and chloride ingress regimes on rate of ingress and redistribution of chlorides in concretes. *Constr. Build. Mater.* **2017**, *140*, 173–183. [CrossRef]
142. Mahdikhani, M.; Bamshad, O.; Shirvani, M.F. Mechanical properties and durability of concrete specimens containing nano silica in sulfuric acid rain condition. *Constr. Build. Mater.* **2018**, *167*, 929–935. [CrossRef]
143. Jin, Z.Q.; Zhao, X.; Zhao, T.J.; Li, J.Q. Chloride ions transportation behavior and binding capacity of concrete exposed to different marine corrosion zones. *Constr. Build. Mater.* **2018**, *177*, 170–183. [CrossRef]
144. Wang, Z.C.; Liu, S.Y.; Tang, G.L.; Wu, K.; Du, G.Y.; Chen, J.F.; Jiang, S.E.; Wang, J.B. Field investigation of ammonia alkali soda residue-based foamed concrete as a subgrade filler. *Acta Geotech.* **2026**, 1–19. [CrossRef]
145. Runci, A.; Provis, J.; Serdar, M. Microstructure as a key parameter for understanding chloride ingress in alkali-activated mortars. *Cem. Concr. Compos.* **2022**, *134*, 104818. [CrossRef]
146. Xu, C.; Tian, W.; Gao, J.F.; Gao, Y. Performance evaluation and lifetime prediction of steel slag coarse aggregate concrete under sulfate attack. *Constr. Build. Mater.* **2022**, *344*, 128203. [CrossRef]
147. Zhao, X.H.; Liu, C.Y.; Zuo, L.M.; Pang, Y.Z.; Liu, Y.F. Experimental research on durability of new grouting materials with soda residue and fly ash matrix. *Ind. Constr.* **2018**, *48*, 31–36.
148. Tosun, K.; Felekoglu, B.; Baradan, B.; Altun, A. Effects of limestone replacement ratio on the sulfate resistance of portland limestone cement mortars exposed to extraordinary high sulfate concentrations. *Constr. Build. Mater.* **2009**, *23*, 2534–2544. [CrossRef]
149. Broekmans, A.T.M. Deleterious reactions of aggregate with alkalis in concrete. *Rev. Mineral. Geochem.* **2012**, *74*, 279–364. [CrossRef]
150. Salim, M.U.; Mosaberpanah, M.A. The mechanism of alkali-aggregate reaction in concrete/mortar and its mitigation by using geopolymer materials and mineral admixtures: A comprehensive review. *Eur. J. Environ. Civ. Eng.* **2022**, *26*, 6766–6806. [CrossRef]
151. Huang, X.F.; Zhang, Q.; Wang, W.; Pan, J.D.; Yang, Y. Effect of carbide slag on removal of Na+/K+ from red mud based on water leaching. *ACS Omega* **2022**, *7*, 4101–4109. [CrossRef]
152. CHhen, X.F.; Bi, Y.F.; Zhang, H.B.; Wang, J. Chlorides removal and control through Water-washing process on MSWI Fly Ash. *Procedia Environ. Sci.* **2016**, *31*, 560–566. [CrossRef]
153. Wu, Y.J.; Li, M.; Fu, D.; Santini, T.C.; Iiang, J.; Hartley, W.; Xue, S.G. Simulation study for the formation of alkaline efflorescence on bauxite residue disposal areas following the phosphogypsum addition. *J. Clean. Prod.* **2020**, *262*, 121266. [CrossRef]
154. Nanayakkara, O.; Gunasekara, C.; Law, D.W.; Xia, J.; Setunge, S. Alkali-activated slag concrete with recycled aggregate: Long-term performance. *J. Mater. Civ. Eng.* **2021**, *33*, 04021167. [CrossRef]
155. Wardhono, A.; Law, D.W.; Molyneaux, T.C.K. Long term performance of alkali activated slag concrete. *J. Adv. Concr. Technol.* **2015**, *13*, 187–192. [CrossRef]
156. Menu, B.; Jacob-Vaillancourt, T.; Jolin, M.; Bissonnette, B. Influence of curing methods on moisture loss and drying shrinkage of shotcrete at early age. *ACI Mater. J.* **2020**, *117*, 95–103. [CrossRef]
157. Zielinski, A.; Schindler, A.K.; Kaszynska, M. Cracking risk of high-performance cement composites due to restrained autogenous shrinkage with and without soaked lightweight aggregate. *Arch. Civ. Eng.* **2023**, *69*, 603–618. [CrossRef]
158. Liu, K.Z.; Yu, R.; Shui, Z.H.; Yi, S.Q.; Li, X.S.; Ling, G.; He, Y.J. Influence of external water introduced by coral sand on autogenous shrinkage and microstructure development of Ultra-High Strength Concrete (UHSC). *Constr. Build. Mater.* **2020**, *252*, 119111. [CrossRef]

 https://doi.org/10.3390/ma19112228

159. Yang, J.; Wang, F.L.; He, X.Y.; Su, Y.; Wang, T.; Ma, M.Y. Potential usage of porous autoclaved aerated concrete waste as eco-friendly internal curing agent for shrinkage compensatio. *J. Clean. Prod.* **2021**, *320*, 128894. [CrossRef]
160. Song, N.N.; Xia, D.T.; Li, B.; Zheng, Y.; Yao, Y.M.; Mou, B.; Yan, W.; Li, S.H. Synergetic shrinkage mitigation and anti-cracking mechanism of SAP and expansive agent in alkali-activated slag/fly ash-based concrete. *J. Build. Eng.* **2025**, *104*, 112238. [CrossRef]
161. Liu, J.H.; Shi, C.J.; Ma, X.W.; Khayat, K.H.; Zhang, J.; Wang, D.H. An overview on the effect of internal curing on shrinkage of high performance cement-based materials. *Constr. Build. Mater.* **2017**, *146*, 702–712. [CrossRef]
162. Al Saffar, D.M. Effect of internal curing on behavior of high performance concrete: An overview. *Case Stud. Constr. Mater.* **2019**, *10*, e00229. [CrossRef]
163. Yang, Y.B.; Yue, X.D.; Yao, D.Y.; Zhang, D.; Guo, W.Y.; Wang, H.C. Effect of soda residue internal curing agent on autogenous shrinkage and early crack resistance of high-strength and high-performance concrete and tts mechanism analysis. *Mater. Rep.* **2022**, *36*, 91–96.
164. Mounanga, P.; Khelidj, A.; Loukili, A.; Baroghel-Bouny, W. Predicting $Ca(OH)_2$ content and chemical shrinkage of hydrating cement pastes using analytical approach. *Cem. Concr. Res.* **2004**, *34*, 255–265. [CrossRef]
165. Pang, X.Y.; Bentz, D.P.; Meyer, C.; Funkhouser, G.P.; Darbe, R. A comparison study of portland cement hydration kinetics as measured by chemical shrinkage and isothermal calorimetry. *Cem. Concr. Compos.* **2013**, *39*, 23–32. [CrossRef]
166. Beltzung, F.; Wittmann, F.H. Early chemical shrinkage due to dissolution and hydration of cement. *Mater. Struct.* **2001**, *34*, 279–283. [CrossRef]
167. Zhao, H.T.; Li, X.L.; Xie, D.S.; Di, Y.F.; Huang, J.; Xu, W.; Wang, P.G.; Zuo, J.Q. Influence of CaO-based expansive agent, superabsorbent polymers and curing temperature on pore structure evolution of early-age cement paste. *J. Cent. South Univ.* **2022**, *29*, 1663–1673. [CrossRef]
168. Gong, J.; Zhang, W.; Zhou, Z. Foam concrete pore structure effect on drying shrinkage and frost resistance. *J. Test. Eval.* **2021**, *49*, 3431–3443. [CrossRef]
169. Xie, Y.D.; He, Z.H.; Tao, H.Y.; Rong, H. Screening binary supplementary cementitious materials for alite-ye'elimite cement: Unveiling diverging synergistic effects on the microstructure and properties. *Carbon* **2026**, *248*, 121184. [CrossRef]
170. Zhu, X.H.; Tang, D.S.; Yang, K.; Zhang, Z.L.; Li, Q.; Pan, Q.; Yang, C.H. Effect of $Ca(OH)_2$ on shrinkage characteristics and microstructures of alkali-activated slag concrete. *Constr. Build. Mater.* **2018**, *175*, 467–482. [CrossRef]
171. Sun, C.; Zhu, Y.; Guo, J.; Zhang, Y.M.; Sun, G.X. Effects of foaming agent type on the workability, drying shrinkage, frost resistance and pore distribution of foamed concrete. *Constr. Build. Mater.* **2018**, *186*, 833–839. [CrossRef]
172. Ning, S.K.; Jiang, X.D.; Li, B.; Shan, L.; Li, H.B. Research on mechanical and shrinkage characteristics of a resource-based cement solid-Waste concrete. *Materials* **2024**, *17*, 177. [CrossRef]
173. Zhang, Y.D.; Liu, X.Y.; Xu, Z.Y.; Yuan, W.G.; Xu, Y.; Yao, Z.B.; Liu, Z.H.; Si, R.Z. Early-age cracking of fly ash and GGBFS concrete due to shrinkage, creep, and thermal effects: A review. *Materials* **2024**, *17*, 2288. [CrossRef]
174. Grishchenko, R.O.; Emelina, A.L.; Makarov, P.Y. Thermodynamic properties and thermal behavior of Friedel's salt. *Thermochim. Acta* **2013**, *570*, 74–79. [CrossRef]
175. Balonis, M. Thermodynamic Modelling of Temperature Effects on the Mineralogy of Portland Cement Systems Containing Chloride. *Cem. Concr. Res.* **2019**, *120*, 66–76. [CrossRef]
176. Tran, V.Q. Temperature-dependent Reactive-Transport Simulation of Chloride Binding in Cementitious Materials with Kinetic Friedel Salt Formation. *Mater. Des.* **2026**, *265*, 116041. [CrossRef]
177. Marty, N.C.M.; Grangeon, S.; Lerouge, C.; Warmont, F.; Rozenbaum, O.; Conte, T.; Claret, F. Dissolution Kinetics of Hydrated Calcium Aluminates (Afm-Cl) As A Function of Ph and at Room Temperature. *Mineral. Mag.* **2017**, *8*, 1245–1259. [CrossRef]
178. Shi, Z.G.; Geiker, M.R.; Lothenbach, B.; De Weerdt, K.; Garzón, S.F.; Enemark-Rasmussen, K.; Skibsted, J. Friedel's Salt Profiles from Thermogravimetric Analysis and Thermodynamic Modelling of Portland Cement-Based Mortars Exposed to Sodium Chloride Solution. *Cem. Concr. Compos.* **2017**, *78*, 73–83. [CrossRef]
179. Li, C.Z. Chloride Permeability and Chloride Binding Capacity of Nano-Modified Concrete. *J. Build. Eng.* **2021**, *41*, 102419. [CrossRef]
180. Mien, T.V.; Nawa, T.; Stitmannaithum, B. Chloride Binding Isotherms of Various Cements Basing on Binding Capacity of Hydrates. *Comput. Concr.* **2014**, *13*, 695–707. [CrossRef]
181. Chang, H.L.; Fan, S.Y.; Guo, Z.K.; Ma, Y.; Li, Z.F.; Zuo, Z.W.; Liu, J.; Yang, L. Chloride binding behavior of calcium silicate hydrates and Friedel's salt of pastes. *J. Build. Eng.* **2023**, *76*, 107318. [CrossRef]
182. Shakouri, M.; Teymouri, M. Investigating the Effects of Testing Protocol Factors on Chloride Binding Capacity in Cementitious Materials. *Constr. Build. Mater.* **2023**, *409*, 134067. [CrossRef]
183. Das, C.S.; Ahmad, M.R.; Zhao, X.L.; Dai, J.G. Influences of Temperatures on the Physical and Chemical Chloride Binding of Calcium Silicate Hydrate and Friedel Salt at Different Chloride Concentrations. *Constr. Build. Mater.* **2025**, *476*, 141303. [CrossRef]
184. Geng, J.; Easterbrook, D.; Liu, Q.F.; Li, L.Y. Effect of Carbonation on Release of Bound Chlorides in Chloride-Contaminated Concrete. *Mag. Concr. Res.* **2016**, *68*, 353–363. [CrossRef]

185. Xu, J.X.; Zhang, C.K.; Jiang, L.H.; Tang, L.; Gao, G.F.; Xu, Y.P. Releases of Bound Chlorides from Chloride-Admixed Plain and Blended Cement Pastes Subjected to Sulfate Attacks. *Constr. Build. Mater.* **2013**, *45*, 53–59. [CrossRef]
186. Yang, C.; Wang, S.G.; Tan, Y.W.; Zhang, C.Q. Influence ofnon-linear chloride binding on the determination of apparent chloride diffusion coefficient for cement iaste with mineral additives. *Constr. Build. Mater.* **2021**, *308*, 104603. [CrossRef]
187. Wu, L.J.; Xiang, Z.Y.; Jiang, H.; Liu, M.W.; Ju, X.L.; Zhang, W.X. A Review of Durability Issues of Reinforced Concrete Structures Due to Coastal Soda Residue Soil in China. *J. Mar. Sci. Eng.* **2022**, *10*, 1740. [CrossRef]
188. Zuo, W.; Zhao, R.B.; Dong, G.H.; Ma, G.L.; Zhou, H.Y.; Song, T.; Tu, Y.; Xie, H.; Wang, X.Y. CO_2-Assisted water-washing process of municipal solid waste incineration fly ash for chloride removal. *Energy Fuels* **2022**, *36*, 13732–13742. [CrossRef]
189. Bouteiller, V.; Tissier, Y.; Marie-Victoire, E.; Chaussadent, T.; Joiret, S. The Application of Electrochemical Chloride Extraction to Reinforced Concrete–A Review. *Constr. Build. Mater.* **2022**, *351*, 128931. [CrossRef]
190. Li, H.; Zhu, S.X.; Yin, W.; Zhu, Z.L.; Zhang, K.; Bai, X.M.; Liu, D.D.; Tang, Y.T. Study on Strength Test and Application of Lime Soil in Pavement Base Modified by Soda Residue. *Adv. Civ. Eng.* **2022**, *2022*, 4887647. [CrossRef]

https://doi.org/10.3390/ma19112228

MDPI AG
Grosspeteranlage 5
4052 Basel
Switzerland
Tel.: +41 61 683 77 34

Materials Editorial Office
E-mail: materials@mdpi.com
www.mdpi.com/journal/materials

www.ingramcontent.com/pod-product-compliance
Lightning Source LLC
LaVergne TN
LVHW070454061126
851858LV00011B/239